Groups:
The Evolution of Human Sociality

Groups:
The Evolution of Human Sociality

Edited by

Kaori KAWAI

Kyoto University Press

First published in Japanese in 2009 by Kyoto University Press. This English edition first published in 2013 jointly by:

Kyoto University Press
69 Yoshida Konoe-cho
Sakyo-ku, Kyoto 606-8315
Japan
Telephone: +81-75-761-6182
Fax: +81-75-761-6190
Email: sales@kyoto-up.or.jp
Web: http://www.kyoto-up.or.jp

Trans Pacific Press
PO Box 164, Balwyn North, Melbourne
Victoria 3104, Australia
Telephone: +61-3-9859-1112
Fax: +61-3-8611-7989
Email: tpp. mail@gmail.com
Web: http://www.transpacificpress.com

Set by Digital Environs, Melbourne.

Distributors

USA and Canada
International Specialized Book Services (ISBS)
920 NE 58th Avenue, Suite 300
Portland, Oregon 97213-3786
USA
Telephone: (800) 944-6190
Fax: (503) 280-8832
Email: orders@isbs.com
Web: http://www.isbs.com

Asia and the Pacific (except Japan)
Kinokuniya Company Ltd.

Head office:
38-1 Sakuragaoka 5-chome
Setagaya-ku, Tokyo 156-8691
Japan
Telephone: +81-3-3439-0161
Fax: +81-3-3439-0839
Email: bkimp@kinokuniya.co.jp
Web: www.kinokuniya.co.jp

Asia-Pacific office:
Kinokuniya Book Stores of Singapore Pte., Ltd.
391B Orchard Road #13-06/07/08
Ngee Ann City Tower B
Singapore 238874
Telephone: +65-6276-5558
Fax: +65-6276-5570
Email: SSO@kinokuniya.co.jp

The translation and publication of this book was supported by a Grant-in-Aid for Publication of Scientific Research Results, provided by the Japan Society for the Promotion of Science, to which we express our sincere appreciation.

ISBN   978-1-920901-78-3

Contents

Figures

Tables

Photographs

Contributors

Kaori Kawai (Introduction, Chapter 7, Conclusion)
Associate professor, Research Institute for Languages and Cultures of Asia and Africa, Tokyo University of Foreign Studies

Kaoru Adachi (Chapter 1)
Lecturer, Ritsumeikan University

Motomitsu Uchibori (Chapter 2, Conclusion)
Professor, The Open University of Japan

Koji Kitamura (Chapter 3, Conclusion)
Professor, Graduate School of Humanities and Social Sciences, Okayama University

Naofumi Nakagawa (Chapter 4)
Associate professor, Graduate School of Science, Kyoto University

Noriko Itoh (Article 1)
Researcher, Wildlife Research Center, Kyoto University

Keiichi Omura (Chapter 5)
Associate professor, Graduate School of Language and Culture, Osaka University

Ikuya Tokoro (Chapter 6)
Associate professor, Research Institute for Languages and Cultures of Asia and Africa, Tokyo University of Foreign Studies

Masahiro Umezaki (Article 2)
Associate professor, Graduate School of Medicine, The University of Tokyo

Hideaki Terashima (Chapter 8)
Professor, Faculty of Humanities and Sciences, Kobe Gakuin University

Toru Soga (Chapter 9)
Professor, Faculty of Humanities, Hirosaki University

Yuko Sugiyama (Chapter 10)
Professor, Faculty of Humanities, Hirosaki University

Wakana Shiino (Article 3)
Associate professor, Research Institute for Languages and Cultures of Asia and Africa, Tokyo University of Foreign Studies

Suehisa Kuroda (Chapter 11, Conclusion)
Professor, School of Human Cultures, The University of Shiga Prefecture

Masakazu Tanaka (Chapter 12)
Professor of Social Anthropology, Institute for Research in Humanities, Kyoto University

Takeo Funabiki (Chapter 13)
Emeritus Professor, Graduate School of Arts and Sciences, The University of Tokyo

Ryoko Nishii (Epilogue)
Professor, Research Institute for Languages and Cultures of Asia and Africa, Tokyo University of Foreign Studies

Introduction

In Pursuit of an Evolutionary Foundation for Human Society

Kaori Kawai

Introduction

Cerebral activity, the single most important key to human evolution, manifests as the ability to develop advanced social relationships. In his work *The Prehistory of the Mind* (1998), eminent cognitive archaeologist Steven Mithen—also an expert in evolutionary psychology—summarizes the modular theory of cognition and intelligence and states that the existence of a cerebral module for highly developed social intelligence was the very thing that made higher primates as such, and is the element that gave rise to hominids and ultimately to *homo sapiens sapiens*. It goes without saying that the ability to develop social relationships, in its most direct form, can be rendered as the ability to exist sympatrically with others—in other words, the ability to form and live in groups—and it is also the ability that guarantees the co-existence within a group of individuals of the same species. In evolutionary historical terms, human beings until very recently (approximately six to seven million years ago) walked the same evolutionary path as great apes such as the chimpanzee and the pigmy chimpanzee (bonobo). In closely examining the nature of human existence and its evolutionary basis, we deem the acquisition of that advanced social ability (hereafter, "sociality") to be the single most important feature in the emergence of the species itself. The aim of this book, therefore, is to theoretically and empirically elucidate sociality from a variety of angles. To that end, we turn to non-human primates for comparative purposes to consider the nature of the evolutionary historical foundations of sociality, as embraced by human beings.

In the context of the above plans, a research group was formed at the Research Institute for Languages and Cultures of Asia and Africa (ILCAA) at the Tokyo University of Foreign Studies in April 2005, under the title "Human Society in Evolutionary Perspectives (1)", to look into the subject of groups (hereafter, the Groups Research Project: GRP) as a joint research endeavor. The research group sought to investigate the concept of groups by placing it in the context of the evolutionary history of human and non-human primates in general, to unpack the concept and investigate the creation and development of groups and group perceptions from an evolutionary perspective. Researchers primarily involved in three fields—primate sociology and ecology, ecological anthropology and socio-cultural anthropology— convened seminars on twenty-one occasions over four years to discuss the shape and extent of variations of groups as sympatric entities, and the evolutionary historical foundations that have led to the orientation of groups in present-day human society. This book is the result.

Group research and evolutionary theory in anthropology and primatology

In this book, specialists on research into human beings, such as socio-cultural anthropologists and ecological anthropologists (hereafter, "human experts"), and those involved in research into non-human primates ("non-human primate experts"), develop the debate about groups in the context of their own fields of expertise, at times in ways that extend beyond the boundaries of their fields. Human beings are primates, but for some reason anthropology is not considered part of primatology, although there are some researchers among non-human primate experts who term themselves "anthropologists".[1] These scholars study non-human primates in order to understand human beings, using methods from primatology to gather and analyze data and elucidate those aspects of humans we hold by virtue of being primates. Through their work they attempt to observe and identify humanity, or its germination. Examples of publications with "humanity" (*ningensei*) in the title include some that start with the term, such as *Ningensei wa doko kara kitaka—sarugaku kara no apurōchi* (The origins of humanity: Approach from the study of monkeys and apes) (T. Nishida 1999) and *Ningensei no kigen to shinka* (The origins and evolution of humanity) (M. Nishida, Kitamura and Yamagiwa (eds) 2003).

As may be surmised from the location of the topic "Kinship groups" in intro- ductory texts on socio-cultural anthropological research into groups, in the past the topic has formed a major and core concern of that academic field. Groups were seen to be the most concrete component elements of a society, and knowing their characteristics was deemed to lead to knowledge of the society's unique mecha- nisms. In recent socio-cultural anthropology, however, such what could be called, "static" structuralist group theories have lost popularity. In their place, community theory,[2] which is at the core of prevailing sociological debate, has been adopted with alacrity (for example, Tanabe and Matsuda (eds) 2002; Oda et al. 2004; M. Tanaka and Matsuda (eds) 2006; R. Nishii and Tanabe (eds) 2006). What has come to be discussed is people's "places of practice", in other words, the existence of places in which the mind, the body, things and activities are constituted in daily practice (Tanabe 2002: 12). In temporal terms, however, in most cases the field—in which debate is furthered using a range of new concepts—focuses on the modern, or the present. In that sense, therefore, while the focus is on gatherings of individuals, just as in our GRP, there is a lack of a long-term, evolutionary historical scale perspective and scope. Conversely, the treatment of these wonderful modern perspectives may be said to be a marked point of difference between discussions of community and the theory of groups dealt with in this book. In this setting, the fact that no-one—even those writers aligned with socio-cultural anthropology— uses the word "community" (albeit that "communal society" is used) is extremely significant, in as much as it characterizes the discussions presented in this book.

At the same time there has been a significant change in ecological anthropology in Japan, which had maintained a close relationship with primatology, and in particular the fields of primate sociology and primate ecology in respect of a single point: evolution. In ecological anthropology, which in the 1970s and '80s saw a frenzy of stimulating debate around re-configuring the ecology and society of early humans, the formulation of a theory of groups reliant on structural similarities to primate society was considered an extremely effective methodology. This was particularly the case in the early days of this academic field, when it was believed possible that the "primitive" form of human beings could be found on the African continent, among the hunter-gatherers of the Congo Basin and Kalahari Desert. Hypotheses and theories such as "pre-band" theory, the theory of the origin of families and the principles of equality and inequality for co-existence emerged from comparative social theory research where data from great apes observable in the wild, such as chimpanzees, bonobo and gorillas, and that on the abovementioned hunter-gatherers, was read alongside the results of ecological anthropology in which the subject was peoples living with greater reliance on nature in a variety of natural environments, such as slash-and-burn farmers in Brachystegia forest (*Miombo* forest) and pastoralists in arid and semi-arid regions (for example, see J. Tanaka 1971; Itani and Harako 1977; Ichikawa 1982; Itani and Yoneyama 1984; Itani and Tanaka 1986). Since the 1980s, however, these societies have been substantially transformed by modernization policies and increasing globalization. Following this lead, ecological anthropology persevered with such keywords as "environment", "ecology" and "adaptation", while gradually distancing itself from the theoretical framework of the biological explanation for humanity—evolution—and allowing its theoretical foundations to slide into the fields of area studies and cultural anthropology[3] (for example, Tanaka, Kakeya, Ichikawa and Ohta 1996).

From a different angle, there has also been a substantial shift in primatology. From the end of the 1940s, sociological research into wild primates was energetically pursued in Japan, with the Japanese macaque as the subject (for example, Itani 2008: Vol. 1, 38–298; M. Kawai 1964). From 1960, the field was extended to Africa with evolutionary research into the ecology and society of the great apes, such as the chimpanzee, bonobo and gorilla, in particular in the context of hominization theory (for example, Itani 1977, 2008: Vol. 2, 5–202; M. Kawai 1977; Kuroda 1982; Yamagiwa 1984). Japanese primatology was nevertheless overwhelmed by the neo-Darwinism of the 1980s; in other words, by socio-biology—also called evolutionary biology or behavioral ecology (for example, Dawkins 1976, 1982; Nishida 1999: 39–58; Strier 2006). This perspective proffered an extreme form of biological reductionism that took the gene as the evolutionary unit and held that individuals would only choose behavior that increased breeding success. Primatology in Japan was further beset by ecological research (for example, Yukimaru Sugiyama 2000), and comparative sociological studies into human evolution gradually waned.

Whereas in the past primatology and ecological anthropology had discussed humanity in the same arena and theorized regarding the evolutionary process, each became increasingly specialized in their separate fields, the former as socio-biology and ecology and the latter as area studies and cultural anthropology. For that reason, the work of teasing out new theories on human sociality by exploring the connection between non-human primates and human beings—continuity and discontinuity in primate and human societies—has lost prominence.

What this book attempts to do is to deliberately go against the tide. At the risk of repetition, our intent is to consciously return to the evolutionary perspective of human society with which primate sociology and ecology and ecological anthropology have persevered with difficulty. In place of the past objective of "reconstructing" the ecology and society of early humans, we instead aim to re-identify the creation and evolution of that which is social, as acquired by human beings, from the new grounds of eluci-dating the evolutionary historical "foundations" of human society, and additionally, to challenge the prevailing theory of groups in socio-cultural anthropology.[4] We are bringing socio-cultural anthropology, ecological anthropology and primate sociology and ecology into the same arena, via the common theme of the phenomenon of groups. At the same time, this work is an attempt by human experts to challenge non-human primate experts—in other words, a challenge from socio-cultural anthropology and ecological anthropology, to primatology—from the perspective of elucidating issues associated with the phenomenon of groups. In the context of this three-way battle we believe it is possible to approach the old topic of groups from a new perspective, and by so doing restore a forum for theoretical and empirical debate on humanity, and in particular, the evolution of advanced socialization.

Groups are a relatively prominent, readily visible phenomena in a variety of social milieu. This is the reason why, when seeking to approach the complex of issues that is the evolutionary historical foundations of human society, we have sought entrée from the phenomenon of groups, rather than from society. What sort of entity is society? There is not necessarily a clear answer to this question; in other words, "society" lacks a definition in high general use. When the words "group" and "society" are used, all one need do is recall the abstractness of "society". For example, without a doubt a legal phenomenon called "society" exists that is integrated with legislation, and it is possible for society to be discussed in the abstract. However, we have not suddenly referred to an abstract entity that is society; we have started from what is its component element and foundational essence, the human group. We are gambling on the concreteness of the group phenomenon. A group is a relatively visible, substantial and real subject, and a subject through which we can observe the behaviors of primates and people and the various events that occur therein. To explain our use of the earlier example, we begin, or seek to begin, from the specifics of a gathering of individuals for which there can be no possible substitute, at the stage before law, or before representation in language—in other words, before losing sight of specific individual faces.[5] I would also like to stress that we endeavor to avoid treating the group as a given, but rather look into the phenomenon of the group as one of our primary foci, in the context of the actions of each of the individuals that are its constituent elements.

The aforementioned community theory in the fields of sociology and socio-cultural anthropology is an understanding common to ours in its goal of honing in on the micro-specifics of society and its emphasis on co-location (sympatry), and in that regard it is difficult to point to a clear differentiation. Ultimately, the differentiation between the two will probably rest on whether or not an evolutionary historical time axis is incorporated into the analysis and inquiry.

Jun'ichiro Itani's evolutionary theory of human society and the Filabanga procession of chimpanzees

Here I must briefly turn to my own academic background. The original cause of my involvement in the type of research discussed thus far was my encounter with the work of my postgraduate supervisor in primatology and ecological anthropology,

Professor Jun'ichiro Itani, and in particular, with his work on the evolutionary theory of social structures in human and other primates. Itani said he was often asked how primatology and ecological anthropology were linked in his work, and his answer, he said, was, "The two fields are essential to considerations of the evolution of human society, and wild primates and peoples who live depending heavily on nature are indispensible givens". He also wrote "It is not that there was first a theory that pervaded both, but that I resonated with the hunch that evolution should link the two, and I think it is safe to say that in that sense I became fascinated by the subject and have continued to walk the path I have chosen" (Itani 2008: Vol. 3, 331). I want to share Itani's instinct. Giving theoretical and empirical form to the idea Itani left when he passed away, having penned the words "I got to this point following a hunch that evolution should link the two", is a topic left for the next generation of researchers to take up. As someone who learnt from Itani and who resonated with his ideas, this is an endeavor I wish to embark upon.

Itani started from sociological research into the Japanese macaque and developed his theory of social structural evolution by studying and comparing both human beings and non-human primates, while at the same time creating, nurturing and developing ecological anthropology. He was of the view that if the aim is to encapsulate the essence of early human societies, which do not tend to remain as a fossil record, primates and peoples who have lived in greater reliance on nature pose difficult subjects. Itani had always argued that sociological study of primate groups (troops) was of fundamental importance in studying human sociality and society. In this book the editor and authors, some two thirds of whom are involved in primate sociology and ecology or ecological anthropology, have to varying degrees directly or indirectly received the benefit of instruction from Itani, a trailblazer in the elucidation of the features of human sociality and society.

Several of this book's chapters focus on themes and topics explored by Itani, and Chapter One by Adachi explains Itani's theory of social structural evolution in primates in detail and examines both structural and non-structural concepts. To avoid duplication, I steer clear of an exhaustive list of Itani's achievements here. Instead, I wish to select one "discovery" that became essential to the understanding of his group (or social) theory, and look at the essence of his theory of social structural evolution in primates.

How we initiated a long-term collaborative research project like Human Society in Evolutionary Perspectives is as described in Chapter Two, but there was no hesitation on my part in identifying groups as the first theme the project should address. The grounds for that was the Filabanga procession, which for me is the most striking description in Itani's work (for example, 2008: Vol. 2, 339–48, 369–76, 391–92, 417–18; 2008: Vol. 3, 419–22). This featured a convoy of forty-three chimpanzees crossing a mountain ridge in a line observed by Itani and Akira Suzuki in Filabanga in western Tanzania in the early days of research into wild chimpanzees. Itani cites data from his observations repeatedly throughout his work, and that is perhaps in part why, whenever I think of a group, the scene that always floats first into my mind is his description.

Research by Itani and his mentor, Kin'ji Imanshi, into the social structure of primates, has been a field unique to Japan carved out since the end of the 1940s. For a long time, overseas researchers in animal behavior studies persisted solely with individuals, and thought of society only as an extension of individual behavior (Itani 2008: Vol. 2, 414). In contrast, Japanese researchers have explored the principles of groups as something dimensionally different than those of the individual. As a result, it became apparent that while a group is a collection of individuals, it has a dynamic structure that cannot be fully understood by simply presenting group composition. Further, there is a profound interrelationship between that structure and the evolutionary lineage of various primates, and a group's structure imposes restrictions on the dispersal of genes within a species that are unique to that species and play an important role in its moderation (Itani 2008: Vol. 2, 415–16). Itani called the group that constitutes the foundation of the social structure of a primate species, the Basic Social Unit (BSU). He claimed that with the sole exception of the orangutan, which at the time was thought to be solitary and to exist without a unique social structure, all anthropoid species have BSUs that are able to coalesce into only one structure (Itani 2008: Vol. 2, 416).

Of all primate species groups, chimpanzee groups caused the most confusion. Jane Goodall, who prior to Itani and his colleagues had blazed a trail in 1960 in the study of chimpanzees in the wild at Gombe Stream in western Tanzania, maintained that she was unable to find any more stable chimpanzee group than that formed by the mother-child bond. While sympathizing with what Goodall felt she was bound to say, having been disoriented by the fission-fusion of chimpanzees, Itani could not bring himself to agree (Itani 2008: Vol. 2, 417). This was because at the time, in studies into Japanese macaques using methodology for individual identification and long-term observation, Itani and his colleagues had confirmed the existence of groups (troops)[6] with established memberships and had identified their social structure. For the Japanese researchers, the existence of larger groups superseding the mother-child relationship in chimpanzees had not only already been predicted, but indeed, confirmed. The sighting of the procession at Filabanga decisively underpinned that belief. The observations had an extremely important influence on the elucidation of chimpanzee social structure, soon inviting corroboration of the existence of groups (troops) with stable memberships.

The procession of forty-three chimpanzees maintained a certain structure. Overall, the procession was lead by a collection of mothers with children, followed by a cohort solely consisting of males. Finally, bringing up the rear was a gathering of sexually active females. Itani called it "effectively, a single crystal, molded and chosen from among the various abilities, norms and biases inherent in the group" and further described it as "not just simple quartz, but a crystal of wondrous clarity" (Itani 2008: Vol. 2, 418). Subsequently, chimpanzees on the Mahale mountains of western Tanzania were successfully attracted with food, and using the method of identifying each single individual that came to the feeding grounds, naming them, and distinguishing the social relationships between individuals, analysis resulted in the delineation of a larger group than the fission-fusion parent group—called a "unit group" by Toshisada Nishida (1981)—and provided proof of its configuration and stability. "However", said Itani, "I want to re-consider the Filabanga procession, that wondrous crystal that bloomed in the bleak wilderness, from all the knowledge we have been able to acquire from this detailed analysis" (2008: Vol. 2, 419).

The discovery of the Filabanga procession directed Itani towards a series of studies on the principle of co-existence among individuals of the same species, in other words, a cluster of problems on the inequality and equality principles for co-existence in primate societies and the structure and non-structure of primate societies, using as a framework for analysis what thereafter came to form the crux of his theory of groups—social structure theory. Itani's interests, however, were not confined to theorizing on groups. Indeed, his primatology must be said to have always taken a wider perspective, and he subsequently analyzed—particularly in the context of the evolution of human society—examined, and theorized a string of wide-ranging social phenomena, including the origin of families and systems, vocal communication and the origin of language, the origin of human equality and distributive behavior and the emergence of value. It is conceivable that this oeuvre began from Itani's work on groups.

Anthropology in a broad enough sense to include Jun'ichiro Itani's primatology has focused on the social structures of groups of human beings and non-human primates as its primary question. Itani's methodology was to use comparative sociological methods to evaluate species and identify lineage relationships. If this is considered in light of subsequent developments in the evolutionary theory of primate social structures, then the social structure of primate species not well understood in Itani's lifetime, and also intra-species variation, become clear. Today, when advances in molecular phylogenics have revealed the detail of lineage relationships between species, then the evolutionary chart that Itani devised now requires several amendments (see Chapter Four of this volume). With the influx of socio-biology, which emanates from the West and seeks to understand variations in social structure between closely related species or within a species as products of environmental selection, then the importance of lineage in relation to social structure, as emphasized by Itani, appears to have abated. Despite this fact, as Naofumi Nakagawa points out towards the end of Chapter Four, both those things together are necessary in configuring theories of social evolution, and in that sense also, it is appropriate to re-affirm the validity of Itani's comparative sociological theoretical structure and methodology.

Contents

I now provide an overview of the contents of this book. Important issues, such as an explanation of terminology used in this setting, will follow the outline below. Between the Introduction and Epilogue are four sections, comprising a collection of thirteen chapters, with an additional three articles and an obituary for the late Professor Hitoshi Imamura. Hereafter follows a brief explanation of content.

In this book we have not adopted the approach of dividing chapters on primatology and anthropology into separate sections, as is the tendency in collections of papers co-authored by writers on non-human primates and human beings. This is because we have recognized a greater point of difference in the theory relating to groups discussed in each chapter than that between non-human primates and human beings. In other words, each paper has been separated according to theme, resulting in a combination that does not distinguish or separate papers on primatology and anthropology.[7]

Part I: The evolution of sociality
The main question that forms the focus of this first section is how to approach groups of human beings and non-human primates and the nature of their evolution. Part I, as an introductory section, presents issues from four perspectives and methodologies in order to delve deep into the phenomenon of groups. Human beings have an extremely sophisticated sociality. As primates, that sociality has been inherited and has evolved from an ancestor common to non-human

primates. Humans have, however, formed a diversity of resulting groups to a degree unobserved in other species. This first section addresses four issues: non-structure seen in mixed species association in Cercopithecine monkeys (Adachi); the possibility of gathering arising from solitude (Uchibori); stable cooperation and decoupled representation (Kitamura); and tendencies to locational dispersal in female primates (Nakagawa).

Part II: The formation of social groups

Here the focus is on the ways in which groups form and are sustained. Even groups identified as biological or social givens, such as family and kinship groups, natal groups or local communities, are not in themselves sufficient for ongoing sustainability. To form and sustain a group requires a variety of contrivances and efforts, and as already mentioned, rather than presuming these conditions are a given, by focusing on the behaviors of each individual we know that there is no shortage of impermanent groups formed, for example, for some particular goal arising from activity, and that these groups are formed for the sole purposes of those activities. In Part II, three case studies are analyzed from this perspective: completely egalitarian distribution and sharing of food acquired among the Inuit (Ohmura); piracy and retaliatory combat among sea-faring peoples of the Philippines (Tokoro); and livestock raiding carried out by pastoral peoples in East Africa (Kawai).

Part III: Creation and development of "we" consciousness

By achieving representation through language, human beings have come to recognize as "us" groups in the present that are founded in relatedness brought about through the conduct of mutual, face-to-face negotiations. At the same time humans have also achieved a "we" group consciousness that on a conceptual level can be manipulated, and in which neither group membership nor borders can be "seen". From this perspective, Part III vividly describes human consciousness in three societies with occupational differences: the band society and its expansion among hunter-gatherers (Terashima); the spectrum from visible grouping (cluster) to invisible cultural category in pastoral peoples (Soga); and the relationship between residential groupings of small villages of slash-and-burn farmers and a kingdom with a population of hundreds of thousands (Sugiyama).

Part IV: Toward a new theory of groups

In this section, we respond to the issues raised for debate in Part I, keeping in mind the highly sophisticated sociality humans have acquired and in light of the nature of the various groups in human society revealed in Parts II and III, and pursue the possibility of the opening of new fields in the theory of groups. In other words, what is proposed are: a mechanism for group creation associated with the excitement that is characteristic of two species of chimpanzee and underlies human groups (Kuroda); a seduction model that contrasts with the conflict model associated with maintaining social order and causal factors in the formation of groups (Tanaka); and the modeling of a forum and structure for human relationships related to the establishment of human groups and discussion of the levels at which human groups disappear (Funabiki).

Following each Part are short papers termed "articles", on the particular themes explored in Parts I through III. Article One (Itoh) confronts fission-fusion in chimpanzees from the perspective of the female, a neglected viewpoint to date. Article Two (Umezaki) extracts the "logic" of frequent conflicts between highlander groups in Papua New Guinea. Finally, Article Three (Shiino) tracks the form and awareness of belonging in groups among the Luo of Kenya, which in the last century have undergone dramatic changes.

The final chapter, the Conclusion, summarizes the sections and articles according to each of the three academic fields of primatology, ecological anthropology and socio-cultural anthropology, and outlines a vision for the next stage in the exploration of the evolutionary historical foundations for human society, while taking a very hard look at the topics of institutions and the ability to represent, imagine and symbolize.

Terminology

It is now appropriate to explain the core terminology used in this book.

Group

The term "group" entails slightly different nuances as used by the authors in each section. It is probably not appropriate at this point to enter into a discussion on intricate differences, but if I were to point to one aspect, it is that the groups we deal with are in the first instance gatherings of multiple individuals, but are in fact closer to a dynamic interactive system (Luhmann 1984) than a static social organization. The approach is not to see the group as a given, but to endeavor to see it from the perspective of the individual behaviors of its constituents where such groups are also characterized by being comprised of people who are able to see each other's faces, with occasion to face-to-face interactions (Naitō 2004:

588–91). In the case of human groups, however, a group identified, for example, by the term "ethnic group" is both a gathering of mutually visible faces living sympatrically and a category of group, for example, "cultural category" in Chapter Nine (Soga), or "reference phase" in Chapter Ten (Sugiyama), and is also a term that can refer to humans who do not actually meet each other as a representational group. This represents an ability (evolutionary historical foundation) that human beings have acquired, but for many groups in socio-cultural anthropology, including in kinship and ethnicity theory, this is where the debate starts.

There are a variety of terms appearing in this book that are strongly associated with groups, but each differ subtly from "group" as such, including: troop, cultural category, ethnic group (category), assemblage, collectivity, grouping (cluster), peer and individuals of the same species. These terms are strictly chosen descriptions selected by each respective author after being oppressed by and grappling with the phenomenon of groups. In order to link each of the terms in the framework of this book while allowing individual definitions, I have taken particular care on at least the following two points. Where a different term is used to discuss the same content, I have sought consistency with a footnote to the effect that "(the term) corresponds to XX used by the author of Chapter XX". Elsewhere I have assiduously sought to avoid depicting different content using the same term.

The reason that particular attention has been applied to how terms are used is because it is very normal—not only in this book—to express things using completely different terms while addressing the same subject. For example, chimpanzees form troops of multiple males and females in a complex group structure, where the normal state is the fission-fusion that, as previously mentioned, perplexed Goodall, but normally they act in small groups of several to a dozen or so individuals, with frequently changing membership. These small groups are sometimes called "parties" and sometimes "sub-groups". Conversely, the troop with stable membership identified by Toshisada Nishida, in other words, an umbrella group that undergoes fission-fusion, he referred to as "Unit Group", but Goodall and other Western researchers call it a "community". This is the true form of what is otherwise termed the chimpanzee Basic Social Unit (BSU).

To give an overarching perspective of the entire book while preserving the independence of each chapter, I locate the groups to which we mainly refer in a very rough hierarchy (Table 0.1). Basically, this means listing groups from small to large, and does not necessarily correspond to ranking according to evolutionary lineage. In other words:

1. Solitary: an individual living alone.[8]
2. Party, sub-group: a temporary gathering of individuals. This corresponds to gatherings of individuals in a troop of primates that display fission-fusion, or to pairs formed only during breeding season among solitary prosimians. In human beings, for example, it may be a gathering of women hunter-gatherers who go gathering.
3. Troop: a gathering of individuals with stable membership. It may include gatherings like those of Japanese macaques that are highly cohesive, or like those of chimpanzees that have low cohesion and display fission-fusion.
4. Band: a gathering of troops, which in non-human primates is observable in gelada baboons, hamadryas baboons and the golden snub-nosed monkey. In human beings it is a word that refers to gatherings of families, and particularly when there is marked fission-fusion, may be called a "camp" and is mainly seen in hunter-gatherer societies in the lower latitudes. In any instance, it is a temporary, impermanent gathering of multiple troops (in non-human primates) or families (in humans) that is linked by relationships that are non-hostile.
5. Herd: among monkeys, a temporary gathering formed by multiple bands in a location like a resting place. Hunter-gatherer bands and camps may also form such gatherings, but they are no more than temporary sympatry and the gathering does not create higher-order social integration. The term herd is also not used in this situation.
6. Local community:[9] used when looking at individuals living in a particular region as some form of union. For primates that form herds, this applies when neighboring herds are viewed as a grouping (cluster).
7. Ethnic group (category), cultural category: a group unit unique to human beings and a gathering that, dependent on representation, may be denoted (for example, see Uchibori 1989).
8. Gathering of persons other than (7): again, a group unit unique to human beings and a gathering based on an individual or group relationship that supersedes that found in (7).

In short, it is possible to depict eight broad stages of semantic hierarchical structures.

Basically, because (1) to (5) are groups with members who act in close sympatric proximity, the members know each other by sight and the groups are "visible". From (6) on, however, the groups include those in which parties may not be known to each other by sight, and where individuals do not exist sympatrically, and the

Table 0.1 Semantic hierarchical structure of groups

	Form of group	Sympatry
(1)	Solitary	-
(2)	Party, sub-group	↑
(3)	Troop	Visible
(4)	Band	
(5)	Herd	↓
(6)	Local community	↑
(7)	Ethnic group (category), cultural category	Invisible
(8)	Gathering of persons other than (7)	↓

groups are thus "invisible" and their actual borders are often vague. In addition, in species that exhibit fission-fusion, like the chimpanzee, for example, the troop (social unit) is "invisible" in that it is very rare for there to be a situation in which all members gather sympatrically. However, if it is likely that as a result of repeated fission-fusion all individuals in the social unit will be diachronically encountered, then it may be more correct to say that this situation is "semi-visible".

The hierarchical structure presented here is one example drawn up in an attempt at a broad sketch of the main groups discussed in this book. In relation to the broader groups of individuals in general, there are probably other methods of classification than these, and in particular for human beings it should be the case that complex group structures involving a wide variety of interim groups (Majima et al. 2006) have been recognized. One point I wish to confirm here, however, is that irrespective of what is normal in human beings, all classes in the hierarchy from (1) through to (8) are inherently possible.

Social evolution and evolutionary historical foundations

In North American cultural anthropology there is a major mainstream theory known as neo-evolutionism. It was proposed as an alternative to nineteenth century social evolution theory by Julian Steward (1958), Leslie White (1959), Elman Service and Marshall Sahlins et al. (1960), who used empirical methods to locate modern industrial society at the peak of complexity in the staged theory of historical development.[10] However, social evolution in the context of the evolutionary historical foundations for human society, which is what we are seeking to elucidate here, does not take into account historical development stages in society, or what cultural anthropology terms social evolution. To begin with, "society" is an extremely abstract concept, with neither substance nor reality, and therefore of itself does not in a real sense evolve. What does evolve are the individuals that voluntarily constitute a society, or more correctly, irreversible changes in physical characteristics or changes in the behavior and actions of each individual. Those changes bring about shifts in relationships between individuals in the society, in other words, in the nature of sociality, and ultimately the structure of the society undergoes change. The subtitle of this book, *The Evolution of Human Sociality*, uses society in that sense. As the term "social evolution" is in this way also ambiguous, insofar as there is no particular need, we do not use it in this book in order to avoid confusion.

What we use instead is the term that has already appeared several times in this chapter, "evolutionary historical foundations". One reason for particularly choosing this phrase was to avoid confusion with the aforementioned "social evolution", but in addition, as has already been discussed, nearly every academic field had become reluctant to attempt to link non-human primates and human beings through evolution. In opposition to that situation, we have settled on the groups that underpin each subject society, and have chosen to focus on their basic nature in order to restore the evolutionary perspective. The nature of groups is something acquired in the process of evolution and may be shared or not, as the case may be, by human beings and non-human primates. In this book the nature thought to have been acquired through the evolutionary process is referred to as evolutionary historical foundations.

Even if we are unable to specifically traverse the evolutionary process and its mechanisms, in the societies at which they have arrived, be they hunter-gatherer, pastoralist, agricultural or the industrial society of modern times, there should be common foundations that have been indelibly handed down from an ancestor common to primates. This is because it has been possible to extract a range of commonalities from the fundamental principles on which both societies stand, in other words, from the sociality of the beings who live therein. As stated at the outset, human beings are creatures who have learned an extremely sophisticated form of sociality. Therein lies the significance of revisiting human and non-human primate groups from a common perspective through the similarities and differences in their sociality, in other words, those aspects that form the evolutionary historical foundations of each.

The presence of Hitoshi Imamura

This may be something that in truth should be written in the Epilogue, but I will here touch upon the fact that professor Hitoshi Imamura participated in our GRP.

While he has not provided a chapter for this book, Imamura had been a member of the GRP since the beginning, and the meeting at the end of the second year was his last before departing this world. I refer the reader to the obituary at the back of this book written in the form of an Epilogue by Ryōko Nishii for an outline of his personality and academic background. However, here I would like to address the fact and the significance of his having been sympatrically present in the GRP.

It goes without saying that Imamura was neither socio-cultural anthropologist, ecological anthropologist nor primatologist, but in fact was a leading social philosopher. It will no doubt come as a surprise that Imamura attended gatherings of types like us (field-based academic disciplines such as primatology and anthropology). For over twenty years, however, he had held a strong interest in socio-cultural anthropology, and had been actively engaged in anthropology, participating in anthropological research groups and visiting anthropologists in the field in Thailand, Myanmar and other South East Asian countries as well as China (for example, H. Imamura 1989a, 1993, 1994a). At the launch of the GRP, Imamura assented to my hesitant and nervously delivered invitation to participate with two words, "Yes please". According to Nishii's obituary, we learn that despite his polite ascension, he attended the GRP with considerable foreboding. Primatologists were also invited to join with anthropologists and elucidate the ultimate question, "How is that which is social created?" In the opening of his last work, *Shakaisei no tetsugaku* (The philosophy of sociality), Imamura asserts, "Social philosophy studies the sociality of humans" (H. Imamura 2007: 5). Imamura pursued the sociality of human beings in the field of social philosophy from a theoretical and methodological base fundamentally distinct from that of anthropologists and primatologists. I do not believe that the encounter between the GRP and Imamura was coincidental. I also wish to believe that there was no greater good fortune than that this encounter occurred.

In as much as Ohmura, Tokoro, Kawai, Terashima, Sugiyama, Kuroda and Tanaka directly quote from Imamura's studies in this book, there is no shortage of members who have been influenced by his social theory. That includes content that conceivably could not have been obtained simply by reading his books. In other words, there was the physical sharing of time and space, debating with the "raw" Imamura as a natural progression in the GRP forums and thereafter in prolonged sojourns at drinking establishments, which were also forums for sparring in stimulating debate hardly distinguishable from academic arenas. In the GRP Imamura would listen to the presentations, at times in apparent comfort, and at times with a tense expression suggestive of him bursting into flame if touched. It was not particularly the case that he often had something to say, but occasionally he would resolutely state his views, or cheerfully make a comment from his immense background of knowledge. At times he would even rage at us en masse. We learnt from him about society and that which is social, and we learnt much about thinking about those things, but overwhelmingly, we were reminded that academic forums are arenas of combat.

It is difficult to find the right words to describe Imamura and his pursuit in the GRP of the question, "How is that which is social created?" In one sense it was naïve, and his intensity and comments reverberated beyond differences between academic fields, or tore down the walls between those fields. Each time he prepared for the GRP as a whole a common battleground was at its very roots. In that sense alone, for the aims of the GRP, his all too sudden death was a cause of intense sorrow.

This book is dedicated to Professor Hitoshi Imamura.

Part I
The Evolution of Sociality

1 The Sociology of Anti-Structure: Toward a Climax of Groups

Kaoru Adachi

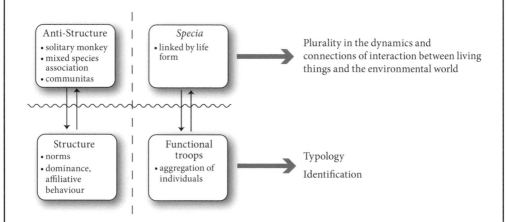

There are parallels between Junichirō Itani's concept of anti-structure and Kinji Imanishi's concept of *specia* (species society). Both step beyond the framework of conventional bio-sociology as defined by gatherings of individuals and their interactions to constitute a unique view of groups that conceives of sociality in the interaction between living things and the environment. The subjects of their consideration are phenomena that belong to the periphery of the group phenomenon, such as solitary monkeys and mixed species association, and they conceptualize a possible new bio-sociology to explain the evolution of groups.

Introduction

When developing his evolutionary theory of primate society, Junichirō Itani, one of the pioneers of Japanese primatology, employed the techniques of social structure typology. In his writings from 1970 to 1980, he developed many theories focusing on structure. From the 1990s on, Itani began to contemplate anti-structure in a binary relationship with structure. The position he afforded anti-structure has an important meaning in terms of the formation of societies and groups in the human evolutionary process.

Primate sociology applies to primates, which form troops that (in comparison to other animals) have particularly strong and stable forms of assemblage, and is a field that has developed to explore the evolutionary historical foundations of human group society, located at the peak of the evolutionary hierarchy. Its primary themes are that the prototype for human society may be found in troops of non-human primates, and similarities and differences with human society can be separated out to hone in on the true nature of the gulf between the two. The grounds for using primates for comparative purposes are close relatedness, and the fact that monkeys are treated as being earlier, in terms of evolutionary history, than human beings. Such perspectives drive living things other than primates beyond consideration as creatures that are, in unspoken terms, evolutionarily "lesser than". In this chapter we use the concept of anti-structure to show that consideration of the sociality of groups may potentially lead to a new bio-sociology located at a certain distance from this type of anthropocentric, or primate-centric, evolutionary perspective.

The specific example of anti-structure that will be centrally referred to in this chapter is the mixed species associations formed by Cercopithecine monkeys in Western Africa (see Photo 1.1). Mixed species associations are troops formed by different species, and diverge from what we envisage when we think of troops— that being groups of individuals of the same species maintaining spatially close contact while interacting in a variety of ways. Precisely because these forms of associations are both unusual and marginal, they are entities that may hold the key to unlocking the true nature of groups.

Photo 1.1 Cercopithecine monkeys in Western Africa (1)

Mixed species associations of Cercopithecine monkeys form frequently in the Taï National Park on the Ivory Coast. The Diana monkey in the photograph is the core species around which the groups are constituted.

The theory of the evolution of social structure and Basic Social Units

Itani's main study of the evolutionary theory of descent is *Reichōrui no shakai kōzō* (The social structure of primates), published as volume twenty in a series on ecology by Kyōritsu Shuppan in 1972. Thereafter, the content of the book became Chapter Three in *Reichōrui shakai no shinka* (The evolution of primate society), published as part of the Heibonsha nature series in 1987 (Itani 1987a). Chapter Eight of the book, translated as "The evolution of primate social structure", was published in the magazine *Man* (1985) as Itani's memorial lecture on the occasion of his receipt of the 1984 Huxley Award, and was again printed, this time in Japanese, in *Sōzō no sekai*, in 1986. In the epilogue to *Reichōrui shakai no shinka*, the author himself says that his theory on the evolution of social structure in primates arising from his comparative sociological approach was largely complete in 1972, and thereafter, he made several amendments and additions based on the results of field surveys from the 1970s to the first half of the 1980s, to arrive largely at completion by 1984.

Itani's evolutionary theory of social structure classifies Basic Social Units (BSUs) typologically, and while it demonstrated that if species are different they will have their own individual BSU structure, at the same time it aimed to elucidate the evolution of primate social structure and the corollary origins of human society by arraying BSU structural typology according to the progress of evolution. BSUs are defined by the sex composition of a group and by patterns of transfer. To provide a bird's eye view of the overall social structure of the order primates, and to attempt to understand it in its entirety, six basic structures were classified typologically and their evolutionary processes conceptualized.

A BSU is defined as comprising both sexes, with a structure that maintains a semi-closed system. In other words, a single *specia* (species society, as discussed below) has a unique type of BSU, and repeated movements into and out of a BSU occur within the *specia,* with individuals transferring among BSUs, solitaries and unstable mono-sex[1] assemblages. This constitutes the emigration and immigration of males and females associated with a BSU, and classifications in patterns that satisfy the definition of a semi-closed BSU system are as depicted in Figure 1.1.

The simplest form is the monogamous type, from which both males and females may immigrate, and it is a structure in which historical succession in the BSU is not guaranteed. Species that adopt this structure include several prosimian groups and New World monkeys. Other BSUs for which succession is also not guaranteed are the polyandrous and polygamous types. The polyandrous type represents a structure into which females do not immigrate, and can be observed in the *specia* of the marmoset family. Polygamy is a structure into which males do not immigrate, and is only observed among gorillas.

There are three types of BSUs in which succession is guaranteed: bilateral, matrilineal and patrilineal. A bilateral BSU is one in which there is emigration and immigration by both males and females. Howler monkeys and the western red colobus in some regions are said to adopt this structure. In these species long-term observations have confirmed bilateral group succession.

In matrilineal BSUs only the males emigrate and immigrate, with the females remaining in their natal group. By this means succession in the BSU is guaranteed through matrilineal kin relations. The structure is seen in both multiple male and single male forms, and within the scope of observations undertaken to date, it is

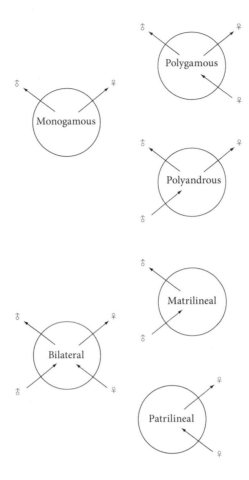

Figure 1.1 Itani's formulation of six types of BSU (with some alterations)
Source: Itani (1987c), revised.

the most commonly observed BSU in primates. In the patrilineal BSU, only females
emigrate and immigrate, and succession in the BSU is guaranteed through the
kin relations of the males, who stay in their natal group. Chimpanzees and pigmy
chimpanzees (bonobo) conform to this structure.

The hypothesized evolutionary sequence of these six basic structures is shown
in Figure 1.2. Solitary individuals that do not belong to BSUs first evolve to a
monogamous social structure, and all other structures are considered to originate
from that form. The monogamous form is one in which a particular male and

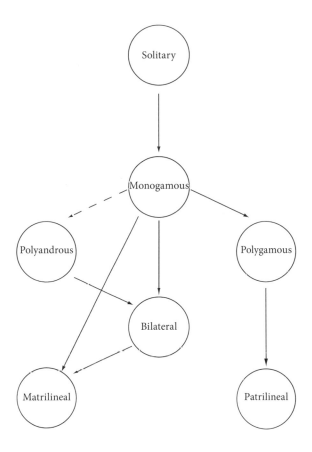

Figure 1.2 Itani's formulation of BSU evolutionary transition (with some alterations)
Source: Itani (1987c), revised.

female have accepted co-existence with one of different sex. The breakdown or re-configuration of the monogamous form gives rise to the polyandrous, bilateral and polygamous structures, and the flow to the left of the polyandrous and bilateral types ultimately arrives at the matrilineal structure. The polygamous gorilla structure is a structural profile unique to apes such as the chimpanzee and pigmy chimpanzee. The evolution of social structures is arrayed directionally from societies without succession to those in which succession is guaranteed. The similarity between bands of hunter-gatherers and the patrilineal chimpanzee BSU points to the likelihood that hominid societies would be classified into the ape profile of evolution to the right.

Once largely completed, the theory of the evolution of social structure had a significant influence on primate sociology thereafter. The results of long-term observations of many primates were collated, and by adding to it knowledge about hominid society from ecological anthropology, Itani's theory was reinforced and modified. The ultimate goal of the theory of the evolution of social structure is to elucidate the origins of human social structure. Refinement of the theory has progressed toward this ultimate goal, and the greatest question has become where to position primordial hominid society in the picture of evolutionary history. For example, Nakagawa in Chapter Four of this volume touches upon the evolution of early human social structures by adding the latest examples from observations in the wild combined with knowledge from social ecology and molecular genetics to Itani's theory of the evolution of social structure.

Structure and Anti-structure

Structure: The theory of co-existence

Itani's theory of the evolution of social structure in primates has as its starting point the fact that the BSU of a species entails a unique structure. The basic social structure is characterized by emigration from and immigration into a BSU by males and females, which when considered from the biological perspective can be construed as avoidance of inbreeding (Itani 1987c). Avoidance of inbreeding (in other words, of mating between close relatives) is achieved by group transfer, that is when either a male or a female, or both, from among individuals who have reached the age of sexual maturity leaves their natal group.

Focusing on the emigration of females divides primate social structure into two broad types. Thinking in terms of a hypothesis (Figure 1.2) derived by inferring the evolutionary sequence for each type provides two major flows, with lineage to the left (leading to a matrilineal society) and lineage to the right (ape lineage, which results in non-matrilineal societies), and it is apparent that both are classified according to whether or not females emigrate from the BSU. The basic structure is determined by whether a female leaves her natal group or not, in other words, the possibility or otherwise of a female co-existing with same-sex individuals.

When the issue of co-existence with same-sex individuals is examined from sociological and behavioral perspectives, there are two types of social behavior that govern the possibility or otherwise of co-existence and that give structure to a society. One is the range of social negotiations in a troop that are based on the principle of inequality, such as superiority and rank order, while the other is affiliative relationships based on the equality principle (Itani 1987c). If viewed from the level of behavioral psychology, we enter the fields of social constraint and social encouragement, with "not allowing" and "allowing", or what Itani calls "social structure creating behavior" (1987d).

Both agonistic behavior (such as strict rank order and superior-inferior relationships) and grooming and other affiliative behaviors work to support co-existence, which in turn forms and sustains social structure. Itani referred to these situations as the "norms" of a troop society. Structure is achieved when individuals associated with a group do not deviate from the norm and behave in accordance with strict group rules.

Anti-structure—what is it?

At the same time Itani began to try to express another aspect of the nature of the evolutionary process from primates through to hominids—unavoidable when considering society—using the term "anti-structure" as distinct from "structure". He published a coherent study of anti-structure in 1991(a) as "Shakai no kōzō to hikōzō—saru, tori, hito" (Structure and anti-structure in society: Monkeys, birds, humans).

According to Itani, if it is primates that form a group, then no matter the society, a single *specia* will be formed with structure and anti-structure. A society of individuals who live a solitary life independently acting without belonging to any groups is an anti-structured society. In contrast to societies with BSUs classified typologically in social structure evolution theory, in other words, societies with groups where BSUs form the structure, independently acting individuals existing at the periphery of the BSUs are called anti-structure. The constituents of anti-structure are individuals of genders that will leave their natal group upon reaching the age of sexual maturity, and these solitary wandering individuals comprise the non-structural portion of a *specia*. Wandering individuals in the process of transfer will eventually again enter a BSU and become a member of a group, returning to structure. No matter how complex the configuration, or large the group, an overall *specia* first becomes feasible when independently acting individuals and groups coexist. When Itani refers to anti-structure, he always emphasizes that it is an important aspect in the composition of a *specia*.

The first entities identified as an anti-structured society were young male Japanese macaques, called *Koen* (solitary monkeys) by Itani. The Japanese macaque BSU is matrilineal, and females remain in their natal group, never transferring to other troops. In contrast, males reaching the age of sexual maturity leave their natal troop and live a wandering life as solitary monkeys, not affiliated with any troop. Their destiny is to ultimately enter a troop other than their natal group, where they will leave children. When a Japanese macaque troop encounters another, both emit prolific warning calls. Encounters with solitary monkeys, however, are controlled in silence. If voice is considered linked to the formation and sustenance of social structure, then an encounter with an anti-structured solitary monkey is conceivably a forum in which some other principle is at work. In a group such as a BSU, the social behavior of an individual is conducted in accordance with group rules unique to the species, and that emerges as the shape of structure. By contrast, the society of the solitary monkey is independent of group rules and freely changes shape in occasional interactions with groups. Itani no doubt conceived the concept of anti-structure by observing such aspects.

Examples of anti-structured society are not limited to independently acting individuals. Even if troops and a group are formed, there are societies that can be called anti-structured. The example that Itani often cites of a group in which there is frequent manifestation of aspects of anti-structure is the mixed species association. A mixed species association is a gathering of troops of different species that act together when ranging and feeding. In birds, during the non-breeding season from autumn to winter it is possible in Japan to observe flocks which are gatherings of more than one species formed from large groups of tits (Parus) and chats (Muscicapinae). On the grass plains of the Serengeti in Africa, multiple species of ungulates, such as gnu, zebra and gazelle, are known to collect and migrate together. All are anti-structured mixed species association societies. Within such groups are those in which there is no apparent structure, other than that they are gathered.

Mixed species association in Old World monkeys (Cercopithecine monkeys)

The phenomenon of mixed species association has long been of interest in primatology. In Africa and South American regions troops of multiple species have been reported as coming together to form mixed species associations, and in relation to this phenomenon, Itani turned his attention to a theory of co-existence that supersedes *specia*. In the case of primates, multiple species that would otherwise form a troop that is a stable single species group do so across species in a way that is maintained over the long term. Among Africa's Cercopithecine monkeys, from a few to up to seven different species may form stable mixed species associations that continue to act together for many days.

Photo 1.2 Cercopithecine monkeys in Western Africa (2)

In a mixed species association of Cercopithecine monkeys, different species of monkey may be observed next to each other in the same tree. To the upper left at A is a diana monkey. To the lower right at B is a red colobus.

In the Taï National Park located in the southwestern region of Ivory Coast in West Africa, seven species, comprising three species of guenon monkey, three of colobus and one species of mangabey, regularly form mixed species associations. A troop of seven different species will maintain separate single species unions while at times amalgamating with troops of different species, feeding and resting together as they range (Adachi 2003). For example, a mother and infant diana monkey may sit on a single branch of the same tree alongside a red colobus male, while on the branch immediately below two olive colobus monkeys are grooming. At the top of a neighboring tree two other diana monkey mother and child pairs may be resting while the diana monkey children are playing nearby with the child of a lesser spot-nosed guenon. Such scenes are commonly observed (see Photo 1.2). The proportion of observation time spent targeting a single species troop in which the troop forms a mixed species association is known as the ratio of mixed species association. The ratio of mixed species association among Cercopithecine monkeys in the Taï Park is very high, and it is rare for a single species to form a single species troop. Troops of seven species of monkey in mixed species associations are nevertheless not always comprised of the same species. Multiple mixed species troops repeatedly coalesce and separate as they move.

For example, if a day is spent observing a troop containing diana monkeys, it will become apparent that as time goes by they will repeatedly form and dissolve mixed species associations with a variety of troops of different species. Mixed species associations formed by diana monkeys constantly precipitate change in the species configuration of a group. For example, as night ends and they depart their sleeping place to begin their activities, the diana monkeys may already have formed a mixed species association with troops of three different species such as red colobus, lesser spot-nosed guenon and olive colobus. As they move, their path may separate from that of the red colobus troop, leaving the diana monkeys in a troop with two other species. In the course of repeatedly feeding and ranging they may encounter black-and-white colobus in a different location, form a mixed species association and thereafter for several hours act in a mixed species association of four species. Again, after ranging, they may at another location separate from the black-and-white colobus, and at that point the troop of lesser white-nosed guenon that had been with them since morning may continue in a mixed species association with the black-and-white colobus, heading off with them and causing the troop to which the diana monkeys are attached to change to a small mixed species association with one other species. Towards the end of the day they may re-encounter the red colobus troop from which they separated in the morning, form a three species mixed species association and head for somewhere to spend the night.

Anti-structure in mixed species association

In mixed species associations in which composition constantly changes in this way, the outlines of a troop are ill defined. In this context it is very difficult to identify something that could be called structure. In a single species troop the constituent membership is stable in the long term, and even though there may be a minority coming and going, it is usually always the same combination of members who are gathered. In contrast, in a matter of hours the constituent members of a mixed species association will change. There are a variety of combinations and continual variation as groups of different species that use the same home range exchange members. For example, when the diana monkeys referred to earlier form a mixed species association of only two species with olive colobus, then in its exchange of calls and movements the troop will at least appear quiet. In contrast, when in a mixed species association of seven species with other guenon monkeys or the colobus or mangabey, they form a rowdy troop in which the repeated movements of the diana monkey are grandiose as they ascend and descend trees making raucous sounds. Repeating these changes over the course of a day, but continuing to constitute a mixed species association overall, is the life of the monkeys of the Taï forest. A group for these monkeys is very different than that for a single species troop.

Something that should also be noted here is that in a mixed species association there is little apparent inter-species social interaction of note. Generally, unity in the society of a single species troop is achieved through affiliative interactions such as grooming and aggressive interactions such as fights based on rank order. There is regularity in the social negotiations—what Itani called the "norms", or what could also be called the rules of the troop—which imbue the troop with a certain order. In a mixed species association, the social negotiations that should impart that regularity are largely not observed. Monkeys that form mixed species associations hardly engage in special social negotiations such as grooming or fighting, they rather coalesce and separate repeatedly in space, and exist in a weak, borderless union that is continuously changing. The mixed species association is neither a close group formed through regular affiliative interactions nor a despotic one controlled by the strict rank order established by repeated aggressive interactions.

In same species troops direct feeding contests are said to be a factor in intra-troop competition.[2] Fights will break out between troop members over limited food resources and the strong will gain the most. As different species of monkeys that form mixed species associations have similar food preferences, it should not be unusual in the above situation for direct feeding contests to erupt, just as they do in a single species troop. Actual research into feeding behavior has revealed that diets largely overlap in terms of fruit preferences. Even when eating the same fruit, however, the location and timing of eating are minutely altered so that the troubles seen in a single species troop are largely not observed between different species. Detailed observation reveals the extent to which such monkeys minutely adjust their behavior in order to avoid fights over food. In large trees bearing abundant fruit, individuals of different species will enter in an unruly fashion and all feed at the same time. In smaller trees, or where trees have only a few fruit, while an individual that arrived first at the location is feeding individuals from other species will stay in surrounding trees and not feed, as if waiting their turn in a reciprocal fashion. When the first customer finishes feeding and moves to the next tree, the monkeys of the other species that had until then been waiting will move in to feed.

Mixed species associations observed in the Taï Park involve troops of different species repeatedly combining and separating, with a constantly changing species composition. United troops with fixed memberships, or groups with rigid organization such as ranking or other systems, seem like highly developed, complex societies. Troops of Japanese macaque, which appear to have strict systems of ranking in which all behavior is regulated, correspond to this image, and the same may be said of human groups identified by ethnic or cultural background. However, a feature of mixed species associations in which membership and group outlines are poorly defined is that they cause us to realize that (it is movement wherein) sociality is ceaselessly created. In an organized group society may be understood to be a bundle of relationships, but in a loose group like a mixed species association, what gives us a sense of society is not a bundle of fixed relationships; there is nothing more than the motion of a series of communications, like the coexistence of monkeys at the feeding location described above. If we focus on the perspective of a series of communications, we realize that a group does not stop with a gathering of visible individuals, but also incorporates leaving somewhere, or separation, and not being somewhere, or absence.

The phenomenon of mixed species association, in which, through repeated coalition and separation, loose bonds are formed in peaceful coexistence, was described by Itani as "indeterminate interest in others that are similar, but not the same" (1995). There is neither adherence to hard rules, such as fighting and rank order, nor powerful centripetal mechanisms, such as grooming or sexual activity, as in a same species troop. Itani had discovered other properties of groups not found in same species troops in the attitude in mixed species associations wherein loose gatherings are continuously maintained, while on the face of it giving the appearance of indeterminate lack of interest. The concept of anti-structure is used to express the looseness of mixed species associations, and of indeterminate gatherings.

Communitas

Itani sees anti-structure as the same as that which Victor Turner proposed as "communitas" (1996). According to Turner, in contrast to structure, which is the state in which a member of a society "is expected to behave in accordance with certain customary norms and ethical standards binding on incumbents of social position in a system of such positions", communitas "which emerges recognizably in the liminal period, is of society as unstructured or rudimentarily structured and relatively undifferentiated". Structure is partial, but communitas applies to the totality; structure is based on the principle of inequality, while communitas is based on the principle of equality; structure leads to sagacity, communitas leads to foolishness; structure is secularity and speech, while communitas is sacredness and silence; and so on. Turner depicts structure and communitas using a variety of approaches. Itani sees Turner's theory of communitas as also holding validity in regards to anti-structure.[3]

Itani acknowledges that the vector of social evolution moves from anti-structure to structure. The usual evolutionary view, even when considering human beings, is to deem that change progresses from elastic groups lacking in outline, to ones that are structured and supported by codes and culture. Despite this, Itani concludes that anti-structure always forms an integral part of groups of primates, including humans, or taking it further, of animal groups. Itani's claim that *specia* always contain both structure and anti-structure means that no matter how structurally complex a society, it will contain within it opportunities for structure to become anti-structure, and new structure to be created from anti-structure, which means that this is the nature of groups of living things. Anti-structure may also be said to pre-date structure. However, that is the picture that is visible when one aspect of groups is severed, and it is certainly not the case that it may end there. In thinking about the evolution of groups, including amongst humans, for as long as the direction from anti-structure to structure was seen as more important, no effort was put into identifying anti-structure in animal groups. By focusing on the diversity of anti-structure, a new approach to the evolution of groups becomes possible. Anti-structure is the flip side of structure, and supports the movement of living groups.

The sociology of life form

Specia—societies without groups

Kinji Imanishi, the pioneer of Japan's primatology and Itani's mentor, was the critic of Itani's theory of social structure evolution who labeled it "biological supremacy" (Imanishi 1976).

Imanishi touted the *specia* (species society) concept and developed his own theory of the social structure of living things (1949). *Specia* is founded in the concept of life form. Life form does not only express shape, but is strongly linked to place of living, in other words, to the environment inhabited by a living thing. Life form must treat the living thing together with the place in which it lives. Life form is not deemed complete unless society as a lifestyle entity has been dealt together with a place of living that enables the lifestyle in the context of the society, and the linkage of place of living and the entity that lives is what constitutes life form. The place of living referred to here is not only a location in space, but includes the various environmental elements that support the phenomenon of a living thing living. For example, feeding and breeding are the most important activities of living things, and all elements associated with these comprise the place of living. What is included therein, among other things, are the non-biotic environment and same and different species individuals. The entity that is the linkage of living places and living individuals is called life form, and a *specia* has been defined as a group of individuals that share the same life form. Another way to put it is that the act of living is the medium for living things, and the linkage between living individuals and environment is life form. By definition, life form is a concept that already includes all kinds of activities essential for an individual's life.

Specia refers to a society formed by living individuals that have the same life form. If life form is different, *specia* is different. Imanishi's famous "habitat segregation" is thought to encapsulate the principle that different species occupy different places of living, but as *specia* is supported by the concept of life form and life form is defined by place of living, inevitably the habitat segregation principle exists between *specia* (Niwa 1993).

What is not taken into consideration in relation to gatherings signified by the word *specia*, however, is whether such gatherings are actually spatially concentrated or dispersed. Whether or not something is functionally recognized as a gathering has absolutely no bearing on whether or not it is a *specia*. In advocating the *specia* concept, Imanishi states: "Rather, here we have arrived at something about which we must be cautious as to whether it is appropriate to use the term group, or communal society. In which case, what then should it be called? It is life form society" (1949: 50). A *specia* is a society comprised of individuals that share the same life form, and all individuals of the same species belong to it. Whether life is gregarious or solitary is an issue internal to the society, and in terms of the definition of a *specia,* either is possible. Imanishi criticized ecologists, saying "The life of a species, or the species as life form, is a social unit in biological nature, and because they have been distracted by the group phenomenon, Wheeler and plant ecologists are the same in having overlooked that point" (1949: 52). Thereafter, he proceeded to construct his own concept of the sociology of living things.

Life form societies and anti-structure

Imanishi charged Itani's theory of social structure with the criticism that it had sunk into biological supremacy, based on the fact that Itani argued about transfer between groups from the perspective of avoidance of inbreeding (Imanishi 1976). Imanishi uses the theory of identification to argue the issues of transfer and paternity, and in that respect his position conceivably differs from that of Itani (Imanishi 1960). In response, Itani appears to have declared his direct opposition, stating he was taking the biological position. The concept of anti-structure proposed by Itani some years after the confrontation between the two and after Imanishi's death, may perhaps be seen to have been a statement of position by Itani that from a different aspect is asymptotic to Imanishi's sociology of nature.

Imanishi's attitude of avoiding an approach that treats society only as a visible group resonates with Itani's perspective of locating anti-structure in solitary monkeys. A solitary monkey in the normal sense is not a gathering, and in a general sense cannot be called a society or a structure. If the definition of a group were to be "individuals collecting spatially", then solitary monkeys cannot be included in groups. What, then, is the reason for deeming solitary monkeys to be an essential element of society? Solitary monkeys as anti-structure, though on the fringe and liminal, are an essential presence in the Japanese macaque *specia*. This is for no other reason than that breeding—which is an essentially inevitable activity of living things—and the existence of solitary monkeys are inextricably linked. When a male reaches the age of sexual maturity and moves to leave the troop into which he was born and transfer to another troop, anti-structure, in the form of a solitary monkey, emerges. The solitary monkey as anti-structure ultimately enters another structure, that of the troop, and is an entity that fulfils its breeding life in accordance with the norms of a troop in a structured world. The pattern of breeding behavior of the Japanese macaque is linked to the non-biotic environment the macaque inhabits, and to the biotic environment of individuals of same and different species, and constitutes the life form of the Japanese macaque. Because the solitary Japanese macaque shares that life form, he is also a member of the *specia*. In Chapter Two of this volume, Uchibori points out in relation to the solitary monkey that in as much as it is incorporated as something that is limited to the life history of the male, it is part of a structured process. In Itani's theory of anti-structure, however, the solitary monkey is perhaps closer to the possible state of "singularity" that predicates on sociality referred to by Uchibori as "anti-structured solitude".

Similarly, in a group that like a mixed species association lacks unity, that continues to loosely change and has no borders, and wherein the composition continues to shift, the concept of anti-structure—which goes to the essence of sociality—is in accordance with Imanishi's view of nature. This perspective configures a *specia* only through the act of living things, and through the thoroughgoing exclusion of the attitude of looking only at the concentration of functional gatherings and overt social negotiations within social groups. Among Cercopithecine monkeys for example, different species with closely similar life forms in terms of feeding behavior create mixed species associations. From among "things that are similar, but not the same" but that consume the same things, overlapping and at times delaying their feeding behavior in their interaction, emerge gatherings that change elastically. Feeding, which is a behavior essential to living things, gives form to a group that is a mixed species association as the result of social interplay between different species. Individuals of the same and different species that share similar life forms create indeterminate gatherings that are borderless, repeatedly gathering at times and separating at others, as the result of a series of actions in which there is interaction with the biotic and non-biotic environments. Conceivably, the anti-structure of mixed species associations is a concept that has been applied to these constant fluctuations of a gathering caused by sequential social interactions.

Looking at evolution from anti-structure

As already mentioned, in Itani's theory of anti-structure social evolution is considered to progress from anti-structure to structure. The theory of the evolution of social structure in the 1980s, which was based on the BSU typology and phylogeny, was in every sense conceptualized in accordance with that direction. While acknowledging the flow of evolution from anti-structure to structure, Itani at the same time points to the fact that the attributes of anti-structure are strongly reflected in the sociality of primates, including humans (1991). What he cites as examples of anti-structure in human society are the naturally occurring singing and dancing engaged in by African pygmies at nightfall, and the great assemblage of people and livestock seen at the time of livestock raiding by pastoralists on neighboring tribes in Kenya. These he deems to be embodiments of anti-structure in human society, and a departure from the principles that daily govern the group societies to which these people belong, as they seek to evade structure and form communitas.

Examples that are not confined within the world of structure but go beyond it and impinge on anti-structure can be seen in a range of places in human society. In evolutionary historical terms, the issue of identity, of who we are, first becomes possible among human beings who have acquired culture and representational ability. As typified by ethnicity, the names of categories of human groups are supported by the logic of the social group, which is the structure to which an individual belongs. By abiding by structure composed of various rules and logics, human beings, who live in highly structured societies, have acquired a sense of identification, or "us". At the same time, however, from among constant acts of daily life human beings continue to create other forms of sociality that have absolutely no connection to the structure that forms the mesh of norms that bind "us".

In a region in which the Kenyan pastoralists, the Samburu and the Rendille, cohabit, Naoki Naitō (2004) has demonstrated that a new group category, the Ariaal, has been created through the daily face-to-face interactions involved in shared experiences related to co-habitation, collaboration and the co-hosting of rituals and weddings. Soga in his paper in Chapter Nine introduces the case of two different pastoralists in Ethiopia, the Gabra and the Borana, in which, through the joint work of collecting water from the same well, one can perceive the outlines of a regional union. Further, in Chapter Ten, Sugiyama tells of loose gatherings seen among women of the slash-and-burn farming Bemba people that are precipitated by a series of mutual on-the-scene negotiations, through the process of sharing food among hearth-hold groups, and claims they are the roots of the human group.

Conclusions

Structure is the basis on which same species troops are formed, and exists as the norms of the troop, absolute rules that must be obeyed by an individual belonging to a troop. In the context of primate sociology, a subsidiary field of biology, it could also be rephrased as something determined by genetics. According to the theory of the evolution of social structure proposed by several researchers, starting with Itani, patterns of transfer and sex composition are largely determined by phylogeny and have conceivably been subjected to the control of genetics in the process of evolution. Hence the chart of primate social evolution based on Itani's social structure has been criticized as overly naturalistic. In response to the charge of naturalism, Itani has displayed a degree of support for the criticism.

In contrast, the concept of anti-structure proposed thereafter takes on aspects of anti-naturalism, or anti-genetic determinism as "anti-structure". Given Itani's statement that in the far distance of the structures unique to each *specia* are worlds of anti-structure, and that in a different dimension not discoverable only from within structure, worlds of what could only be called freedom and fun exist, the impression is given that anti-structure is something that is free of genetics and not determined by adaptive evolution (Itani 1995). As was the case with Imanishi's *specia*, Itani's anti-structure comprises a theory that does not fit into the usual biological categories. In addition, when we realize that human groups strongly reflect the attributes of anti-structure, we are forced to the awareness that anti-structure is something else, at a distance from genetics and the theory of adaptive evolution.

I am nevertheless of the view that there is great significance in continuing to consider anti-structure by limiting it to biological foundations. Therein rests the intention in having semi-forcibly linked Itani's anti-structure with Imanishi's *specia* via life form. Sociality that transforms constantly and freely while manifesting the totality of what anti-structure or communitas express is not limited to functional groups that are mere spatial concentrations of individuals, but also includes the phenomena of separation from the group and singularity, dealing with the sociality of groups in its broadest sense.

At its foundation are the life forms of living things, which are an aggregate of interactions between the individuals and their environment through the various aspects of their activities. The environment referred to here is not just that which exists physically around an individual, but the *umwelt* that Jakob von Uexküll and Georg Kriszat (1934) revealed as an element indivisible from the action of an agent, and environment in the sense of affordance[4] (Sasaki 1991). For individuals that share the same life form, the manner of linkage between one action and another, which rest in the constant repetition of actions essential to living, can be none other than anti-structure. We should be able to come closer to the diverse modes of anti-structure through the biological analysis of foundational life form.

Primate sociology has sought the evolutionary historical foundations of human society and pursued the course of evolution from anti-structure to structure. That pursuit has given rise to many outcomes, and continues to reveal the characteristics of groups unique to humans in the context of achieving structure. The boundary that crosses humans and what predates them has been investigated from a diversity of aspects, at times with apes (or primates in general) such as chimpanzees and gorillas having been explored as a guide to that boundary. At the same time, Itani claims that there is a characteristic of groups known as anti-structure, in groups of both humans and what predates them, and notes that on the human side of the boundary its function is particularly important. He goes even further to suggest the possibility that it has power sufficient to cancel out the dichotomy of human beings and what predates them.

In his studies of anti-structure, Itani does not limit himself to human beings and other primates, but extends his interest to birds and ungulates. Anti-structure is particularly apparent in highly evolved ungulate society, and he points to the possibility that they may open up new horizons in social evolution, since they are species that have been more domesticated through the process of interaction with human beings. It is possible to expand the scope of the evolutionary theory of social evolution, which originally applied only to human beings and primates, to taxonomic groups that to date have been thought to be unrelated to group and social evolution, without being misled by the apparent degree of structure in the form of complex social negotiations or sophisticated intellect or the composition of large groups containing many individuals. By incorporating Imanishi's *specia* into Itani's theory of anti-structure, it may well be possible to conceptualize a new bio-sociology. That endeavor will be underpinned by a vivid sense of reality, and may possibly develop concepts that embrace not only non-human primates, but all groups of living things.

2 Assembly of Solitary Beings: Between Solitude and "Invisible" Groups

Motomitsu Uchibori

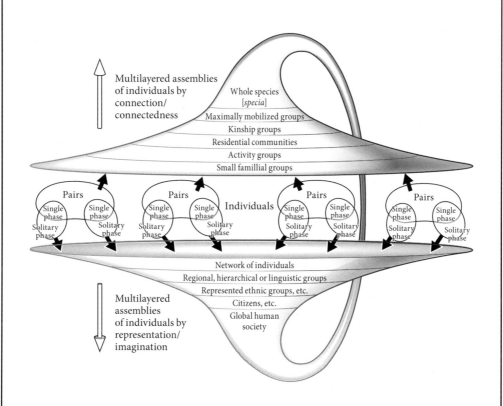

Individuals seem the most concrete entity. However, they are only concrete in the sense of something that exists between two collectivities. One is a collectivity based on the connection of individuals and emerges from the phase of singularity and bigeminality of individuals; the other is a represented collectivity based on what might be termed the imagined solitary phase of individuals. Both have a multilayered structure and are linked at the top tier, remote from the concreteness of individuals.

From solitude

In order to think about the collectivity of humans within an evolutionary framework, it seems necessary to view the very opposite state of it. Humans are able to be solitary; or, in other words, they have a strange inclination towards solitude. This might serve as a major but potential impetus driving humans to herd together, and forms the incipient concept underlying this discussion.

Perhaps the term "single" should be used rather than "solitary", as far as it is referred to at a behavioral level and the psychological phase is omitted from the discussion. However, the term "solitude" shall be used initially in order to indicate that situation where an individual is specifically solitary as against, or even within, a "group", rather than as an existence among multiple solitary beings.

The common image of solitude seems to refer to the concrete and self-conscious existence of human individuals, who came to be referred to as egosyntonic in modern society. Undeniably, there is an affiliative relationship between solitariness and modernity. However, the central point of the current discussion is the proposition that the basis of the development of such affinity is something inherent in human assemblage, and in fact characterizes the existence of human groups in crucial aspects within an evolutionary framework.

Being within a group and being solitary are seemingly opposite states. However, assemblage and solitude are not mutually exclusive concepts, and the latter is at least logically premised on the former. It is possible that individuals sequentially fluctuate between these opposite states. This is observed not only in humans but also in other primates. In the case of Japanese macaques, which are known for the existence of solitary individuals among them, individuals in the solitary state are always male and such an existence seems to be a rather structured process (see Yukimaru Sugiyama 1990). It should be noted that the solitary state of humans discussed in this chapter is, on the contrary, not part of a structured process but their ability or inclination as a *potentia* and hence not an inactively occurring regular state. What actualizes from such a *potentia* is something that can be referred to as "solitude outside structure", in the sense that it is not foreseeable from structure.

Assemblage and sociality

There is some kind of ambiguity in the concept "group". Above all, it is too broadly applicable. In order to narrow the range of its relevance, its position among a set of similar concepts such as "troop", at one extreme, or "society", at the other, has to be clarified. At first, it has to be decided as to which is more appropriate to the pursued discussion: classifying these concepts in layers or giving each a separate position according to the individuality of perspective. This will be followed by the question as to which concepts are optimally applicable when discussing human assemblage within an evolutionary framework. This is tantamount to asking the point of the question itself, because the applicability of a concept depends on the question posed in relation to it. This proposition shall be discussed in the most formal terms.

Imagine the ties of all relationships between individuals of the same species. When the margin of a distributional space shows a certain degree of discontinuity, this can be regarded as a border and the space delineated by that border can be viewed as a "society". The largest of such entities is equal to what Imanishi tried to express in the term "*specia*" (as discussed in Chapter One of this volume). Societies defined in this way entail diversity in the size of population, the closeness of relational ties and temporal persistence, often in a multilayered manner.[1] When seen in this way, the number of layers is finite and perhaps not so large. It should be noted that inclusive layers do not form a unilinear ladder or chain of order. Instead, each layer below a *specia* consists of multiple societies, each positioned in a polythematic manner, and it is thus possible to employ more than one classification according to the individuality of perspective even in the same layer at the same time.[2]

Such multilineal multilayeredness dwindling downward from a *specia* indicates nothing more or less than the existence of divergent and diffusive classifications. Although even societies are thus diversified, individuals existing therein are indivisible entities and therefore there exists a point, close to the base societal layer, where multiple divergences essentially coalesce. This can also be regarded as a nodal or convergent point at which societies become identical to an aggregate of individuals, no matter which classification is used. The term "troop" shall be used to indicate such a point where multiple divergent classifications re-converge. A troop as an aggregate of individuals in this usage is hence a physically and substantially coexisting state that is the largest unit seen from the point of the base societal layer.

Then, what should be actually placed in the base societal layer? Naturally, this would be the smallest plural number (i.e. pair) as far as more than one individual is associated with the term "society". However, a solitary individual in a physical sense can be regarded as the lowermost social form by departing from this view because, in reality, even though a solitary individual can be physically alone at one point in time, he/she cannot be merely solitary during the passage of a certain period of time, and he/she is inevitably in relationships with other individuals as far as the term "relationship" does not only signify direct interactions.

When the above usage of "society" and "troop" is accepted, the collective existing state of multiple individuals, for which the term "group" is applicable, can be found in any society in layers above that of a pair and below that of a *specia*. Depending on how borders are drawn between societies, the whole society in one layer is regarded as a single group or the whole society (society as a whole) is regarded to consist of multiple groups. A group, which is a kind of society, is formed based on a relationship as a certain commonality among its elements. There is no need to question whether such a commonality is real or imagined. Even though individuals as such elements possess a degree of cognitive or discriminatory ability, they do not have to feel a sense of belonging to the group. A group emerges primarily depending on an external perspective. As far as it depends on such, at times groups come to closely resemble categories. A wide variety of academic and intellectual efforts have been dedicated to distinguishing these concepts, but in many cases such efforts have gone unrewarded.

Here, I am talking about groups that can be referred to by the term "ethnic group", for example. The border surrounding such a group is commonly "invisible", particularly for individuals within it. When groups are classified based on the commonality among constituents, they do not really differ from "ethnic category". If there is any meaning in asserting that something surrounded by a physically unrecognizable border is not a mere category but a group, it is because the commonality among individuals is perceived and related as if it concerns some sort of physical linkage among individuals. Such a perception is a product of a very specific multifaceted representative ability—so-called "imagination".[3] Realistically speaking, when it comes to the issue of human society, extended groups created through this ability (i.e., fictitious groups) are of great significance since specific substance in relation to the evolution of human sociality in comparison with other primates is thought to exist in the basis of the development of fictitious groups. This chapter, stopping short of historical investigation, focuses on the logically underlying basis that immediately precedes and prepares for the emergence of fictitious groups rather than their actual concrete development in the historical sense.

The evolution of primate society has been thematically propounded from two perspectives. One is a macro perspective in which basic social units (BSUs) in primates' *specias* are identified and the phylogeny of human society is investigated in interspecies comparison; the other is a relatively micro perspective in which the formation of structural characteristics of groups is investigated by disaggregating interactions between primate individuals into patterns. Apparently, it is possible to elaborate the discussion within a primatological context from both perspectives. Focusing on what immediately preceded the moment of the emergence of specifically human society is to determine the very limit for the necessity of thinking the content of such a context where there is no room for fictions/constructions.

It has been a matter of course to talk about the evolution of human society through reference to hunter-gatherers. This tendency is especially pronounced in the anthropology of Japan, evident in the studies of the Mbutis in forestlands and the San/Bushmen in dry areas of Africa. However, is it really justifiable to privilege societies of the hunter-gatherer stage, present or past, simply because it is presumed close to "nature" or it occupies "most of human history"? As for the whole *dynamis* of human sociality, all developmental stages including modernity should be equally taken into consideration. Although sociality at the hunter-gatherer stage might have served as a basis for "subsequent" change, what is important is to see possible bases in it for the emergence of various "subsequent" forms, not such substantial continuation itself. Otherwise, "subsequent" forms would be referred to as if they baselessly emerged from zero potentiality, or, to put it in other words, sociality at the hunter-gatherer stage would be considered to have been largely severed from subsequent *energeias*. That is the opposite but exact error by which the modern family was placed at the top of the evolutionary hierarchy in the vulgar social evolutionism of the nineteenth century.

On reflection on such past errors, though it might be a hazardous attempt, it is still possible to discuss human sociality in terms of evolution while examining the themes of socio-cultural anthropology and using the terminology of that field. This is because there is no choice but to start with various actual *energeias* to detect possible bases for the emergence of various "subsequent" forms in past stages. In other words, though it is hardly a new concept, modern families and even nation-states can be considered an *energeia* in human society equally along with various forms of kinship or cohabitation groups (e.g., the Mbuti, the Nuer, the Yolngu or the Kachin), rather than a temporal destination point in a social evolutionary sense or in the context of "evolutionary history". From that perspective, the actual succession of the diachronic formation of modern families and nation-states does not have to be disregarded, but yet the history moves out to an explanatory position as secondary in importance.

One, a pair and the human troop

In biological terminology, there is a view that humans are a gregarious animal rather than a social one.[4] In a strict biological sense, sociality is thought to indicate the existence of a strictly divisional/specialized social structure accompanied by physiological and morphological differences between individuals, as seen in social insects such as bees and ants. It is probably inappropriate to be bound by such specialist terminology, yet enough attention should be paid to the fact that human gregariousness inherently lacks a fixed division of labor or unbalanced positioning between individuals based on their characteristics.

Needless to point out, social/sexual division of labor or the differentiation of status hierarchy based on various criteria in human society emerges in a social layer other than that of gregarious assemblage.[5] According to the discussion in the previous section and taking it one step further, a species-specific troop is defined as the largest physically and substantially coexisting state that is *universally* observed somewhere upward from the base layer of a *specia*. The "universality" is meaningful in the sense that, if a society has groups larger than a troop in layers above the latter, they are not physical or substantial, and, as an additional condition, in the sense that a society consists of either more than one such group, each of which contains troops, or of just a singular group of the sort; in other words, of such a total group-society that is identical to an entire *specia*. What forms of troops in this sense are seen in humans?

To come right to the point, there is no species-specific troop that exactly meets the above criteria in human society. Based on common sense or reliable ethnographic references, however, there seems to be two kinds of groups in terms of troop-like coexisting states in humans. One is the relatively long-term cohabitative or residential group, and the other is the short-term activity group that forms for specific purposes. Neither perfectly meets the above universality criterion. Human assemblage does not, therefore, always emerge in the form of a troop with a single social positioning.

This further indicates two facts in relation to the specific characteristics of human assemblage. One is that human society is dichotomously diversified in terms of the presence/absence of specific troop-like groups or units; the other is that the size of units tends to disperse from large to small or to the individual level, which certainly does not constitute a troop. This may sound contradictory; the internal composition of gregarious coexisting states in humans is yet unexpectedly uniform, regardless of the diversity in human society in terms of the presence/

absence of troop-like units. Such inherent contradictions (i.e., diversity vs. uniformity; assemblage as a basis for the dispersion tendency) may provide us with a clue to think concretely about human groups as coexisting states.

The dispersion tendency shall be considered in particular here. A solitary individual, certainly not a part of a troop, is an existing state that only has meaning in relation to assemblage. Being solitary is not a preexisting state. It emerges as dispersion from a certain coexisting state, sometimes as a result of, more commonly in the transitional phases of such dispersion. This is the way of being human as a simple given. The problem is in what circumstances does such dispersion incorporated into the way of being human become visible. Seeing this as a social process involving a solitary being, dispersion is a divisive, disintegrative or eliminative function of a group, or an act of withdrawal, isolation or independence for an agent in the process. These terms, which express concrete phases of dispersion, also articulate various moments and motives in relation to the parting phases of human group realignment that includes alternate fission-fusion. Such moments and motives are often identified in primate society of several species, including humans, and form a subject of independent investigation. It should be noted that the dispersion of human groups is a basic tendency inherent in a group and a collective manifestation of the individual's natural inclination to be of solitary existence or to form a pair. It is in this sense that individuality is said to emerge from assemblage.

The characteristic of the dispersion tendency of human groups is not that either male or female leaves a troop but that both male and female can leave a troop-like unit, if any, separately or together as a pair. In that sense, the characteristic of human sociality is that there is a small or little difference between sexes in terms of their relationship toward a group. Humans and gibbons are known to have less physical difference between sexes among large primates. The reputed strong tie in gibbons' heterosexual pairs may suggest a similar tie in human pairs. In gibbons, its heterosexual pair would not face any larger visible group (i.e. troop). In humans, on the other hand, its heterosexual pair appears to face a group of a certain size as a solitary pair under specific conditions in the parting phases, or as a pair as a component part of the group under ordinary conditions. In that sense, the formation of a human pair is regarded as occasional while that of gibbons is considered naturally structural.

A human pair is not inseparably assimilated into a group even under ordinary conditions where it emerges as a component of the group. Humans can and often do exercise their relative autonomy within a group because they always carry an inclination toward dispersion, even as a pair within a group. In plain terms, a manifestation of such autonomy is the formation of something familial; in other words, a human family is an extension of a human pair and is based on the reproduction of an individual by such a pair. It may sound a most non-cultural-anthropological expression, or sound like denying the diversity of human society from an ethnographic perspective; however, the diversity nurtured in historical times must remain within its limits as far as it is discussed on an evolutionary plane.[6] It should be noted that the extracted emergence of a human pair precedes the reproduction of an individual in relation to something familial and that such extraction is an end in itself and does not initially aim for the reproduction of an individual. It may be possible to discuss the formation of a pair in close association with the reproduction of an individual through a purely evolutionary biological (or socio-biological) inferential procedure. However, loss-and-gain motives behind this are not always clear for both males and females. In fact, the characteristics of a human pair are thought to be inherent in such ambiguity.

This is also associated with the fact that the relationship of a human pair is not always stable and strict. Whether it is possible to compare with that of gibbons or not, the relationship of a human pair is basically casual. There is an anthropological assertion that human pairs naturally form polygamous relationships. What this means is that, in humans, multiple pairs can be simultaneously formed with one male as a hub. This does not deny the stability of a single pair, but carries the seeds of its successive instability.[7] The problem resulting from such instability is the pair-child relationship, or more precisely the male-child relationship. To put it simply, there is a male-female pair and an inevitable consequence of their self-awareness as a pair is the emergence of paternity. This enables, as a remote consequence, the formation of a group based on paternity to varied degrees, although, needless to say, it does not signify organizational or relational patrilineality in itself.[8]

Effects of "assemblage"

In human society there is no troop in a fixed position in layers or that is socio-structurally inevitable. However, in reality, individuals or pairs spend a great deal of time coexisting with other individuals or pairs in the same place at the same time in a certain coordinated manner. This is a given matter of fact, and there is no point in doubting this. Nevertheless, talking about evolution is to question the grounds for such a simple given situation: What does coexistence in a place mean and what are its effects or adaptive value?

Photo 2.1 An Iban woman feeding pigs before dawn with her husband watching

Among the Iban, people usually use teknonymy, that is, people (even spouses) refer to others who are parents by the names of their children (i.e., 'X's mum' or 'X's dad'). This couple already had a married daughter at the time of the photograph, but they often still called each other by their real names. As a bystander, it sounded to me as if they were still lovers.

Source: Taken by the author on a hill in Sarawak, a western part of Borneo in February 1982.

As a matter of course, what has been called a physically and substantially co-existing state actualizes in a certain place at a certain time. Multiple individuals coexisting in the same place at the same time shall be collectively referred to as an "assembly" for the sake of convenience. This term enables us to indicate a group resulting from assembling with a view to the motives for and the processes underpinning such an action. When it comes to the issue of human assemblage, the first question is: what does an assembly face with the effects of assembling? Is that another human assembly or an environment mainly consisting of non-human animate and/or inanimate entities? As far as the ultimate effects of assemblage are concerned, though it is obviously not a question with two choices, these should be distinguished when we focus on the question as to what humans directly face by assembling. The former is directly concerned with the competition between human assemblies, while the latter hones in on the effects of assembling for individuals within an assembly. As for the former, the characteristics of the internal structure of assemblies would determine the degree of their effects to the competition between assemblies, although arguing such effects premised on assembling is not to delve into questioning the grounds for such a situation. As for the latter, the effects of assembling itself would be firstly recognized, and then the effects of positioning the inherent properties of individuals within the internal structure of an assembly would be evaluated. If there is any meaning to questioning the necessity of groups, this is because there are grounds for discussing the effects of assembling itself.

This chapter advocates that there are such grounds. One possible effect is that by assembling solitary beings or pairs this unfailingly maintains their individuality as a consistently realizable *dynamis*. Though there are several perspectives to express such maintenance, a familiar comic pseudo-slogan, "If everyone crosses against the red light together with others, there is nothing to be afraid of", shall be taken up and explored for its logical composition. In short, the elements of the pseudo-slogan are: who everyone is; what they are going to cross; and what they are afraid of. You may go a step further, if you wish. The slogan's queer profundity partly results from the fact that it is (or is thought to be) largely false, and thus the falsity in itself should be discussed at a meta-level.

On the surface, however, what the pseudo-slogan indicates is the way of action taken by multiple individuals in response to a certain hazardous environmental condition. In this case, assembling is a response action as crossing is a premise. Due to the effects of assembling, the degree of hazard involved in crossing may be actually reduced or, in a happier case, the individuals are made to believe that there were no hazards facing them. This way, either objective safety or psychological security or both are ensured. It should be noted that these are the initial effects of assembling when talking about the characteristics of human society in an evolutionary phase. In this regard, human society may not inherently entail high differentiation or organization but rather exist upon the base property of a fluid assembly where similar individuals gather to avoid hazard and then disperse once the danger is diminished. Such fluidity seems to match the image of "everyone" in the pseudo-slogan.

The view that humans assemble because of the existence of hazard is more specifically argued by advocators of the "Man, the Hunted" hypothesis (Hart and Sussman 2005).[9] Hazardous environments include open spaces such as savannahs or the outer edges of forests. Theories as typified by the "hunting" hypothesis seek the grounds for human sociality in their active pursuit of the "possibility of eating" animal meat as "hunting apes" in such spaces as a background, whereas in the "Man the Hunted" hypothesis the most significant moment is an avoidance of the "risk of being eaten". Again, if safety and security (i.e., the sense of "fearlessness" in the above slogan) prompt the most significant moments in terms of individuals' responses to environmental conditions, this leads to the assertion that this response constitutes their *adjustment* to the environment or their passive ingratiation through the action of assembling, rather than the collective *adaptation* of a structural group to the environment through the utilization of its fixed structure. Constant realignments with fission-fusion marked in human troop-like groups are perhaps a direct indication of this.

There is something worth questioning here as a hypothetical situation. This is the question of whether individuals need to be together or not if they do not "cross the road"; in other words, whether or not a simple coexisting state, not being merely an individual or a pair, produces any effects at all other than the avoidance of hazard. It is possible to associate the effects of assembling with the states expressed by such terms of varying connotations as "sense of being together", "state of coexistence", "state of collective excitation" or even *"communitas"*. However, it is not in the scope of this chapter to explore this issue in detail, as it has been discussed extensively elsewhere.[10] I shall limit myself to stating that these effects can be seen as an extension of the abovementioned passive effects of assembling (i.e., safety and security), and that these effects can only be produced in an inherently unstructured assembly and are incompatible with the close-knit internal organization or stable membership of a rigidly-structured group, as also discussed by Kuroda (refer to Chapter Eleven of this book, or Kuroda 1999).

Group and network

As far as the initial stage of evolution is concerned, the above manner of assemblage can be regarded as the characteristic underpinning the collective existence of humans. However, when it comes to all possible states of human sociality, our quest is not yet over. Attention must be paid to various groups as well as approximate social organizations that evolved historically, all of which maintain, at base, the initial states of human sociality in evolutionary terms. From an ultra *longue durée* perspective, there is no doubt that even the potentiality of the historical formation of approximate social organizations was fostered in an evolutionary process. Two matters shall primarily form the focus here from such a perspective. One is the evolutionary positioning of extremely small groups typified by the nuclear family; the other is the characteristics of "invisible" groups that are discussed in association with the representative ability and imagination of humans.

As for extremely small groups, there is no need for further analysis in this chapter. As noted above, they should be regarded as something that emerges as an extension of pair formation on an evolutionary scale. Although the variations emergent in historical space-time include interesting elements, the width of such differences is actually not particularly large. More interestingly, in relation to the multilayeredness in human society, there seems to be a force operating to disrupt the naturalness of such groups based on pair formation from certain directions throughout historical times. This force comes from various directions and can be a real coercive power or an ideological influence; at any rate, it can only operate in relation to invisible groups.

The assembly discussed in the previous section consists in principle of solitary individuals but, in reality, it is made up of solitary individuals as well as pairs of coexisting solitary beings. It seems there is already a multilayered structure between small familial groups and such an assembly, yet it is an assembly of mutually visible individuals and pairs, among which the aforementioned disruptive force does not operate. Why, then, does this force only operate on the interface between such an assembly and invisible groups? To answer this question is to think about what represented, imagined or, if it may be said, merely denoted groups are, particularly in comparison with a direct assembly of individuals and pairs; is it also based, at a profound level, on an assembly or troop-like unit or essentially not?

The answer to the above question is inevitably an eclectic mix of affirmation and negation. By definition, an invisible group almost invariably manifests itself as an aggregate of constituent individuals as well as a superior "core/entity" facing individuals. The core/entity is an existence that gives the appearance that it possesses independence and concreteness in the same sense as constituent individuals, and that can be counteractive against the existence of the latter. It can be an organization (organism) with an internal structure or a more ambiguous community of various kinds. The force, which operates to collapse small natural groups, may emerge as the counterforce of the core/entity. Examples of invisible groups which have emerged throughout approximately 10,000 years of historical space-time include unilineal descent groups as judicial-political organizations (corporate groups), modern business corporations, private territorial states, ethnic groups and nation-states. The attributes we see in the existence of these bodies are entirely dissociated from those of the form of assembly which has been universally seen in the initial stage of evolution and that has not yet completely disappeared from human society. Conversely, representation, which makes such an invisible community hypothetically visible, is often of a small familial group as an extension of a pair and an assembly of such groups. In other words, when an invisible community becomes "visible", representation, which has made it visible, tries to establish that it is rooted in, or based on, what is part of the near initial appearance of sociality in evolution.

On the other hand, the invisibility of an invisible group results from the sheer fact that its external edge is remote from individuals or is out of their operative control. The external edge of such a group is also that of a network that extends with individuals' recognition of others as its nodes and often forms a limit beyond the reach of such recognition. Represented and imagined groups are therefore not a mere vacuous aggregate but are based on the network of mutual relationships or mutual recognition between individuals. It should be noted that such basic bi-directionality of the network is premised on solitary beings; in other words, the nodes of the network are always individuals. Solitary beings are transitory in themselves. However, as far as such transience is unquestioned, this latent vector toward individuality is considered to be a real basis of networks as well as invisible communities. Hence, highly paradoxically, what holds the latent force to collapse invisible groups is the vector toward purified individuality and the solitariness of solitary beings.

From assemblage to institution

The current discussion shall be concluded here. Overall, it focused on formal logic throughout. There remain many obstacles regarding whether this discussion is shadowed by actual proof or not, yet I think I was able to show my own direction.

The following is a summary of the main points outlined in this chapter.

1. Humans become gregarious by assembling; however, they also often leave a troop and form a new one. This results from a circumstantial and occasional behavioral human process, rather than their pre-programmed essential process.

2. Real human sociality exists, like a pendulum balancing between the vector toward individuality/solitariness and that toward the "gregariousness accompanied by excitation", as discussed extensively by Kuroda in Chapter Eleven of this book.

3. Both individuality/solitariness and gregariousness are matters of direction, not actual states manifesting themselves. The inherent "non-structural property" (i.e., that which is structurally malleable) and fluidity of human groups seem to emerge from here, and so does the state of solitary existence as the potential of individuals within such groups

4. On an evolutionary scale, an assembly of solitary beings and pairs has passive effects as safety and security for individuals, but individuals are still its basis. It seems that as a result humans are or can be solitary at times to varying degrees.

5. The abovementioned "non-structural property" of a group directly appears as a lack of structural order and minute differences between individuals. A certain degree of sexual division of labor and the guardian-ward relationship governed by age are regarded as part of the few structured relationships between individuals in a group.

6. An invisible group developing on a historical scale takes the form of a community of representation and imagination. It is also an extended network between individuals with varying degrees of real basis, which is denoted and substantiated as a community by naming and representation. The solitariness of solitary individuals may have the latent force to collapse such a community.

Taking the endpoint of the current discussion as a starting point, our next quest is institutions as an *energeia* of the representative ability and imagination exercised in society and various fictions/constructions as its secondary products. The specific characteristics as the representative ability and imagination, which humans acquired during the latest evolutionary stage, differ, and the latter is premised on the former. In the next phase of our quest, the social development of imagination needs to be discussed as a crucial moment for human evolution.

3 From Whence Comes Human Sociality? Recursive Decision-making Processes in the Group Phenomenon and Classification of Others through Representation

Kōji Kitamura

Transition from monkey stage to human stage in the group phenomenon.

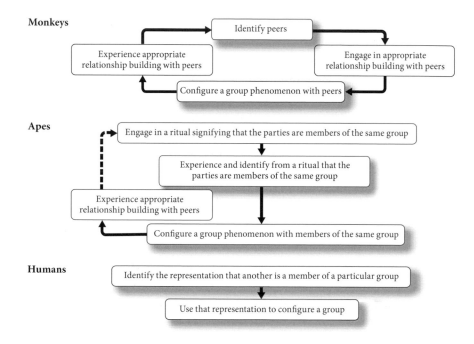

The decisive difference apparent in the transition to the human stage is whether each individual as a member of a particular group has become the "meaning" perceived by the third person; in other words, the emergence of decoupled representations identified through the mediation of language. Chimpanzees, which should correspond to the interim stage in the transition, express the decoupled representation that the parties are members of the same group through particular ritual display, but chimpanzees may be considered interim, because this includes the meaning "same", as perceived by the parties.

Introduction

The aims of this chapter are to demonstrate that it is possible to understand the group phenomenon in human society from a basis common to the societies of non-human primates, while demonstrating the possibility of a new understanding of groups. By at the same time looking at specifically which interim stages might act as a bridge for the transition from monkey to human in the group phenomenon, the intention is to reveal factors that could conceivably give rise to the differences between them in the context of evolution. We will nevertheless take the position that the basic activity an individual of a certain animal species needs to carry out in order to survive is building relationships with its surrounding environment. On a related concern, the core topic of interest in this chapter is how the individual—in the process of opening up routes to forming relationships with another individual of the same species—turns their hand to the very activities that engender something that could be called the group phenomenon. In this context, the phenomenon associated with relationships between groups could be positioned at the higher logical level of the group phenomenon, but that issue will be left to other chapters of this book.

In this chapter, when a difference is identified that suggests major change to a particular characteristic among multiple phylogenically closely related taxonomic groups, the process that brings about that difference is described by the term "evolution". Where there are clusters of phenomena addressed herein that should be linked to evolution, while it is possible to identify a common basis attributable to close relationships, there are clear differences resulting from some factors, and an interim stage is identifiable that suggests that the change that created the difference materialized in stages over a certain period of time. Consequently, we will be seeking to understand, in evolutionary historical terms, the group phenomenon in the phylogenic cluster of primates that includes humans.

Let us next clarify what it is that we seek to consider through the terms "group" and "group phenomenon". In this chapter, because we are aiming to understand the human group phenomenon from a common basis to that of animals other than humans, we will not adopt a definition of groups that applies only to human societies. Conversely, in as much as we aim to thoroughly evaluate changes arising in the process of evolution to human society, we must not exclude phenomena unique to human society. To that end, we use the term "group phenomenon" to indicate the overall subject of consideration. In relation to factors associated with the definition of groups that give rise to the group phenomenon, we will as far as possible use generic concepts capable of addressing the proposition that individuals of the same species that share the same life form[1] and occupy the same ecological niche have a high probability of engaging in the same relationship-building process toward the same objects. In addition, by using the term "group phenomenon" we are endeavoring to allow for inclusion of the broadest possible phenomena that can conceivably be linked to the concept of groups.

The group phenomenon discussed herein specifically signifies the following. First, the most typical scenario is the phenomenon associated with some degree of sustainability as a state at a certain point in time, which translates to multiple individuals being in spatial proximity and stable co-existence. A phenomenon that allows a slightly more dynamic understanding is that in which multiple individuals in spatial proximity are engaging in the same activity together. It is not always the case that it is possible to clearly distinguish between these two phenomena, and it could be that it is not particularly necessary to do so, but where a troop[2] forms a group with stable membership, which can be considered a more developed form of the primate group phenomenon, it is conceivable that the two phenomena exist simultaneously. In other words, when members of a group are always in spatial proximity and seek to co-exist stably by avoiding hostile confrontations, they are said to share a broad framework of activity and to be acting together (this point will be discussed in detail below). The group phenomenon in later human society, therefore, could have the attribute considered typical of a group, stable membership, but at the same time may not always have to satisfy the condition of spatial proximity.

In the case of human society, the group phenomenon is formatively reliant on information and information processing that enables the building of relationships based on classifying others as "members of a group", and "not members of a group", given the existence of something that should be called a group. Conversely, a group can be independent of the group phenomenon and can form just as a group. In other words, a group member could initially be identified by information obtainable from a third party perspective—for example, information that identifies the child of a particular man—irrespective of whether or not group members are in mutual spatial proximity, indicating that membership is to a certain degree stable. A reflection of this is that the condition that a group comes into being in spatial proximity becomes unnecessary, while, in the group phenomenon, where some or all group members seek to create social union out of the development of order in mutual interaction, or seek to create social union out of sharing a pre-set framework for activity, then the group phenomenon is recognizable in each. In both cases, however, the former example of group phenomenon includes the state in which group members are in stable co-existence while in spatial proximity, while the latter includes the phenomenon of engaging in the same activity together while in spatial proximity. Both may therefore be considered to incorporate the two ways monkey troops form, and to have a common basis.

This understanding of groups and the group phenomenon is dealt with in other chapters of this book, which address claims that the human "troop" was originally non-structured (Chapter One by Adachi), and while humans are troop-like, that they are capable at times of becoming solitary to varying degrees (Chapter Two by Uchibori). The group phenomenon emerges here—not only in human societies, but also in the societies of other animals—through the activity of seeking to create social union in the course of building a relationship with the environment at that time, or, in other words, through the activity associated with achieving balance between appropriate relationship building with the environment and appropriate relation-ship building with peers. For that reason it originally emerges through an optional activity engaged in as needed. In the monkey stage, however, it is modified by the fact that relationship building with peers is sufficiently important as to affect the nature of developing relationships with the environment at that time, and is a way of doing things that enables the stable re-emergence of a troop where members of the troop are at all times in stable co-existence in spatial proximity while engaging in the same activity together. The group phenomenon in the monkey stage, in other words, is something prescribed by the structural factors in troop formation, and at all times seeks to create social union. In contrast, in human society it is sufficient to choose social union that addresses issues occasionally needing to be looked at as appropriate to circumstances, taking cues from existing group frameworks.

Let us summarize the key points of the discussion thus far that invite the fol-lowing considerations. The first thing that must be noted relates to what might almost be called an allergic reaction to research that seeks to deal head on with groups and the group phenomenon. Such research is strongly criticized for hy-pothesizing—in the absence of evidence—an "animal instinct" that seeks to create social union, and a "human nature" strongly desirous of sociality. Such criticisms are leveled by both biological research, that adheres to the dogma that individuals that survive are those that have a high likelihood of doing so by pushing others aside, and by social studies of modern civil societies, which laud the freedom of the individual. The thinking in this chapter, however, is that such debate that calls for an either/or decision between whether such instincts or nature exist or not, is unproductive. Because, as has already been discussed, there is no more than a possibility that certain activity will be adopted where necessary to create social union in the course of engagement with the environment.

Second, rather than the activity of relationship building between individuals of the same species, what is important is activity that gives rise to a group phenomenon of that type that is associated with achieving a balance between appropriate relationship building with the environment and with peers. In response to uncertainty that becomes apparent when seeking a balance between these two types of relationship building, there is an endeavor to pursue relationship building with peers, which is easier to address, and to create social union in relationship building with the environment.

The biological basis of the group phenomenon

The group phenomenon we consider here may be identified in the domain in which two types of activities intersect: relationship building with resources in an environment and that with individuals of the same species. In other words, there is a quest to create social union in the course of intentional adjustment to realize a particular result where behavior to build relationships with resources is modified, cued by avoidance of agonistic interactions with individuals of the same species, and there is a striving to create social union in the course of autonomous choice where relationship building with the environment—which anyone can take on— is adopted, prompted by sharing a framework for relationship building with the environment with individuals of the same species. In order for this type of group phenomenon to form, in building relationships with the surrounding environment that each individual of a certain animal species engages in to survive, there must be a situation in which individuals of the same species are at least differentiated as something special, that is, they are differentiated from other things (resources) as objects of relationship building.

At a certain stage of biological evolution, individuals of the same species become something special as objects of relationship building by an individual, because they become a "symmetrical other" at a level of high probability that they will engage in the same relationship building toward the same objects, in other words, they become peers. What is meant by "peer" in this case is an entity that provides clues to the choice of framework for relationship building when the range of options about what type to engage in is so wide as to be confusing, because in particular circumstances there is a high probability they will engage in the same process of building relationships with the same objects. Similarly, they are an

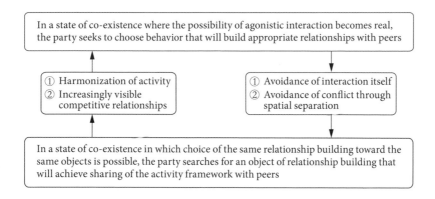

Figure 3.1 Recursive decision-making processes for building relationships with peers

entity that will become the "other party" in interactions that should be avoided in order to survive, because there is a high probability they will engage in the same relationship building toward the same objects and therefore a high likelihood that individual behaviors will inevitably come into contact and hostile clashes will become more visible. As for each party the other is in a symmetrical role, however, in such relationship building with peers, by seeking to make a choice about their own behavior dependent on choice of partner, both parties fall into a double bind in which they are unable to make a choice by themselves (Luhmann 1993), and in that sense a decision becomes impossible to make.

The most basic response to undecidability is to opt to think of the situation confronted as something specified, in which each individual that will become a party responds with a particular approach, where the following two specific types of approach are conceivable.

The first is to deem the situation as a state of co-existence in which it is possible to make co-locational, simultaneous choices with peers to build the same relationship to the same objects, and to adopt the approach—for the moment—of choosing to build relationships with the environment according to one's own motivation, but also to seek to realize a state of co-existence by combining one's motivations with those of surrounding peers in harmonious activity. In this case, however, each must simultaneously take the approach of avoiding direct relationship building with peers at any time (Figure 3.1 ①).

The second is to deem the situation confronted as a state of co-existence in which competitive relations with peers around the same objects are increasingly

visible and the possibility of agonistic interaction becomes a reality, but by avoiding clashes where one more strongly excludes and another withdraws, a state of
co-existence in spatial separation is realized (Figure 3.1 ②).

Of these, the former corresponds, for example, to fish swimming in a school,
while the latter relates to what is seen in the societies of many species, where an
individual establishes a territory in which they pass their life.

Even before such divergence comes into being, it is certainly the case that
physical equality in individuals of the same species will be grounds for the
phenomenon of harmonization of activity and a state of co-existence in spatial
separation by competitive relationship building with peers. In that case, however,
the phenomena are no more than highly localized, brought about by each
individual deciding impulsively to act or not act, and at that stage, the activity of
each that would seek a harmonized state or a competitive one is thought to form
a state of equilibrium in which others are mutually negated. In other words, by
the activity of seeking harmonization on the one hand, the likelihood of agonistic
interaction increases, and by the activity of seeking competition, through the state
of co-existence in spatial separation, the activity of seeking harmonisation would
increase. Thereafter, indivisible from a state of divergence of relationship building
with individuals of the same species from activities to build relationships with
everything else in the environment, relationship building that already eliminates
the possibility of one of the two types of approach to forging relationships
with individuals of the same species comes into being and disrupts the state of
equilibrium, conceivably creating a situation in which only the other becomes
pre-eminent.

In that context we must here confirm that in the process of relationship building
with the surrounding environment engaged in by an individual, the same type
of undecidability can be found as in that with individuals of the same species. In
other words, in the context of building relationships with an object in the environment by relying on meaning identified in an object when seeking to choose
relationship building behavior appropriate at a particular time, the meaning is
in fact what the particular behavior that seeks to realize some particular value
through relationship building at that time should find in the object appropriate to
the relationship building process. Therein is the same recursive decision-making
process as that discovered in relationship building with peers, and that is thought
to be what causes undecidability (see Figure 3.2 below). The undecidability emerges
in a form in which uncovering an appropriate object in the surrounding environment will not necessarily be possible, even if the result that should thereafter be

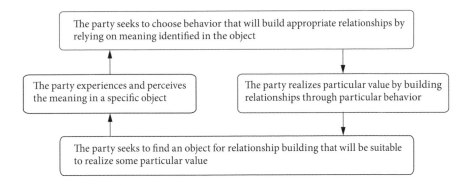

Figure 3.2 Recursive decision-making processes for building relationships with the surrounding environment

realized is clear. In addition, it emerges in a form in which achieving appropriate relationship building with a particular object, which is a pre-condition, will not necessarily be possible.

However, from among resources that form the ecological niche for an animal species, in relation to identifying the meaning of major resources essential to the survival of an individual of the species, and in relation to behaviors that use them, through the working of natural selection to choose and adjust behaviors that can identify the meaning of objects and realize particular values by using them, the process of relationship building is thought to be treated as if it were possible to make a decision at the level at which all individuals in a species can always participate in (Kitamura 2008). Even so, much relationship building with the surrounding environment could be said to be limited to things about which there is undecidability, but the normal response of many animals is simply to dispense with such a process as something unnecessary.

If it is assumed that relationship building with the environment engaged in by an individual in an animal species in order for it to survive is as described above, then identifying the meaning of resources essential to survival and choosing behavior that will realize particular value for survival can be hypothesized as an ability shared by all individuals of the species. If that is the case, in the sense that individuals of the same species become others in a symmetrical role at a level where there is high probability that they will engage in the same relationship building process with the same objects, or they become peers, it is likely that as objects of relationship building they are special entities.

The group phenomenon, descent and living a gregarious lifestyle

With a few minor exceptions, nearly all species in primate society, including humans, adopt a gregarious lifestyle, in the sense that multiple individuals of the same species engage in the same activity together while co-existing stably in close spatial proximity. In mammalian societies in which society is based on a gregarious lifestyle, normalization of relationship building with peers is based on realizing a state of co-existence in the form of harmonized activity through sharing a framework for activity with peers now and then and on seeking to realize a state of avoidance of agonistic interaction with peers. Relationship building with peers is therefore governed by two approaches: social encouragement in the sharing of an activity framework, and social restraint in agonistic interaction (Kummer 1971), and group phenomena with the following characteristics thereby come into being.

In the social encouragement of shared activities, it is not simply that there is an endeavor to share motivation with surrounding peers to build relationships with the environment. It is through each party seeking to engage in the same specific activity as their peers that sharing of an activity framework is encouraged, which is thought to make it likely that harmonization of activity will be more certain. At the same time, in the social restraint of agonistic interactions, it is not simply that there is a seeking to avoid direct interactions with peers, rather, it is feasible that agonistic interactions with peers will become the reality that causes each party to choose and adjust their own behavior to realize a result where that is avoided. This is thought to make it probable that avoidance of agonistic interactions will be more certain.

An extension of these types of approaches is the formation of a group with established membership, which corresponds to the Japanese macaque troop. In order for a troop that is a group with established membership to form, segmental assembly in the spatial distribution of members of a species society first provides a clue, wherein the possibility of using a classification system in which the categories are "peer" and "not peer" is thought to be essential. Then, stable reproduction of the social union known as the troop becomes possible through sharing an activity framework with others who are deemed peers, and simultaneously through modifying their behavior to avoid a state that would lead to hostile clashes from mutual interactions. In other words, the peers of this classification are not simply individuals of the same species, but individuals of the same species that are highly likely to share the motivation to seek to engage in the same activity together, and highly likely to share the goal of achieving a social order that avoids agonistic interaction. Conceivably, therefore, it is more certain that a group with established membership will be reproduced when there is an endeavor at all times to engage in the same activity with such peers, and a simultaneous striving to achieve social order to avoid agonistic interaction.

As already indicated, the harmonization of activity based on the symmetry of individuals of the same species is believed to come into being as a direct outcome of first eliminating the other possibility of relationship building based on the symmetry of individuals of the same species, which gives rise to some hostile interactions. The partner is therefore one that is identified through the classification of others on the biological basis of both being individuals of the same species, and the process of relationship building is the repeatedly chosen and fixed one. In contrast, there is no biological basis for the group phenomenon founded in troop reproduction activity, where classification of others into partners and not partners provides cues for activity. It is thought to be possible on the grounds that a social union, of which framework corresponds to segmental assembly in spatial distribution of the members of a species society, has come into being at a particular point in time. In other words, this classification of others is thought to be both a cause of activities that give rise to the social union known as the troop, and a result of those activities. Further, seeking to harmonize activity with peers, or to avoid agonistic interactions on the basis of such classification, are not activities that are biologically regulated or that must always be engaged in. It is exactly because they become necessary when there is an attempt to achieve balance between relationship building with the environment and relationship building with peers, that they are thought to be activities chosen as appropriate to time and place.

In the first instance, it is not the existence of a rigid relationship with another in which both are peers that drives these kinds of activities to be chosen at a time or place. In that sense, attention turns to the nature of the relationship, in that for each individual the possibility of engaging in mutually harmonizing activity with others or avoiding agonistic interactions with others can become greater or lesser. What is then thought to happen is that there is a seeking to confirm the nature of the relationship with a view to engaging in relationship building with a partner about whom it is hoped there will be greater such possibilities (Kitamura 2008). Secondly, it corresponds to activity involving picking and choosing between possibly engaging if it is a member of the same troop, and not engaging at all if it is an individual of the same species who is not a member. This is an approach that is important in ensuring the reproduction of a group with established membership. Thirdly, a more important matter is that in a state of co-existence, where it is possible to make co-locational and simultaneous choices about the same relationship building process to the same objects, even if by choosing to share the same activity framework with peers a state is realized in which activity is harmonized, the choice is problematic. This is because simultaneously a choice is being made that creates a situation in which the likelihood of competitive relations with peers will become more visible. Agonistic interaction becomes concrete as a result, despite the fact that in seeking to enable co-located and simultaneous choices about the same relationship building with the same objects, there is deemed to be a need to act to avoid agonistic interaction. If activity to avoid such interaction is actually chosen, then a state of co-existence is prepared in which co-located, simultaneous choice is possible, and it becomes possible to respond to the need to share the pre-condition framework for activity with peers at that time.

Herein lies the recursive decision-making process for two types of mutual activity essential to the reproduction of a troop: harmonization of activity and avoidance of agonistic interaction. In other words, these two types of activity are related, in that each are both another cause and a result, and it is thought that the group phenomenon first becomes possible in this type of society when the two come into being simultaneously (Figure 3.3). Specifically, in the first instance they are activities in which there is a seeking to choose an approach to relationship building with the environment that anyone can accept, by consistently seeking to share the approach with peers specified in the framework that corresponds to the form of segmental assembly that has resulted from previous activity. At the same time, in seeking to realize a result by relationship building that anyone can accept by adjusting behavior to build relationships with the environment such that social order is achieved that avoids agonistic interaction with peers with which the activity is being harmonized, the social union that is the troop is stably reproduced.

Collective activities to build relationships with the environment in humans

At the level of human society, the nature of the group phenomenon undergoes a complete change. In the case of human society, for activities which bring together people who mutually have not seen and do not know each other, where common attributes of some sort that enable such collectivity of activities between such people are hypothesized, such activities should also be labeled group phenomenon. In that sense, the group phenomenon in human society is thought to pre-suppose the existence of something that should be called a group that is believed to come

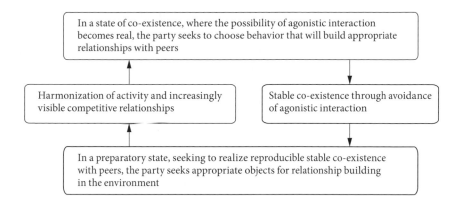

Figure 3.3 Recursive decision-making processes between
two types of activities needed for the reproduction of a troop

into being in dependence on something that prescribes the group framework. Irrespective of whether or not they are in mutual proximity in a particular place, however, as the hypothesis is that members of a group in human society are identifiable from information that can be acquired from a third party perspective, where a portion or all of the members of a certain group act to enable mutual tolerance and coordination in the same relationship building to the same objects, and even when the activity is engaged in a situation that is neither co-located nor simultaneous, that may also be called a group phenomenon. For that to happen, however, the relationship building process, which is neither co-located nor simultaneous, must be shared exclusively with the members of that group, which is in contrast to what is normally termed the group phenomenon, which is thought to correspond to co-locational and simultaneous engagement in activity that would enable mutual tolerance and coordination to be achieved among members in the same relationship building process to the same objects.

Despite the fact that in such human societies the group phenomenon on the face of it appears to be very different from what is evident in other primate societies, its nature is thought to be absolutely unchanged. In other words, where it is necessary at the level of the individual to achieve balance between relationship building with the environment and that with peers, the group phenomenon comes into being, and in the case of human society, the special feature of choosing as appropriate to need is clearer. People seek to create social union by adjusting relationship building with the environment with a view to a result that anyone can accept, which is achieving order of mutual interaction, or they seek justification for such activity as something any of one's peers can accept by sharing with others judgments about relationship building with the environment and engaging in relationship building in some form of social union. In a group of primates other than humans with stable membership, specification of group members through classification of others, thought to be essential to the creation of this type of group phenomenon, is dependent on the results of the very activity of forming a group in particular places, and harmonization of activity and avoidance of agonistic interaction— activities essential to the reproduction of a troop—always have to come into being simultaneously. In contrast, human society is different, in that classification of others is already provided, and if that is taken as a cue, it is sufficient to engage in a response that emphasizes one or other of the two types of relationship building as appropriate to the issue at hand.

For the group phenomenon—contextualized by troop formation—among Japanese macaque, as classification of others into peer and not peer that prescribes the group framework is both a cause and a result of the group phenomenon, activities essential to reproducing a troop—harmonization of activity and avoidance of agonistic interference between individuals for whom others are mutually classified as peers—must always simultaneously come into being. In contrast, in human society the criteria that enable such classification of others are already provided as something not dependent on activity in a particular place, and group members are identified by processing information obtainable from a third party perspective using the criteria. In seeking to respond to the issue of achieving order in a mutual interaction, and to the issue of justifying relationship building to the environment, there is thought to be an attempt to address the issues through collective activity, where the attributes shared by group members provide a clue to the choice of a response appropriate at a particular time and place.

Let us take a look at the foregoing situation from the opposite direction. In the case of the Japanese macaque, when we presumed that there is a classification of others that prescribes the group framework, the reason we said it is necessary for the interpretation that the group phenomenon arises from classification of others to be justified is because this conceivably makes a choice to build relationships with peers that leads to more certain troop reproduction. In addition, for mutual relationship building of that type to actually be realized, and when there is such appearance, either of the parties to the relationship building process must be experiencing the relationship building process as something appropriate to replicating that order. Consequently, the more fundamental process that is enabling the group phenomenon here is not relationship building based on the classification of others itself, but rather, it is likely the experience in which the parties realize they are striving towards harmonization of activity and avoidance of agonistic interaction. Starting one's thinking from this point also makes some cases of human society easier to understand. In other words, people will engage in collective action that brings the group phenomenon into being when there is that need and there is a partner with whom such relationship building can be hoped for. In the case of humans, however, in seeking to find such a partner, third party information about the affiliation of people to social groups and social categories (collections of people who share socially important attributes) is essential.

Let us conclude with a specific example. We will look at food sharing among the Central Bushmen hunter-gatherers of southern Africa (refer to Kitamura 2008 for details) as a group phenomenon configured in response to the issue of achieving order in a particular place of mutual interaction at a particular time. Among Bushmen peoples, a restrained, humble attitude is thought to be required in food sharing from both the person distributing food and the recipient.[3] In reality, agonistic interaction with peers is avoided in the food sharing setting through adjustments to achieve the goal that the food is transferred from the person distributing the food to the recipient. The activity in such a setting may be understood as something that is mutually adjusted with a view to realizing a result that either side will accept, in that, as each of the parties interacts, the appearance of eliciting an autonomous reaction from the other, without forcing onto the other specification of either's goals, is created. In other words, in the sharing of food in the context of such adjustments a group phenomenon based on the sociality of both parties is configured in that place which leads to a result that anyone can accept.

As another group phenomenon configured in response to the issue of justification of relationship building with the environment in a particular place such that any member of a group will accept it, let us consider healing rituals for illness among the Turkana herders of eastern Africa. As a final option for dealing with illness, the Turkana undertake a healing ritual involving the offering of livestock in sacrifice. A diviner is consulted in the presence of the gathered relatives and neighbors of the patient, and it is believed that by conducting a healing ritual in accordance with the diviner's instructions the illness will be cured. The belief that illness will be cured thereby arises because everyone accepts that the method of addressing the issue is as per directions from a supernatural being present through the divining, and in that sense it has been socially justified. Illness is a misfortune brought on by the environment that threatens a person's life, and is also a classic example of a problem that is difficult to resolve through normal means. When seeking to address such a problem, a relationship building process is justified when members of the same group engage in the problem through collective action, and can result in the experience of the misfortune being resolved (Kitamura 2007).

Conclusions: Bridging the divide between monkeys and humans

From the discussions so far, what gives rise to the divide between the group phenomenon in monkey society and that in human society is conceivably some difference that exists between the classification of others into peers and not peers, and into members of a certain group and not members of that group, that should prescribe the group framework. What can first be identified as a difference between these two classifications of others is that the former is one from the perspective of a particular party, and individuals of the same species other than that individual may be classified into one of the two. In contrast to it being a relative classification, where if the individual at the centre of the perspective is different then the classification of the individuals is also different, the latter is an absolute classification, in the sense that a decision as to whether or not a certain individual is a member of a group is enabled on the basis of third party criteria. When, however, there is an endeavor to make it possible to engage in the same relationship building to the same objects between oneself and an individual deemed to be a member of the same group by taking the latter classification as the cue, the same relationship building as used in the former case comes into being. It is in relation to this point that in the consideration so far we have maintained that there is no intrinsic difference in the group phenomena of the two societies.

As already indicated, however, mutual relationship building based on these two classifications of others could conceivably become decisively different things. We humans have acquired absolute classification of others from a third party perspective, and together with that which previously existed and must be called relative classification of others from the perspective of the parties, are able to use both in perfect freedom and therefore do not normally notice the difference between the two. Understanding that difference would be decisively important in considering what might be important among the events that happened in the course of evolution to human.

Relative classification of others into peer and not peer from the perspective of the parties in monkey society, in the context of building relationships with individuals of the same species in their society, corresponds to the classification of others adopted when a party to relationship building seeks to choose an appropriate partner to achieve the result that should be realized from the relationship building activity at the time. Mutual relationship building based on the classification of others is thought to be enabled by each of the parties to relationship building seeking to achieve it by choosing a partner with which they have previously built that type of relationship and with which they have always confirmed relationship building to be appropriate. In contrast, in absolute classification of others from the third party perspective in human society, the general nature of the mutual relationship—rather than reliance on what type of interactions it is that will be achieved using the classification—becomes the issue. The nature of the mutual relationship is specified by third party information about affiliation to social groups and social categories, and in response to an issue at the time the group phenomenon is configured through relationship building interactions using the specification.

How mutual relationship building based on these two classifications of others can become decisively different should become clearer with an understanding of the nature of the interim stage that bridges the two. The interim stage that bridges the transition from monkey to human in the group phenomenon should correspond to what is known as "groups that undertake frequent fission-fusion"[4] in the group phenomenon of the genus *Pan* (chimpanzee and bonobo). In particular, what becomes an issue is what happens when, after a long period of separation, temporarily separated sub-groups (parties or sub groups) re-unite. In the society of genus *Pan*, despite frequent fission-fusion, groups (as social units) that regularly form gatherings in spatial proximity are thought to exist (T. Nishida 1968; Kuroda 1982), and when such a group is taken as a reference point in the classification of others into members of the same group, it cannot possibly be the same classification as a peer in Japanese macaque society, where an individual has been confirmed as an appropriate partner because for a long time they have created the same groups together. Conversely, and despite which, it is inconceivable that third party information that in human society could be used at any time to enable classification of others into members of the same group is prepared in this way.

In the society of the genus *Pan*, when sub-groups of the same group have been ranging separately for a time and encounter each other again, the following typically eventuates. The chimpanzees exchange ferocious calls as they gradually draw closer, and in an excited state, including mutually driving each other ahead, combinations of individuals between those of the sub-groups change one after another in stylized interaction until they arrive at a state of tranquil co-existence (see Chapter Eleven of this volume). The enormous ruckus of the occasion is thought to be a state of excitement brought on by the possibility of the encounter developing into a hostile clash, despite which they have approached, but it is unlikely that they are seeking to realize some particular, specific result by so doing. Rather, individuals of the sub-groups engaging in stylized interaction are thought to be confirming the nature of the relationship by mutual relationship building. That is not, however, simply an endeavor to confirm that the various individual combinations are in relationships in which non-hostile interaction is possible, but it is thought that each of the parties, with the common attribute of being members of the same group (or, not members of different groups) (refer to Article One, for this topic), are in a sense staging a sham to confirm the possibility they are in a stable state of co-existence and are together able to share activities in the same gathering.

It is a type of ritual conducted on re-encounter that imprints the re-encounter, and by conducting such ritual it is thought that those involved are able to mutually confirm they are parties qualified to participate in it, and the general nature of each other's relationship is that of members of the same group. Despite the fact it is not the case that they have to do what they do and they want to do it, they do not seek to avoid the contact. In addition, by going out of their way to approach and mutually connect their behavior, they are therein expressing that each is identifying the others as members of the same group, and by collectively engaging in that expression, they are able to confirm that each deems the nature of their mutual relationship to be that of members of the same group. Taking as a cue the nature of the mutual relationship thus confirmed, in responding to the issues of balancing relationship building with the environment and that with peers which they can thereafter be expected to encounter, they will be able to configure a group phenomenon by sharing the activity of relationship building with the environment, or avoiding mutually agonistic interaction (see Key Ideas on page 60).

From the foregoing consideration, the decisive difference that arises in transition from the monkey to human stage is the emergence of what should perhaps be called a "decoupled representation" (Uchibori 2007) that is identified through the mediation of language. In other words, at the same time that (being a member of) a group becomes a meaning identifiable by a third party, it has become, by being generalized further, something that is not experienced as motivating specific behavior, is itself identified as meaning, and is able to be used as meaning when some sort of choice about behavior is needed. What fulfils a function similar to that of language at the chimpanzee stage to display decoupled representation is the stylized interaction performed when seeking to confirm the mutual relationship of participants in a ritual. Certainly, the group thus displayed corresponds to a group not specifically present as a phenomenon in that place, and it is detached from behavior that gives rise to the actual group phenomenon. At the same time, within the meaning is incorporated an element attributable to the perspective of the parties seeking to use the meaning as a clue to choice of behavior that give rise to a group phenomenon, that of (being a member of the) same (group), and in that respect also, the interim nature of the phenomenon is clear (see Key Ideas on page 60).

By theoretically re-configuring the nature of the interim stage that bridges the transition with a view to positing the factors creating the difference between the two, the route to understanding the group phenomenon—in evolutionary histori-cal terms—in the phylogenic cluster of primates that includes humans is revealed, and based on that understanding, we believe we have been able to advance con-siderations of human sociality. This progress emphasizes that in human sociality (being a member of) a group becomes a decoupled representation, and that cue gives rise to a new route, in the form of collaboration with peers. While on the one hand there is a possibility it will be posited as something more excessive, be-cause this type of representation does not directly specify specific corresponding behavior in the sense that it permits a more flexible response that conforms to the issue at the time, it is possible it will be posited as something more restrictive and respecting of individual freedom. It could be said that it has therefore ceased to be something that stipulates that human sociality must be a particular way, despite which, a single individual of a species of the animal, hominid, seeks to somehow survive by building relationships with the environment in the ecological niche of that species and seeks to create a stable state of co-existence by making relationship building with co-located peers. Conceivably they seek to respond to an issue in collaboration with peers neither because they have to, nor because they want to, but because they should enact what any of their peers would think is an appropriate response.

4 The Function and Evolutionary History of Primate Groups: Focusing on Sex Differences in Locational Dispersal

Naofumi Nakagawa

Evolutionary history of primate groups

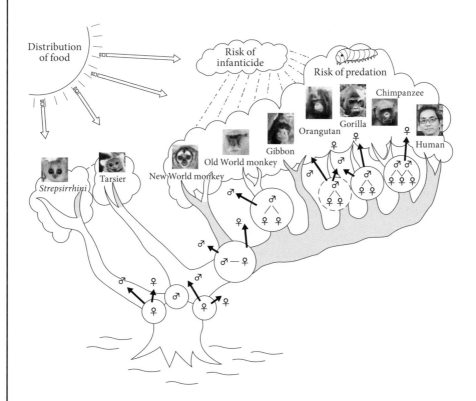

This diagram depicts the phylogenetic relationship of each taxonomic group as a phylogenetic tree, drawn with a focus on the social structure of the ancestral type of each taxon, and in particular the sex composition of, and dispersal sex from, groups. The solid lines linking male and female depict a mating relationship, while the length of the arrows showing dispersal from the group reflects dispersal distance. Primitive primates—among the small solitary primates of which males disperse further—form one male-one female groups. Through to early humans, such monkeys have had monogamous mating systems, and have maintained the characteristic of both male and female offspring departing the group on reaching sexual maturity, with the males demarking territory close to that of their parents (female-biased dispersal). As a result of the interplay between various environmental selection pressures such as food, risk of predation and risk of infanticide, groups have at the same time evolved to have a variety of sex compositions and patterns of succession (see endnote 1 for photographic credits).

Introduction

Thanks to advances in molecular phylogeny, the descent of humans from apes is incontrovertible. Humans parted company, first from the gibbon, then the orangutan, the gorilla, and finally the chimpanzee and bonobo, and thereafter evolved as primates with a unique erect bipedal form of locomotion. Researchers vary slightly on the era of divergence, but divergence from the chimpanzee and bonobo is said to have occurred some six to seven million years ago (Saitō 2007).

Meanwhile, palaeontology is not to be outdone. Since the discovery in 1974 of our brethren of approximately four million years ago, *Australopithecus afarensis*, there have been a succession of discoveries that have threatened its firm status as the oldest humans, starting with *Ardipithecus ramidus* in 1992. The estimated era of divergence of *A. ramidus* is 4.4 million years ago, and that of *Australopithecus anamensis* 4.2 million years ago, but *Orrorin tugenensis* was six million years ago, *Ardipithecus kadabba* 5.6 million years ago and then *Sahelanthropus tchadensis* six to seven million years ago (Suwa 2006). The momentum is sufficient to necessitate amendment of the era of divergence from the chimpanzee and bonobo. The appearance of early humans is being revealed from both sides: human and ape.

There is nevertheless an aspect that will never be revealed by the research outcomes of molecular phylogeny or palaeontology alone. That is, of course, ways of life. What did early humans eat, what sorts of groups did they form, what mating systems did they have and how did they raise children? Did they have something approaching language, what about their culture, and so on, and so on? Their way of life is no doubt engraved genetically, but cannot be distinguished from them. Nor is there a fossil record, and anthropologists have taken on the challenge of unraveling the knot these problems pose using two broad approaches. One is the ecological anthropology approach. Its techniques look for hints in the way of life of present-day peoples living the hunter-gatherer lifestyle that early humans probably led. The other is the field of primatology. Its methods seek clues in the way of life of extant primates, and in particular apes, as close relatives of human beings.

Since Kinji Imanishi, Japanese anthropologists have also adopted one or other of these approaches, but one of Imanishi's immediate apprentices, Junichirō Itani, began his work in primatology and subsequently switched careers to ecological anthropology. What might be called Itani's magnum opus as a primatologist is his study, *The Evolution of Primate Social Structures* (Itani 1985, 1987b,

2008a: Vol. 3). Itani held that primate groups had basic structures unique to each species, and combining the prevailing knowledge about the form of groups of the various primates with the then understanding of genealogic relationships, he asserted the importance of phylogenetic inertia, in other words, the historical flow of evolution in the development of primate social structure (for more detail, see Adachi's paper in Chapter One of this volume).

As will be illustrated later, locating various evolutionary traits, such as social structure, on a phylogenetic tree for closer examination is nowadays a commonplace technique for exploring them. This also allows for the need to probe phylogenetic inertia in each characteristic. In the West at the time of the initial publication of Itani's study, however, primatology dealt with social structure as a product of adaptation to the environment, and the study of socioecology, which explores the selection pressures that cause social structure to evolve, was about to mature. The timing corresponded almost exactly with the middle of the period between the publication of Carel van Schaik's socioecology model (van Schaik 1989, 1996; Sterck et al. 1997), which is today the most influential in the field and will be discussed further below, and that of its prototype, Richard Wrangham's model (Wrangham 1980). Itani hypothesized that there was significance in the fact that with at least one or the other sex emigrating from their natal group before reaching sexual maturity, groups were avoiding reduced vitality through inbreeding (inbreeding depression). This was the only selection pressure considered, however, and probably for that reason it went largely unnoticed. It was in that context that Yukimaru Sugiyama enthusiastically introduced Wrangham's model to Japanese primatologists. Unlike Itani, who thought that each species formed groups with their own unique basic structures, Sugiyama took the position that even within a species, basic structures could be changed by the environment (Yukimaru Sugiyama 1987).

In this chapter we will first introduce a typology of social structure, from the perspective of primatology, that is somewhat different from that used by Itani, and pose a hypothesis from socioecology attributable to the West that was absent from his hypothesis that relates to selection pressures that advance the evolution of social structure. As most primates form and live in groups that have a variety of structures, what in effect we will be doing here is to develop a functional theory of groups. By further focusing on sex difference in locational dispersal, which is becoming better understood as a result of recent advances in primatology, we will again point out the importance of phylogenetic inertia. Finally, we will use that result to pose a new hypothesis in relation to early human society.

The typology of social structure

Let us get an understanding of the typology of primate social structure to be used in this chapter. The structure of primate society is in the first instance classified according to whether or not a group is formed, and if it is, whether or not the adult members are comprised of one or multiple males or females; in other words:

1. a solitary monkey, be it male or female
2. one male-one female group
3. one male-multiple females group
4. multiple males-one female group
5. multiple males-multiple females group.

The terms monogamy, polygyny and polyandry as used by Itani, are normally used to describe mating systems. Generally, the one male-one female group is thought to correspond to monogamy, the one male-multiple females group to polygyny and the multiple males-one female group to polyandry, but this is not always the case. For example, even for a solitary monkey, if the territory over which a single male ranges and that over which a single female ranges almost exactly overlap, then if mating is observed only between those two, their mating system could be called monogamy. If the territory of a single male incorporates that of multiple females, and he does not allow overlap with the territory of another male, then it may be called the polygyny.

The second typological criterion is succession in groups, which Itani also considered important, and which is categorized into four according to whether or not there is succession in a group, and if there is, which of either sex are the successor:

1. no succession
2. matrilineal
3. patrilineal
4. bilateral.

The third criterion is whether group cohesion is high or not. In contrast to a group with high cohesion, where most of the members are always together, a group with low cohesion is one in which some members change their involvement from time to time; basically, the group repeatedly undergoes fission-fusion. Finally, the fourth criterion is the issue of whether a group tends to be mono-level or multi-level. Multi-level refers to multiple groups that are not necessarily affiliative, but are at least joined by non-antagonistic relationships to form a higher level of group structure. In reality, many primate groups have high levels of cohesion and a mono-level structure, so that with some exceptions, the first and the second criteria are sufficient.

Current socioecological models

Selection pressures determining the number of females in a group

Let us introduce the most well-known current socioecological model. Its core is van Schaik's model regarding to the following determinants (Figure 4.1):

1. formation of a multiple female group with high cohesion, including one or more males
2. female emigration and immigration
3. dominance hierarchies among females.

According to this model, such ecological factors as food distribution and predation pressures give rise to a group way of life amongst females, and determine the main competitive regime to which females are exposed. Differences in competitive regime influence linear and other dominance hierarchies among females, whether or not there is alliance from kin-related individuals in fights between them (nepotism), and whether or not there is immigration and emigration from a natal group by females. As a result, primate groups with multiple females may be classified into the four following categories:

1. dispersal-egalitarian
2. resident-egalitarian
3. resident-nepotism
4. resident-nepotism-tolerant.

The criteria for the typology of social structures dealt with in this chapter relate only to group composition, succession, cohesion and mono/multi-levels, and the characteristics associated with the social relationships of the constituents, such as whether or not there is a linear dominance hierarchy or nepotism, are not included in the typology criteria.

We will now explain only the part that relates to selection pressures that cause multiple females to form a matrilineal group.[2]

First, the greatest selection pressure that gives rise to female gregariousness is conceivably that from predation. If females are in a group, then the number of eyes available to warn against predators is increased, which means that the vigilance ability is enhanced accordingly. Also, there is a "dilution" effect in a group. If alone when found by a predator then the target is, of course, the individual. If a female is in a group of ten, however, at least when there is one predator, there will

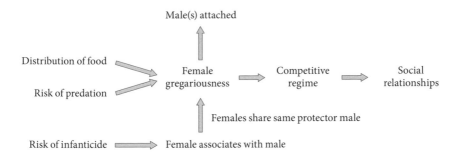

Figure 4.1 Flow chart of basic thinking in Van Schaik's socioecological model
Source: Adaptation of Fig. 6 in Sterck et al. (1997).

only be one from among ten animals that becomes prey. The effect is therefore one of simple probability: for a single attack, the probability of being preyed upon is one in ten. Another possibility may be that by group members cooperating for protection, they have the advantage that they are able to defend themselves from a predator that could not be resisted alone.

There is another element that group members should cooperate to protect: food resources. The single most important resource for females could be said to be food. That is because food is a resource that is directly tied to reproductive success. With more females in a group, the weight of numbers will make it possible to chase smaller groups away from food resources, which means the amount available for feeding will increase accordingly. When competition between groups associated with antagonistic interactions over food (between-group contest) arises, the group with the most females will probably be superior. Food is also a negative selection pressure that gives rise to group living amongst females. Even if a large group is victorious in a between-group contest and acquires a large food patch,[3] because there are more mouths to feed, the amount available per group member may end up being less than for a member of a smaller group. This is also a form of competition over food, called a "within-group scramble". This factor is not given much weight as a disadvantage of group living, but competition over food, known as a "within-group contest", could be considered a disadvantage. Within-group contests are competitions that reach the stage where high-ranking individual group members interrupt feeding of low-ranking individuals, causing a decline in the food available to the latter group.

As may be understood from this, factors that contribute to an increase in the number of females in a group are the advantages of avoiding predators and be-tween-group contests, and factors that reduce females are the disadvantages of within-group scrambles and contests. The number of females in a group is deter-mined by a balance between the two sides. The intensity of either competition, however, differs with the size, density and the distribution pattern of food resources.[4]

Selection pressures that form matrilineal, multiple female groups

From the perspective of group succession, multiple female groups are divided into matrilineal and non-matrilineal (bilateral, patrilineal) groups. We will now look at factors involved in the formation of one or the other. In terms of competition over food, matrilineal groups form in environments in which within-group contests, between-group contests, and both, have a strong effect.[5] Equating this to the pattern of food distribution, in an environment that includes food patches that are of high quality and dispersed with low density, but that do not contain much food, within-group contests will increase, and if kin-related females then aid each other in fights, hypothetically the amount of food they acquire will also increase. Conversely, in an environment that includes patches that are also of high quality and dispersed with low density, but have abundant foods, between-group contests will increase, and if individuals with a high degree of kin-relatedness cooperate compared to other groups, hypothetically the amount of food acquired as a group will also increase.

In terms of patterns of competitive regimes over food, in environments in which both within- and between-group contests are weak, non-matrilineal groups will form.[6] Equating that to the pattern of food distribution, in an environment in which food is either dispersed or concentrated on a very large patch, within-group and between-group contests hardly have any effect, and only within-group scrambles have a strong impact. Because within-group contests are weak, dominance hier-archies either will not exist between females, or even if they do exist, they will not be linear; in other words, they will be egalitarian. Even if kin-related females assist each other during fights, there is no need for kin-related females to be in the same group, because it is not advantageous. If a female emigrates to another group in which she has no relatives and becomes low ranking, she does not experience much disadvantage and the dispersal cost is therefore relatively low. On the other hand, because within-group scrambles are intense, if a female immigrates from a large group to a small one, the benefits from fewer scrambles is significant. For this reason, females will readily disperse.

Selection pressures that attach males to multiple female groups

Where an explanation of the composition of multiple female groups is next needed is on the topic of males joining primate groups. In this model also, the most important resource for a male are the females that relate directly to their reproductive success, and we will follow the precedent of the classical view that male dispersal is determined by the temporal and spatial distribution of females; in other words, the more females in a group, the more males. It is nevertheless not uncommon in primates for there to be species in which mating is limited to particular periods, and compared to non-seasonal mating species—even if the number of females in a group is the same—a tendency has been confirmed in seasonal mating species for the number of males to be greater. That is because, compared to the former, more of the latter females will come into oestrus simultaneously.

As a principle for explaining why males form a group with females in spite of the non-mating season, it is, however, self-evident that the above explanation alone is inadequate, given seasonal mating species. This is because such species also form bisexual groups, even in the non-mating season. The most likely selection pressure for the formation of bisexual groups is infanticide committed by males. Yukimaru Sugiyama was the first to discover infanticide among Hanuman langur (1965, 1980, 1993), and shortly thereafter the phenomenon was theorized by Sarah Hrdy as a male reproductive strategy (Hrdy 1977). Generally during lactation in mammals, at least for a certain period, there is a period of non-oestrus (amenorrhea) when the female will not fall pregnant. Thus, a male will kill an infant sired by another male to speed up female oestrus and sire its own infant. Females form a group with males to receive protection from infanticide by males, and multiple females sharing the same protector male is thought to bring multiple female groups into being (Sterek et al. 1997; Nakagawa and Okamoto 2003; Shima 2004; Yamagiwa 2008).[7]

Selection pressures that form societies other than multiple female groups

Van Schaik's model outlines the formation of multiple female groups with high cohesion, and in terms of selection pressures, may be considered to also broadly apply to several other social structures. Solitude, where food is dispersed and the risk of infanticide is low but predation pressures are high, is a social structure that evolved in small species that avoid predation by concealment.

One male-one female groups form when food is dispersed and there is a high risk of infanticide. Even where predation pressures are high, small species are able to avoid them through concealment (van Schaik and Dunbar 1990; van Schaik and Kappeler 1997). In small species generally, the weight of the infant is heavy compared to that of the mother, which means that the mother alone is unable to safely raise an infant. Conceivably, this structure therefore evolved to bring about infant care by males.[8]

Multiple male-one female groups, or multiple male-multiple female groups that entail polyandrous mating relationships with one reproductive female are a form that evolved in twinning small species, and because the mother's infant care burden is thus increased, the fathers of each of what are fraternal twins both help.[9]

Selection pressures that form patrilineal societies

As the final topic in discussing selection pressures and group formation, let us look at patrilineal societies. Van Schaik's socioecology model presents a detailed theory of selection pressures that give rise to female dispersal, but there is a need for caution in that it does not address the social pressures that lead to the formation of patrilineal groups. In the female dispersal-egalitarian society, there is no distinction in bilateral types between those in which both females and males disperse and patrilineal types in which males basically do not disperse. In reality, patrilineal groups all have low cohesion and manifest what is called fission-fusion, and for that reason they are not included in this model, which deals with what from the outset are highly cohesive, multiple female groups.

Furuichi (2002) summarized the theory of patrilineal group formation in chimpanzees developed to date by western researchers, primarily led by Wrangham et al., and additionally proposed his own idea to fill the gaps. Prevailing theory could be called "the food dispersal hypothesis", he said, and may be summarized as follows:

1. in order to enhance the efficiency of fruit feeding in small, dispersed patches in tropical rainforests
2. females disperse and range without forming large groups
3. males cooperate to herd the dispersed females
4. a patrilineal society results in which males remain in their natal group and conversely, females emigrate.

The doubt Furuichi harbored regarding the food dispersal hypothesis was the question of whether, even if females adopted a dispersed way of life in order to increase feeding efficiency, would that actually lead to males cooperating to herd multiple females. If females were dispersed, could not males also disperse and herd them, and they establish territories as pairs? This scenario certainly makes sense. Furuichi's thinking was that it is difficult to explain cooperation between males and the formation of patrilineal societies using the dispersal of females in adaptation to fruit feeding alone.

What Furuichi proposed instead was "the operational sex ratio hypothesis". The following is an outline.

1. the inter-birth interval in ancient apes becomes longer and females enter oestrus only rarely; in other words, the number of adult males per female in oestrus (the operational sex ratio) increases
2. competition over females in oestrus between males becomes fiercer
3. maintaining male-female pairs becomes difficult
4. as a result, males come to remain in their natal groups to facilitate cooperation in resistance against other groups.

The bonobo, however, studied by Furuichi et al. for an extended period of time, revealed that the above hypothesis absolutely does not apply. Bonobo have a strong tendency to enter oestrus, even during periods when pregnancy is impossible, and despite a long inter-birth interval their operational sex ratio is much lower than that of the chimpanzee. Perhaps for that reason, male behavior to protect females from other groups could not be identified. The bonobo nevertheless also form patrilineal groups.

Compared to the gorilla, the subject of his own study, Yamagiwa (1994) focused on the fact that the bonobo of course, and also the chimpanzee, manifest swelling of the sexual skin and have prolonged oestrus. He was of the view that having acquired this characteristic males aggregate oestrus females and form multiple male groups, and kin-related males cooperating together to protect oestrus females gives rise to patrilineal groups. Yukimaru Sugiyama (2002) linked the fact that males among the subject of his study—the chimpanzee of Bossou in the Republic of Guinea—do emigrate and that there are no adjacent groups, and he deemed joint protection of oestrus females by kin-related males to be a factor in the formation of patrilineal groups.[10]

There are also views that the objects of shared protection by males in patrilineal groups are not just oestrus females, but also infants and their mothers (van Schaik 1996; Williams et al. 2004).

The evolutionary history of patrilineal society and locational dispersal

The primate molecular phylogenetic tree

In the discussion thus far we have examined how various social structures seen among primates have formed and looked at the selection pressures involved, while explicating the prevailing mainstream theory; in other words, the functional theory of groups. This point is not considered at all in Itani's theory posited in *The Evolution of Primate Social Structures*, but we will here re-focus on the phylogenetic inertia emphasized by Itani, or what is otherwise called the historical flow of evolution. In particular, we will focus on the formation of patrilineal societies. In that context, what I have incorporated as a new perspective is sex difference in locational dispersal, and what has made this possible are research advances in the following three academic domains:

1. molecular phylogenetics that reveals the genealogical relationships of each species of primate, which are the basis for analysis
2. research into the social structures of New World monkeys from field studies
3. molecular genetic research into the locational dispersal of males and females.

The advance of molecular phylogenetics has not only revealed the genealogical relationship explicated at the outset and the era of divergence between we humans and the apes, but has also systematically revealed the same between each species of primate (Figure 4.2). Starting with the oldest and through to humans, the following is an explication of the genealogy and eras of divergence. Strepsirrhini, comprising Lemuroidea and Lorisoidea, diverged sixty-three million years ago, followed fifty-eight million years ago by Tarsiidae.[11] Then Platyrrhini (monkeys that inhabit Central and Southern America and are therefore usually termed "New World Monkeys") diverged forty million years ago.[12] Next, Cercopithecidae (monkeys that inhabit Asia and Africa and are therefore usually termed "Old World Monkeys") diverged twenty-five million years ago. The Old World Monkeys and Hominidae are included in the classification Catarrhini, but what diverged eighteen million years ago from this common ancestor were the lesser apes, in other words, Hylobatidae. As stated at the outset, in this chapter we use the taxonomic name Hominidae to refer to great apes and humans, and orangutans, gorillas, chimpanzees and humans are deemed separate genera. The more detailed genealogic relationships described here will be addressed sequentially as necessary.[13]

Social evolution model of New World monkeys: The New World monkey ancestral type

Based on knowledge of the social structure of New World monkeys amassed to that point, Theresa Pope (Pope 2000) made an attempt to illustrate their social evolution (see Figure 4.2). What is novel about her model is that it looks not just at social dispersal, but also at locational dispersal, and has been configured as a model that addresses their continuity while at the same time distinguishing between them.

Thus far the issue of male-female dispersal has been addressed from the perspective of group succession. In other words, as a matter of course, males in a

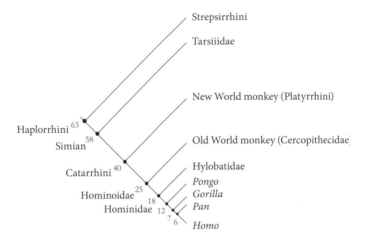

Figure 4.2 Molecular phylogenetic tree of extant primates and estimated era of divergence (units are in millions of years)
Source: Adapted from Fig. 3 in Matsuzawa et al. (2007).

matrilineal group will disperse further than females, and females in a patrilineal group will disperse further than males. In a bilateral group where both sexes disperse from their natal groups, or in one male-one female groups where there is no succession, does either sex then disperse widely? Even if they emigrate from their natal groups, do they then only disperse to a distance not far from the home range? In contrast to dispersal from a natal group, which is that from familiar society and should therefore be called "social dispersal", relocating a long way from the natal group's home range is a coming of age process and involves dispersal from familiar country, and should be distinguished by being termed "locational dispersal" (Isbell and van Vuren 1996; Yukimaru Sugiyama 2002). Conversely, not dispersing and instead remaining in natal groups is called philopatry.

Pope deemed the ancestral type to be territorial one male-one female groups, where the mating system is monogamous societies as identified in extant *Aotus* (Cebidae), *Callicebus* and *Pithecia* (as above, Pitheciinae in Atelidae). In each of the genera, the father not only carries the infant, he sometimes feeds it. Because they are one male-one female groups, both females and males disperse socially, but geographically the females disperse further. Males stay longer in their natal group than females, and sometimes reach sexual maturity there. When sons form new groups with single females, they typically locate in the periphery of the parent's

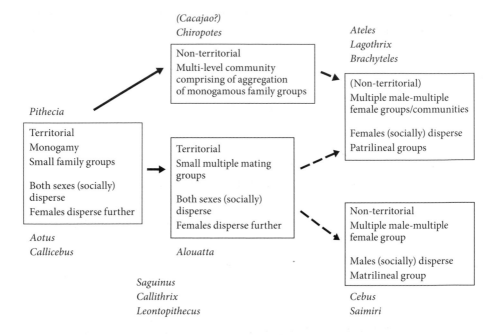

Figure 4.3 Transition of mating systems in New World monkeys, assuming territorial monogamy/small group polygamy as ancestral type

Note: There is no intention to suggest phylogenetic relationships.

Source: Adapted from Fig. 19.4 in Pope (2000).

territory. Neighboring groups will from time to time eat peacefully in the same feeding tree, and it is likely these are father-son groups.

Chiropotes in Pitheciidae is the genus in which the gathering of multiple groups with strong male philopatry is more normalized, and it is likely that *Cacajao* is the same. Their monogamous one male-one female groups dispense with territoriality, forming multi-level communities (regional groups) in which collectives have developed. Males completely abandon emigration from the community, and if mating systems become promiscuous, they become patrilineal multiple male-multiple female groups of the kind seen in Atelinae in Atelidae.

Conversely, in ancestral one male-one female groups, if territories are crowded, sexually mature sons will remain in the natal group as helpers, forming the monogamous multiple male-one female group. Females may also immigrate to the group, but oestrus in these females is suppressed, they do not engage in reproduction, and by playing the role of helper, are tolerated by the group's female. The mating system then becomes that of a monogamous multiple male-multiple female group. Males may also immigrate, in which case the mating system becomes that of a polyandrous multiple male-multiple female group. Reproductive females will give birth to fraternal twins with different fathers, requiring infant care by the two males. These are all social structures identified in Callitrichidae, and fathers not only carry infants, but also feed them.

Other than the mantled howler, which typifies red howlers, *Alouatta* (Atelinae) form small, territorial, polygynous one male-multiple female groups. Between twenty and thirty percent of females remain in the natal group, where cooperative behavior based on kin-relatedness between females is highly developed; for example, relatives will cooperate to evict other relatives from the group. In that respect the groups are like matrilineal groups, and may have evolved into the matrilineal multiple male-multiple female groups of *Cebus* and *Saimiri*. Conversely, while nearly all males emigrate from the natal group prior to reaching sexual maturity, rare examples are known of a male remaining as a helper, ultimately beginning to reproduce, and succeeding to group leader on the death of their father. Males that do leave typically demark a territory adjacent to that of the natal group, and their locational dispersal distance is therefore shorter than that of females. The mantled howler is territorial, but creates multiple male-multiple female groups. The females all emigrate from the natal group, and because each immigrates to a group independently, there is no kin relationship between females. In contrast, because many males remain in their natal groups, the groups are similar to patrilineal groups and may have evolved into the patrilineal, multiple male-multiple female groups seen in Atelinae. Because a portion of emigrated males also demark a home range in neighboring locations, locational dispersal distance in males is short.

Pope herself is self-effacing about her model, saying it was not intended to take into account phylogenetic relationships, so we will apply these social structures to a modified version of Purvis' (1995) phylogenetic tree (Figure 4.4). Just as she hypothesized, using the Maximum Parsimony Method,[14] it is clearly appropriate to deem the territoriality and the monogamous, one male-one female group mating system identified in *Aotus*, *Callicebus* and *Pithecia* as the New World monkey ancestral type. What is also clearer than anything else is that New World monkeys are "patrilineal", in the sense that since the ancestral type era, males have consistently had stronger philopatry than females, and the matrilineal groups seen in *Cebus* are a derived trait.

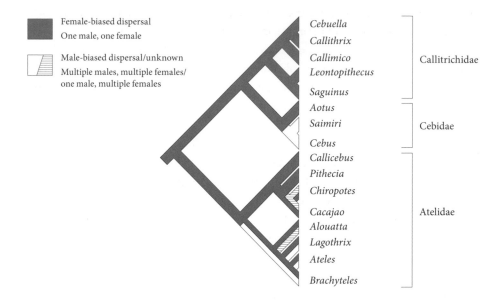

Female-biased dispersal
One male, one female

Male-biased dispersal/unknown
Multiple males, multiple females/
one male, multiple females

Cebuella
Callithrix
Callimico
Leontopithecus
Saguinus
Aotus
Saimiri
Cebus
Callicebus
Pithecia
Chiropotes
Cacajao
Alouatta
Lagothrix
Ateles
Brachyteles

Callitrichidae

Cebidae

Atelidae

Figure 4.4 Social evolution of New World monkeys

Because we have placed Pope's (2000) information into the phylogenetic tree without amendment, it does not reflect the male-biased dispersal in common marmosets resulting from Table 4.1. Emulating van Schaik and Kappeler (2003), we also deemed Callitrichidae to be one male-one female with a polyandrous mating system, emphasizing the fact that there is one reproductive female.

Source: The phylogenetic tree is an adaptation of Fig. 4.1 in van Schaik and Kappeler (2003) and Purvis (1995 amended version).

New World monkey ancestral type: Male-biased dispersal, one male-multiple females

Old World monkeys (Cercopithecidae) that emerged after divergence from the ancestral simian include such well studied genera as the Japanese macaque (*genus Macaca*), the baboon (*genus Papio*) and the vervet monkey (*genus Cercopithecus*). Because all of these form matrilineal multiple male-multiple female groups, there is a tendency to think they are representative of primate social structure. New World monkey researcher Anthony Di Fiore investigated inter-species similarities in a variety of social traits; not just those addressed so far in this chapter—the composition of groups, succession patterns, mono/multi-levels, mating systems, dominance hierarchies and infanticide by males—but also care by females of the offspring of others, infant care by fathers, grooming between the sexes and antagonistic interactions between the sexes, and found markedly higher levels of similarity among Old World monkeys than compared to other taxons (Di Fiore and Rendall 1994).

Di Fiore and Rendall (1994) reached the conclusion that given the remarkable magnitude of change in social traits from ancestral primates to ancestral Old World monkeys compared to that in the ancestral type of other taxons, Old World monkeys display derived traits. Prior to them, the doyen of New World monkey research, Karen Strier (1990), had also asserted that as female (social) dispersal is seen in New World monkeys and apes and even in Old World monkeys, and seems in fact to be the majority, then female philopatry must be a derived trait in Old World monkeys. Di Fiore and Rendall (1994) quoted this as well, and particularly cited social dispersal by itself, stating that female philopatry is a derived trait seen in Old World monkeys and several other species, and that female dispersal seems to be a primitive trait in primates in general.

It would be irresponsible to decide, however, that simply because something is seen in the majority it is a derived trait, and it would appear impractical to apply a conclusion of derivation determined from a global evaluation of social traits, without amendment, only to the particular trait of social dispersal. There is no doubt, therefore, that male-biased dispersal is the Old World monkey ancestral type, but as Strier (1999) subsequently identified, it is not possible to decide whether it is the simian ancestral type or a derived trait of Old World monkeys. As will be discussed below, just on the basis of the evidence discussed in this chapter so far, the former would appear to be the fairer view. It is clear that the group composition of one male-multiple female groups that in two subfamilies manifests in most Colobinae, and manifests in most of the in Cercopithecinae, is the ancestral type.

Selection pressures that lead to sex differences in locational dispersal

Let us now for a moment return to the topic of selection pressures. To date, broadly there have been four hypotheses proposed about selection pressures that bring about sex differences in locational dispersal (sex-biased dispersal) in animals not limited to primates. The first is the local resource competition hypothesis. This hypothesis, proposed by Paul Greenwood (1980), focuses on the fact that in contrast to mammals—in which males generally disperse further—in birds it is typically females that disperse over the greater distance, with the benefit of the ability to search for local resources. In birds, most species have a monogamous mating system and males do the infant care. The males of these species secure information about resources important to the survival of offspring, such as suitable food or nest sites, and the females choose their mate according to the value of resources

Photo 4.1 A group of Japanese macaque in Kinkazan Island

A group of Japanese macaque, in which individuals affiliated with several family lines, have gathered and are resting (Kinkazan Island, Miyagi Prefecture). The genus *Macaca*, which includes the Japanese macaque, are the most studied taxon in primates, and because they and the genus *Papio* form matrilineal, multiple male-multiple female groups, there is a tendency to consider that these groups constitute the mainstream in primate social structures. The reality is that they are more likely an offshoot.

within a territory, so that also from the perspective of enhancing mating success, there are advantages in remaining in familiar country. In contrast, there are many mammal species with a polygynous mating system and where the male hardly engages in infant care. In such species, because the female establishes a territory to protect food resources—the most important resource—remaining in familiar country is advantageous. The second, the local mate competition hypothesis, was proposed by Stephen Dobson (1982), who investigated the few mammal species where there is monogamy and focused on his inability to identify a tendency for females to disperse further. In species with polygyny, competition between males over mates becomes intense because males mate with multiple females, and to avoid competition between relatives males disperse widely. In contrast, in species with monogamy, we can predict that sex-biased dispersal will not be identified because there is no sex difference in competition over mates. The third is the in-breeding hypothesis (Wolff 1992, 1994). According to this hypothesis, the sex in which the risk of inbreeding is highest should disperse. In species where there is polygyny, male offspring (sons) are at the highest risk of inbreeding because only the mother lives in proximity to the offspring, and the males should therefore disperse. In contrast, the prediction that sex-biased dispersal will not be identified is upheld in species with the monogamous pattern. The final hypothesis is the local resource enhancement hypothesis (Perrin and Lehmann 2001; Le Galliard et al. 2006). According to this hypothesis, the focuses on resource protection and infant care are the advantages of cooperating with kin-related individuals, and the more advantaged sex remains. The hypothesis of factors in formation in primates of matrilineal multiple female groups, as mentioned above, and patrilineal multiple female groups, equates to this thinking.

Research into locational dispersal using molecular genetic techniques

Even when addressing the aforementioned locational dispersal, when one sex socially disperses from a group and the other sex remains, as in matrilineal and patrilineal groups, the resulting outcome is the same, and should be apparent just by careful observation of individuals affiliated with a group. An individual that has exited a group is nevertheless difficult to observe, and for that reason, where both sexes will disperse socially from a group, such as in bilateral one male-one female groups, and among solitary monkeys, it is not easy to investigate which of the sexes has dispersed further.

What has overcome this difficulty is research into the social structure and genealogical structures of primates using molecular genetic techniques, which have become very popular of late (see the comprehensive explanation in Di Fiore 2003). Sustained research projects using these techniques accord with the basic principle of population genetics, that dispersal brings about genetic flow. For example, where male-biased dispersal occurs, because males disperse uniformly within an appropriate range and then reproduce, we can predict there will largely be no localized identification of particular genotypes in the Y chromosome DNA uniquely inherited by male mammals. In contrast, we can predict there will be such identification in the mitochondrial DNA uniquely inherited by females. We can also focus on the diversity of mitochondrial DNA. The diversity of mitochondrial DNA in individuals that live in a group or in a certain region can be compared between males and females. If there is male-biased dispersal, we can predict that male genotype diversity will be greater. Conversely, sex-biased dispersal can be identified by investigating the distribution of these genotypes and their diversity in a particular region that includes multiple individuals or groups, and if it is identified, it is possible to investigate whether it is male- or female-biased.

The simian ancestral type is the solitary monkey of male-biased dispersal

The personal theory that I will explicate from the next paragraph will take us one step back, but I wish to review evolutionary history with the aim of covering the entire evolution of primate society, starting from the New World monkey ancestral type. In Strepsirrhini, Lorisoidea (Galagidae, Lorisidae) are nearly all solitary primates, but Lemuroidea (Lemuridae and others) typically have structures that are similar to one male-one female groups (Figure 4.5). Varied pairs manifest a variety of group types, including multiple male-one female groups and multiple male-multiple female groups, and while one male-one female groups are the most common, no more than ninety percent of all species correspond to this form. These are differentiated from uniform pairs, however, to which over ninety percent of species that form one male-one female groups correspond. Even in uniform pairs, dispersed pairs are known in which each sex engages in solitary feeding and movement, but each of their territories overlaps exactly and they often share the same nest. This is different from the one male-one female groups described in this chapter, in which males and females are nearly always together. As shown in Figure 4.5, there are no Strepsirrhini species that are thought to have evolved directly from solitary primates to one male-one female groups, and it can be seen that they have emerged via the form known as the varied pair. In addition, one male-one female groups are very much in the minority, and the Strepsirrhini ancestral type would have been solitary.

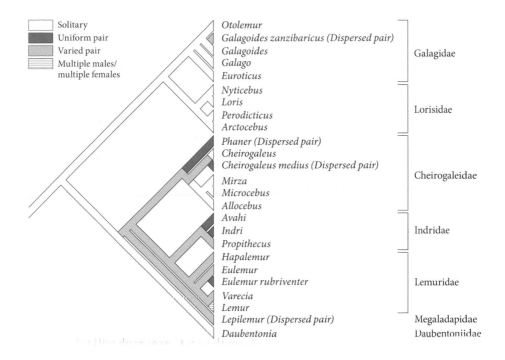

Figure 4.5 Social evolution of Strepsirrhini
Source: Phylogenetic tree is an adaptation of Fig. 4.1 in van Schaik and
Kappeler (2003): adaptation of Purvis (1995).

Let us take a look at Table 4.1, which summarizes the results of research into
the locational dispersal of males and females using molecular genetic techniques.
For both one population of the solitary Coquerel's dwarf lemur (genus)[15] and three
populations of grey mouse lemur (genus), there is molecular genetic evidence of
male-biased dispersal. For that reason, at this point in time it is surmised that the
Strepsirrhini ancestral type was male-biased dispersal. In contrast, in dispersed
pairs of fat-tailed dwarf lemur (genus), it was not possible to confirm a tendency
to sex-biased dispersal in a particular population, but evidence was obtained from
the population of another region that indicated female-biased dispersal. In the
latter population, while fathers do not carry infants, it has been reported that they
do play a babysitting role, moving and sleeping in hollows with infants (Fietz and
Dausmann 2003). Further, in Scandentia, formerly classified as a prosimian and
even today positioned as a separate order most closely related to primates, form

Table 4.1 Molecular genetic studies of dispersal patterns in non-human primates, excluding matrilineal Old World monkeys

Common name	Scientific name	Group composition	Dispersal patterns	References
Fat-tailed dwarf lemur	*Cheirogaleus medius*	dispersed pair	no sex-bias	Fredsted et al. (2007)
Fat-tailed dwarf lemur	*Cheirogaleus medius*	dispersed pair	female-biased	Fietz et al. (2000)
Coquerel's dwarf lemur	*Mirza coquereli*	solitary	male-biased	Kappeler et al. (2002)
Grey mouse lemur	*Microcebus murinus*	solitary	male-biased	Radespiel et al. (2001)
Grey mouse lemur	*Microcebus murinus*	solitary	male-biased	Wimmer et al. (2002)
Grey mouse lemur	*Microcebus murinus*	solitary	male-biased	Fredsted et al. (2004)
Gentle lemur	*Hapalemur griseus*	varied pair	male-biased	Nievergelt et al. (2002)
Red-fronted lemur	*Eulemur fulvus*	varied pair	male-biased	Wimmer and Kappeler (2002)
Verreaux's sifaka	*Propithecus verreauxi*	varied pair	male-biased	Lawler et al. (2003)
Common marmoset	*Callithrix jaccchus*	varied pair	male-biased	Nievergelt et al. (2000)
Common marmoset	*Callithrix jaccchus*	varied pair	no sex-bias	Huck et al. (2005) in Di Fiore (2009)
Wedge-capped capuchin	*Cebus olivaceus*	multiple males-multiple females	male-biased	Valderrama Aramayo (2002) in Di Fiore (2009)
Mantled howler monkey	*Alouatta palliata*	multiple males-multiple females	no sex-bias	Ellsworth (2000) in Di Fiore (2009)
White-bellied spider monkey	*Ateles belzebuth*	multiple males-multiple females	female-biased	Di Fiore et al. (2009)
Lowland wooly monkey	*Lagothrix poepigii*	multiple males-multiple females	no sex-bias	Di Fiore et al. (2009)
Hamadryas baboon	*Papio hamadryas*	one male-multiple females	female-biased	Hapke et al. (2001)
Hamadryas baboon	*Papio hamadryas*	one male-multiple females	female-biased	Hammond et al. (2006)
Siamang	*Symphalangus syndactylus*	one or two males-one female	female-biased	Lappan (2007)
Sumatran orangutan	*Pongo abelii*	semi-solitary	no sex-bias	Utami et al. (2002)
Bornean orangutan	*Pongo pygmaeus*	semi-solitary	no sex-bias	Goossens et al. (2006)
Bornean orangutan	*Pongo pygmaeus*	semi-solitary	male-biased	Morrogh-Bernard et al. (2011)
Sumatran orangutan/ Bornean orangutan	*Pongo* spp.	semi-solitary	male-biased	Nietlisbach and Krutzen (2010)
Western Gorilla	*Gorilla gorilla*	one male-multiple females	female-biased	Bradley et al. (2004)
Western Gorilla	*Gorilla gorilla*	one male-multiple females	some males dispersed much further	Douadi et al. (2007)
Bonobo	*Pan paniscus*	multiple males-multiple females	female-biased	Gerloff et al. (1999)

Bonobo	*Pan paniscus*	multiple males-multiple females	female-biased	Eriksson et al. (2006)
Common chimpanzee	*Pan troglodytes*	multiple males-multiple females	female-biased	Morin et al. (1994)
Common chimpanzee	*Pan troglodytes*	multiple males-multiple females	no sex-bias	Goldberg and Wrangham (1997)
Common chimpanzee	*Pan troglodytes*	multiple males-multiple females	no sex-bias	Mitani et al. (2000)
Common chimpanzee	*Pan troglodytes*	multiple males-multiple females	female-biased	Langergraber et al. (2007)
Common chimpanzee	*Pan troglodytes*	multiple males-multiple females	female-biased	Inoue et al. (2008)

dispersed pairs, and while infant care by fathers is not known, there is molecular genetic evidence of female-biased dispersal (Munshi-South 2008). In dispersed pairs that are not one male-one female groups, but similar, it should be noted that a tendency common to that of New World monkeys has been confirmed—identification of female-biased dispersal and infant care by fathers.

We will now return to the phylogenctic tree and take a look at Tarsiidae (see Figure 4.2). The Tarsiidae ancestral type also includes the solitary primates. The dispersal distance of solitary Tarsiidae is unknown, but the varied pair eastern tarsier manifests male-biased dispersal (Gursky 2007).

It is becoming increasingly difficult to deem territorial, female-biased dispersal in monogamous one male-one female groups as the simian ancestral type. Because the prosimian ancestral type, including Tarsiidae, is male-biased dispersal, Old World monkeys have inherited the trait, and because of this it would seem appropriate in accordance with the Maximum Parsimony Method, to in fact deem female-biased dispersal in New World monkeys as a derived trait.

Female-biased dispersal in almost all apes is attributable to the simian ancestral type

We have finally arrived at what may be called the interesting stage of this chapter. In this section and the next we will incorporate recent knowledge of locational dispersal to verify whether it is possible to deem the New World monkey ancestral type—territorial, female-biased dispersal in a monogamous one male-one female group—that we said could not at that stage be judged the New World monkey ancestral type, to be the simian ancestral type. Because this means that the Old World monkey ancestral type—the matrilineal one male-multiple female group—

is a derived trait, it indicates that the simian ancestral type is identifiable all the way through to the ape ancestral type. As you read, refer to Figure 4.6 and Table 4.1, which depict the social structure of each ancestral type of each genera of Hominidae, Hylobatidae, Old World monkeys, New World monkeys, Tarsiidae and Strepsirrhini.

Hylobatidae, which was the first of the apes to diverge, basically takes the one male-one female group,[16] and the tendency for male offspring to demark a territory closer to the parent is known in multiple species. Among siamang, there is molecular genetic evidence indicating female-biased dispersal. Despite the fact that there is conceivably little need for fathers to engage in infant care, because gibbons are large, it has been reported that siamang, and albeit anecdotally, black crested gibbon fathers will carry infants (Fuentes 2002).[17] In the present-day, it is usual to categorize gibbons into four genera, but among them, genus *Symphalangus* is the next after genus *Nomascus* to be said to have this ancestral trait (Roos and Geissmann 2001; Takacs et al. 2005), and conceivably, infant care by fathers at the very least existed in the gibbon ancestral type. The fact that infant care by fathers, for which there is conceivably little need, is known only in these two genera, also suggests that the characteristic is a legacy of the simian ancestral type, and therefore it may be appropriate to think that it was subsequently lost for this reason.

Let us now track this flow, and look according to age order at eras of divergence of extant greater apes from the human lineage. The next to diverge after Hylobatidae was the orangutan. The orangutan has for a long time been thought to be solitary, but the Sumatran orangutan female in particular has been revealed as slightly collective, and currently is termed "semi-solitary". Molecular genetic evidence indicating sex-biased dispersal has not been identified in either the Sumatran or Bornean orangutan, thought to be the result in recent years of loss of places for males to go due to rapid deforestation. Very recently, however, there have been announcements of evidence of male-biased dispersal in both species.

Of further interest is the gorilla, which was next to diverge. A variety of group configurations that change according to number of years since group formation have been identified among mountain gorillas living on the Virunga Massif, located on three national borders; that of the Democratic Republic of the Congo, Ruwanda and Uganda. These form a very small number of one male-one female groups, roughly seventy percent one male-multiple female groups and just under thirty percent multiple male-multiple female groups. One male approaching a female from another group can be the trigger for the female to immigrate, forming

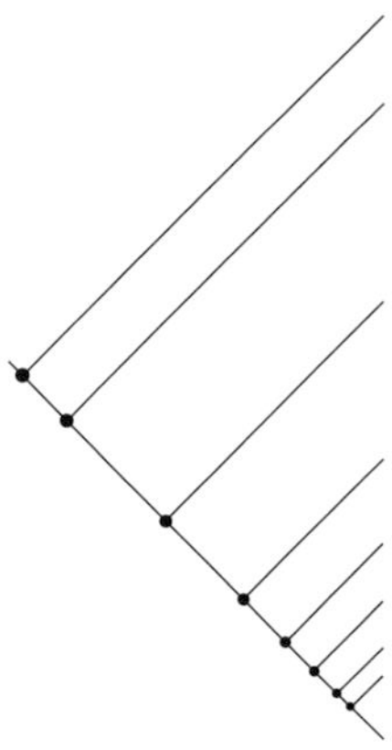

	Locational dispersal	Social dispersal	Group composition	Infant care by males	Territoriality
Strepsirrhini ancestral type	male-biased	male/female	solitary	×	○
Tarsiidae ancestral type	male-biased	male/female	solitary	×	○
New World monkey ancestral type	female-biased	male/female	one male-one female	○	○
Old World monkey ancestral type	female-biased	male	one male-multiple females	×	×
Hylobatidae ancestral type	female-biased	male/female	one male-one female	○	○
Pongo ancestral type	male-biased	male/female	semi-solitary	×	×
Gorilla ancestral type	female-biased	male/female	one male-multiple (one) females	△	×
Pan ancestral type	female-biased	female	multiple males-multiple females	×	△
Human ancestral type	female-biased?	male/female?	one male-one female?	○?	○?

Figure 4.6 Archetypal social structure in each genera of Hominidae, Hylobatidae, Old World monkeys, New World monkeys, Tarsiidae, and Strepsirrhini

a one male-one female group, and gradually thereafter the number of females increases, forming a one male-multiple female group. Later, when the male has become elderly, in the multiple male-multiple female groups thus formed, sons will emerge that will stay in the natal group, even having reached sexual maturity. Even in groups where the father has died, there have been reports of some females remaining in their son's group. In other words, there are instances where a group is succeeded along patrilineal lines (Yamagiwa 2005).[18] Males will not carry infants like the siamang, but they do fill a babysitter-type role, tolerating children nearby (Yamagiwa 1994). In the eastern lowland gorilla, mature sons remain and multiple male-multiple female groups are formed, but sons who have exited reportedly form their own groups very close to that of their father. Molecular genetic evidence has been added to these reports. In Mondika in Central Africa, the western gorilla appears to manifest female-biased dispersal. Multiple male-multiple female groups are not seen in western gorillas; they form one male-multiple female groups, but both males and females disperse socially before sexual maturity, which is thought to result from males forming their own groups in neighboring territories.[19] As a matter of interest, between-group relationships in this population are comparatively affiliative, because the silverbacks in these several groups are in a patrilineal kin relationship; in other words, they are thought to form a patrilineal community (Yamagiwa 2006).

While the two species of the *Pan* genus—the chimpanzee and the bonobo, which diverged most recently from the human lineage—are exceptions, they are basically patrilineal and form groups with a multiple-male multiple female composition and high fission-fusion. In accord with results from behavioral observation of patrilineal groups, molecular genetic results of female-biased dispersal have also been obtained from chimpanzees in Tai in the Côte d'Ivoire and Mahale in Tanzania and bonobo in Lomako in the Democratic Republic of the Congo, and studied across a broad area of the latter country.[20]

As indicated, female-biased dispersal has been identified in all apes other than orangutans, and may be considered the Hominoidea ancestral type. That is truly in contrast to the fact that, in Maximum Parsimony Method terms, it is impossible to explore the ancestral type in relation to such traits as sex differences in social dispersal, group composition and the presence or absence of infant care by the male parent, because there are so many variations. In this way it has been possible, at least from sex differences in locational dispersal, to seek out the ancestral type from extant apes. Viewed in Maximum Parsimony Method terms, it can at last be said that the likelihood that male-biased dispersal is the simian ancestral type,

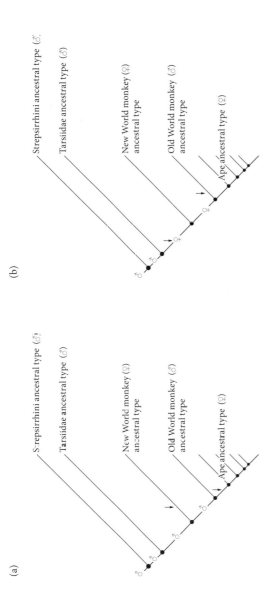

Figure 4.7 A diagram to enable consideration of whether male-biased dispersal is the simian ancestral type or a derived trait of Old World monkeys, using the Maximum Parsimony Method

Because the ancestral types of both Strepsirrhini and Tarsiidae are male-biased dispersal (shown by ♂), then through to the Haplorrhini ancestral type is male-biased dispersal. Thereafter, if it is assumed that the simian ancestral type was also male-biased dispersal, then a change to female-biased dispersal (shown by ♀) occurred twice in the New World monkey lineage and the Hominoidea (apes and humans) lineage (a). On the other hand, if it is assumed that a change to female-biased dispersal (shown by ♀) occurred at the simian ancestral stage, then additionally a change to male-biased dispersal happened in Old World monkeys, also indicating that a total of two changes occurred (b).

or a derived trait of Old World monkeys, is finally equal (Figure 4.7). If we were to incorporate the claims of Di Fiore and Rendall (1994), as discussed above, then male-biased dispersal is even more likely to be a trait of Old World monkeys. As may be understood from the story so far, their claim that female-biased dispersal is a primitive trait is clearly wrong, and the primate ancestral type is definitely male-biased dispersal. At the simian ancestral type stage, however, it became female-biased dispersal, and thereafter can be thought to have been progressively inherited, at least through to the ape ancestral type.

We would now like to surmise, albeit daringly, that not just female-biased dispersal, but also the one male-one female social structure were successively inherited from the simian ancestral type through to the ape ancestral type. There is one reason for this, and that is that the one male-one female trait matches very well with female-biased dispersal and always appears to evolve in concert. It is also apparent that this trait is linked to the monogamous mating system, infant care by males and territoriality. As already indicated, territoriality, monogamous one male-one female groups and infant care by males were all seen in the ancestral type of both New World monkeys and Hylobatidae. While dispersed pairs rather than true one male-one female groups, we would ask the reader to recall that in Strepsirrhini the same tendency was seen as a derived trait.

Selection in sex-biased dispersal: Female-biased dispersal, one male-one female groups

Why then, do female-biased dispersal, one male-one female group formation, monogamy, infant care by males and territoriality evolve in concert? This issue cannot be resolved without scrutiny of selection pressures that bring about sex-biased dispersal, outlined as a hypothesis earlier in the chapter.

Dobson, in his thesis, proposed the local mate competition hypothesis of sex-biased dispersal, which had the shortcoming that the number of mammalian species he studied was small. Handly and Perrin (2007) added new data, including some from molecular genetic evidence discussed earlier, and re-reviewed the tendency for sex-biased dispersal in mammals. As a result, they confirmed a tendency towards female-biased dispersal in the monogamous pattern and male-biased dispersal in the polygynous pattern, but because there are exceptions, at the same time took a hard line on Greenwood's (1980) local resource competition hypothesis, arguing it was overly simplistic. The red howler, the western gorilla and the mountain gorilla all manifest female-biased dispersal as well as the polygynous pattern that Handly and Perrin deemed the exception, and all certainly have female-biased dispersal and clearly do not match the hypothesis.

There is a certain impossibility in seeking to treat the polygynous pattern of groups formed by many primates and the solitary polygynous pattern typical of mammals equally. This is because, as theorized in socioecological models, by staying in a group there is a group function that arises involving the shared protection of female resources by kin-related males, and of food resources by kin-related females. The advantage of remaining in familiar country, of obtaining information about food resources that are also applicable to the solitary primates theorized in the hypothesis of local resource competition, is dimensionally very different.

In contrast, it is clear that the two examples deemed exceptions for manifesting male-biased dispersal, despite a monogamy pattern, were incorrectly quoted. Irrespective of whether or not they can be explained by the local resource competition hypothesis, at least for primate species that form monogamous one male-one female groups, it would seem that identification of female-biased dispersal is certain.

Why is this the case? If we recall Greenwood's (1980) local resource competition hypothesis mentioned above, it is very obvious that female-biased dispersal, one male-one female groups, monogamy, infant care by males and territoriality are all connected. Unfortunately, however, it is difficult to conceive that this explanation applies without amendment to one male-one female primate groups. The one largest difference is that despite the identification of infant care by males, compared to birds, it is uncommon for males to collect food and give it to both infant and mother. No doubt there is the fact that carrying an infant over familiar country must be convenient, but it would be far more convenient for females that give infants nutrition by suckling. There is also no evidence, as is known in birds, that females choose a male for a mate on the basis of the value of resources within the territory demarcated by the male. In primates that form one male-one female groups, there is at least evidence of staying in the natal group for a while, even having reached sexual maturity, to act as helpers by protecting territory and helping with younger siblings, and most typically it is the male offspring that stay the longest. Given this, it may be that because competition between males over females is weak in the monogamous pattern, there are few disadvantages for a father if a son demarks a territory nearby, and even after the son has demarcated his own territory, in some cases there may be an advantage in father and son mutually tolerating each other and cooperating to address unwanted intrusion by others. On the other hand, a mother is in competition with a daughter over food. It may be the interaction of both factors that brings about female-biased dispersal. Ultimately of course, because the significance of sex-biased dispersal is avoidance of inbreeding, we imagine it may be the result of local resource competition, local mate competition and local resource enhancement at work. Pleasing everyone, and speaking frankly, it must be recognized that there are elements about which there is insufficient evidence, but the reason why there is a connection between such traits as female-biased dispersal and one male-one female groups, and so on, probably lies therein.

In conclusion: A new hypothesis of the social structure of early humans

Finally, therefore, let us follow Itani's lead, but, using a different basis for our argument, let us also turn our thoughts to the social structure of early humans. The synopsis has already been established. In a continuous succession from simian ancestral type to ape ancestral type, it may be that a series of social traits—female-biased dispersal, one male-one female groups, monogamy, infant care by males and territoriality—were also inherited by early humans. Proposing a rather daring hypothesis like this given that extant apes have lost nearly all these traits aside from female-biased dispersal does have some grounds, albeit somewhat shaky. That is because the reason that a series of traits demonstrated by the ape ancestral type have been lost by extant apes is clear. As bodies became larger, the weight of infants became lighter relative to the mother's body weight, and for that reason offspring grew up, even without involvement by the father. For that reason the male was released from the investment of energy in rearing a single infant, and was able to turn that energy toward acquiring many females. That led in turn to the formation of polygynous semi-solitary primates, or one male-multiple female groups and promiscuous multiple male-multiple female groups.

Let us return to the topic of early humans. At the outset I wrote that we cannot capture our way of life from paleontological evidence, but in fact, there is one single, but important piece of information it provides. This is the sexual dimorphism of bones and teeth, which readily remain as fossils. Compared to species such as gorillas that form one male-multiple female groups in which there is fierce competition between males over females, in species that form one male-one female groups in which competition between males is weak, such as gibbons, sexual dimorphism is known to be significant, and in this regard, *Australopithecus* had been thought to form polygynous mating systems. However, a result that overturns this hypothesis has recently been obtained by increasing and re-analyzing samples to find that *Australopithecus* formed monogamous mating systems (Reno et al. 2003). In the very earliest of humans, *Australopithecus ramidus*, sexual dimorphism in the canines has been found to be relatively small (Suwa et al. 2009). The isotopic composition ratio of radioactive strontium in the molars of *Paranthropus robustus* and *Australopithecus africanus* as compared with the same in flora and fauna in the vicinity of their excavation, were similar in a large tooth thought to be that of a male; but by contrast only half were similar in small teeth, therefore suggesting female-biased dispersal (Copeland et al. 2011). Considering the high degree of connectivity in the series of traits, could it be a vain hypothesis to simply write off the depiction of early humans as having demarcated a territory, living in one male-one female groups with monogamous mating systems, and having manifested infant care by fathers, as a pipe dream?[21, 22]

In this chapter we have explored the evolutionary history of primate groups from the Strepsirrhini ancestral type sixty-three million years ago to early humans six to seven million years ago. Small, solitary primitive primates evolved to form one male-one female groups for some twenty-three million years. Like the extant titi and owl monkey, they had a monogamous mating system, and the male engaged in infant care to protect the offspring of his own group from infanticide. Both male and female offspring would emigrate from the natal group on reaching sexual maturity, but male offspring would demark territory near the parent and cooperate with them against invaders. It would appear that this series of traits was preserved as they increased in size to early humans, and we looked into the fact that there was an established historical flow to evolution. At the same time, the reason groups evolved that had a variety of sex compositions and succession patterns, such as one male-multiple female, multiple male-one female, multiple male-multiple female, and matrilineal, patrilineal and bilateral forms, most certainly related to the action of a variety of environmental selection pressures such as food, risk of predation and risk of infanticide (see Figure 4.1). Itani emphasized phylogenetic inertia, and Wrangham and Schaik highlighted environmental selection pressures. Some twenty years after their initial theories were posited, it became clear that both were essential elements in exploring the evolutionary history of primate groups. There nevertheless remain many mysteries to be clarified in the evolutionary history of groups itself. No doubt these will be resolved in the coming time by the strong ally of molecular genetics.

Acknowledgements

I would like to express my gratitude to Eiji Inoue for reviewing the first draft of the manuscript and identifying inappropriate turns of phrase, and Juichi Yamagiwa and Shinichirō Ichino for the invaluable photographs they provided. This text has been slightly revised through the addition of results of research subsequent to the publication of the Japanese version.

Article 1

A Group of Chimpanzees: The World Viewed from Females' Perspectives

Noriko Itoh

This is a story about the wild chimpanzees that inhabit the forest of Mahale (See Photos A1.1, A1.2 and A1.3). There are many kinds of chimpanzees, but in this article I pay particular attention to females. When I first began studying chimpanzees, I was told "Females must be boring because they are only interested in nurturing and eating". Regardless of what they are interested in or not, they certainly appear to be very indifferent in the way they move away from others they meet without any interaction, even after spending time only with their children, or when they unexpectedly move away from others even after intimate grooming has taken place. However, this inclusion of *partings* as an ordinary state is one of the characteristics of chimpanzee groups. This article discusses the aspects of their groups elucidated not only by meetings but also through partings.

Series of <events>

The term "group" evokes various images. There are countless forms of groups, such as: those in which members unite to work toward a common goal; those in which

Photo A1.1 Mahale Mountains National Park

The Mahale Mountains chain is on the left with Lake Tanganyika on the right. M Group chimpanzees, the subjects in this article, inhabit the undulating forest developed between the mountains and the lake, known as one of the most humid areas in Tanzania.

Both photos taken by Michio Nakamura.

Photo A1.2 Chimpanzee male

Bonobo (adult male on the right) is looking at Wakusi (adult female on the left) eating a piece of fruit. Bonobo was just watching and Wakusi kept eating. Wakusi seemed uneasy and gradually turned her back on Bonobo.

members have clearly distinct roles; those filled with patterns of domination and subjugation; and those in which members gather more loosely without any clear or common reason. I would like to start with the image of the group to be discussed in this article.

The image of the group in question here is a matter that will be ruined as soon as it is termed "group". The act of verbalization is nothing but an act of distinction. Non-human primates do not name their groups. Instead, they create <events> by connecting one's actions with others', and by repeatedly creating <events>, they form a group (see Itoh 2003 for details).

Imagine a situation involving two individuals, A and B, who can see each other. If A moves in one direction and keeps moving while B remains, A and B will eventually move out of each other's sight. Conversely, if B moves in the same direction as does A, A and B will continue to be able to see each other. The change caused by B's act is a state which I call an <event>.

Such moment of <event> creates aggregation, which is distinguished from the others, and is observable. Given that the <events> are repeated, if B moves in the same direction as A, it is described as the act of *following*, while if A stops moving until B starts moving, it is described as the act of *waiting*. Either movement is one of various possibilities that are actualized as an act depending on the other's movement.

Photo A1.3 Chimpanzee female
Ako (an adult female) carrying a child on her back is just about to cross a river. Some were moving
ahead, and others were falling behind while still eating.

The process in which an <event> is created and each <event> creates other
consecutive <events> is in question. Perhaps, such a process is a basic phase of groups
in the sense that it is common in any group situation.

From the sociology of integration to that of fission-fusion

According to the above depiction, there is no guarantee that the group thus formed
will continue to exist. It may last only for a moment or for an hour, half a day, a day
or longer. As the basis of sociological analysis in primate studies, the focus has been
on the longest succession of each primate species. This succession is called a Basic
Social Unit (BSU) (see Chapter 1 of this volume). In principle, a BSU firstly needs
to be determined based on individuals' acts, but once it has been established, the
individuals determined as the members of the same group play a leading role in

analysis regardless of their actual behaviors, as if individuals are an element of a BSU which is supported on the premise of the BSU. In other words, these individuals will be reconnected according to blood relations or a dominance hierarchy or the like, and will be portrayed as a single structured *whole*.

In the case of chimpanzees, however, this is problematic regarding the existence of females. Past studies have highlighted only some males who are diligent in ascending or maintaining their social status in the dominance hierarchy (e.g., de Waal 1998). Such behaviors have as yet been rarely observed in females, which in turn makes it difficult for researchers to locate females in the concept of a structured whole. Moreover, females have been regarded as "asocial" due to their few marked social interactions in comparison with males. This conclusion is in part a result of the fact that researchers have evaluated female behaviors on the premise of a structured group; a picture extracted from the frequent and particular types of interactions. Junichiro Itani, who studied primate society from the direction of "integration" as a social structure, states as follows:

> 'Meetings' and 'partings' are equivalent as a social phenomenon, however past studies,
> early ones in particular, have regrettably disregarded 'partings' and been devoted to
> investigating the internal structure of troops as a social unit, in other words, *the sociology
> of integration*. (Itani 1987b: 37–38, emphasis added)

Each independent chimpanzee individual leads their life in a cycle of meetings and partings. To approach how they experience "group", I would like to deepen my understanding of such fission-fusion not through the "sociology of integration" but rather through the "sociology of fission-fusion". Looking at this from the perspective of females who are considered asocial should shed new light on chimpanzees' sociality, or their group life.

Wholeness created by fission-fusion

Chimpanzees live a nomadic life moving from place to place in wide forestlands. They continuously meet and part with other chimpanzees during this process. When they meet other chimpanzees, they may stay with them for a while, participate in grooming, or simply move away without any interaction, which is the most common outcome. If one continues to observe whom a chimpanzee encounters, the number of individuals will peak after a certain period (for example, it was around sixty in the recent study of M Group in Mahale.). The observer will determine it as a BSU at

Photo A1.4 Chimpanzees moving along the observation trail

When we meet chimpanzees in the forest, they are usually with a few others as shown in this photo. If only the young and the adults are taken into account, the number of such an association will be three to four on average (Itoh and Nishida 2007). The constituent membership of this association continuously changes, again and again, at a bewildering rate. This is the phenomenon called fission-fusion. It is difficult to predict who is about to arrive and from where.

this point; although, in Mahale no matter how long one may observe, there will be little chance to witness this *whole* (BSU) in one panoramic view (see Photo A1.4).

For instance, when Gwekulo, an adult female, was followed, at times she came across several other chimpanzees together while on the move, and at others several chimpanzees moving together passed through where Gwekulo stayed alone. Sometimes, chimpanzees coming from different directions arrive and meet at one location. Importantly, they repeatedly meet and part with the same individuals in the course of their daily lives.

Although the above was from Gwekulo's perspective, the same kinds of <events> also simultaneously happen to other individuals. In other words, an individual who concludes one <event> with Gwekulo will create another with a different individual at a different location. Thus, the concluded <event> will lead to many other <events>. When this succession continues it can be examined from afar, enabling the observer to extract a BSU as a *whole*. But again, each <event> cannot provide a panoramic view of the *whole*.

As we have seen, there is no *whole* that one can observe in one view, and yet a kind of *wholeness* in the fission-fusion process is perceived beyond the limited time and space of the "here and now", while remaining within such time and space. This is simply because partings create an <event> in the here and now and the subsequent

Photo A1.5 What is approaching from the direction of gaze?

Visibility is very poor in the forest of Mahale. Though the observation trail is wide enough for one person to have relatively good visibility, the range of view is easily obstructed due to its twists and turns. This is a photo of Ako, a mother chimpanzee (behind on the left) with her two children. They were sitting along the trail but, no matter how hard they looked, nothing could be seen beyond the bend in the trail ahead. Subsequently it was literally 'confirmed' that other chimpanzees were coming. The baby looking into the camera seemed curious about me/the camera within his range of sight, rather than the approaching creatures beyond his sight.

partings in this created <event> are perceived as acts leading to create subsequent <events>. I suggest similar perceptions exist for chimpanzees, as discussed below.

As visibility is very poor in the forest of Mahale due to the density of vines and bushes, it is impossible to know who/what is approaching until it comes within a very close range. In this environment, chimpanzees try to see (=meet) others even before they actually meet. As you can see from Photo A1.5, Ako is trying to see who is approaching. I, who am following her, am also trying to see what is ahead. Ako and I continue to gaze intently in the direction of the faint sound of approaching footsteps on fallen leaves (it can sometimes be a subtle smell wafting in the air that alerts one to the approach). We "know", at least, that it is a chimpanzee that is approaching even though it is still out of visibility (though I am not sure whether they can distinguish "who" is approaching). It is sometimes difficult for me to judge whether the animal is some unknown creature or a chimpanzee. At those times, I always look at the chimpanzee nearby and based on their attitude I can tell that it is a chimpanzee and not another animal approaching.

Such prediction is possible probably because *wholeness* is created by connecting <events>, and this process is stable enough to be referred to by chimpanzees. Human observers inadvertently assume an encircled *whole* because of such stability, but the *wholeness* continues to be a dynamic process which is always re-distinguished from the others by concrete <events>. The following is one of the simplest examples of such *wholeness*.

Wholeness and shimmering boundaries

As previously noted, if one continues to accumulate the number of participants in <events>, the "group" will peak after reaching a certain point. In the case of females, however, this number may completely change once it has reached its apex. Most males stay where they were born and raised, meeting and parting with familiar individuals, whereas a majority of females, at a certain point in their lives, come to meet and part with totally unfamiliar individuals. This phenomenon is called an intergroup "transfer" in general terms. Local research assistants call such females involved in the process *mgeni* ("guest" in Swahili). These females may continue to stay or suddenly move out. As discussed below, we are able to catch a glimpse of the shimmering boundary of *wholeness* from the behavioral patterns of guests and other individuals.

One day, I entered the forest accompanied by another researcher as well as local assistants. We located chimpanzees on some trees, guided by their vocalizations. They were on a treacherous slope but we were able to observe them in tree crowns almost at eye level. It appeared to be difficult to follow them as a deep valley led to a steep rise to the east. Thus, we decided to split up to at least identify the chimpanzees in the area. As soon as we began observing, we noticed an unfamiliar female nearby. Surrounding chimpanzees seemed to direct their attention toward her. Perhaps that is why we were able to notice her quite quickly.

Fuji, an adult female who was in the same tree as the guest and closest to her, displayed threatening behavior, waving her hands in front of the guest's face as if to shoo her away. Then, in the next moment she stretched her hand out toward the guest. Again, she displayed the threatening action but then started grooming the guest. The guest was looking around restlessly, sometimes grimacing at the threats of Fuji but also letting her groom her without resisting. These guest's behaviors were also, if described from the binomial confrontation between hostility and affinity, both hostile and friendly without remaining one or the other.

Previous studies have emphasized hostility between guests and other females, but what was observed above was not something that simply highlighted either hostility or affinity. Such apparently incomprehensible ambivalence did not indicate that there was something occurring between chimpanzees' groups; as has been generalized, "chimpanzees' intergroup relation is hostile". Even so, it still differed from encounters observed before that time.

Threatening behaviors displayed by Fuji created an *unusually* tense atmosphere, because females rarely fight each other in comparison to males, and also because it was displayed under the gaze of surrounding chimpanzee observers. The guest was not behaving in a particular way to draw such attention, but her reaction also helped create the atmosphere. She was just there, neither leaving nor displaying dissatisfaction by screaming or barking nor retaliating, but just patiently continued to stay under Fuji's threat. Because of that, Fuji's hostile behavior did not settle into a mere antagonistic interaction. The situation changed in the next moment when Fuji stretched out her hand toward the guest. The guest again continued to stay, which rendered the interaction amicable.

The repetition of such ambivalent negotiations between females is *unusual*. They do conduct alternate interactions of fighting and reconciliation, but if it had conformed to *usual* behavior, it would have not alternated repeatedly over such a short period. Such interactions, which seem to be ambivalent in terms of continuation, indicate that what is happening is NOT a phenomenon of previously bounded intergroup relations. Rather, in this instance the shimmering boundary of *wholeness* makes a brief appearance.

At some point, the guest may come to participate in the creation of *normal* <events>. She will initially follow and stay close to one particular individual, as the repetition of meetings and partings is not as straightforward as it may seem. At times, the guest may change her precursor, but her specific behavioral pattern will continue for a while. This pattern enables the guest to obtain information about the new environment. However, it is not enough for the guest to meet and part with others in such a wide forest (ca. 30 km^2 in Mahale) only to source the location of food. This is possibly because the *wholeness* in *usual* fission-fusion will be realized only when each individual's concerns (e.g., when/where to part, where to go next and what to eat) are connected with <events> which, whether visible or invisible, are created and concluded everywhere.

From existence to act

As far as the collective behavior theory based on the male-specific hierarchical structure is advocated, chimpanzee females will continue to have no place to settle. To be resolved, females have been excluded from BSUs, or ranked under all the males to fit into the hierarchical structure. But even when excluded from BSUs, they associate with other males and females. Even locating them as subordinate to males, they do play with them if they so desire, ask them for food if they want it and can send them away rather coercively if they wish.

Females act more autonomously in comparison to those males who are dependent on strong relationships with other males. Females are thus not submissively locked into the hierarchical order. In this regard, females certainly do seem to create rifts in the "group", on the premise that a group is a single whole. Conversely, their apparently indifferent behavior in meetings and frequent partings create the *wholeness* of fission-fusion by connecting <events>, no matter how superficial or inconspicuous they may seem. This *wholeness* is not clear, but is in fact a concrete process.

While remaining in this concrete process and by doing so, females also bind such *wholeness* in the course of their lives. This boundary is defined based on neither complete rejection nor acceptance. Both of these continuously appear in a cyclical fashion and by such ambivalent interactions, *wholeness* is bounded.

Sometimes, completely hostile interactions between males of *different groups* are thought to indicate groups and the intergroup boundary. However, the same violence is also directed *inward* (de Waal 1998). If one male in one "group" uses violence against another male in another "group", researchers regard this as an *intergroup* conflict, whereas if one male uses violence on another male in the same "group" they view this as an *intragroup* political struggle. It is, however, nothing but an interpretation of male violence based on the "group" (more accurately, a BSU as a whole). Alternatively, females create *usual wholeness* through the continuation of creating <events>, and establish boundaries through the heterogeneity of patterns of negotiation itself.

Chimpanzees have become infamous as violent apes (de Waal 2005a), but from female behaviors, different aspects that may be expressed in such terms as autonomy, flexibility and tolerance for ambiguous interactions emerge. Such aspects are directly connected with the images of their groups. Apart from the established depictions of their groups as well-defined, orderly and previously bounded, new images that are ambivalent and yet concrete and dynamic seem to appear.

Female behaviors discussed in this article are not limited to females, and not all females are the same. The common images of chimpanzees seem to be based on descriptions of males' behavior, but those who have been the subjects of investigation in fact only comprise a minority of all males and, even for this minority, only certain periods of their lives have been highlighted. In terms of their long lives, studies of wild chimpanzees have just begun. Different phases of the Mahale group may be unveiled through continuous observation of these long-living creatures with marked individuality.

Part II
The Organization of Social Groups

5 The Ontology of Sociality: "Sharing" and Subsistence Mechanisms

Keiichi Omura

The structure of reciprocity and the formation of groups

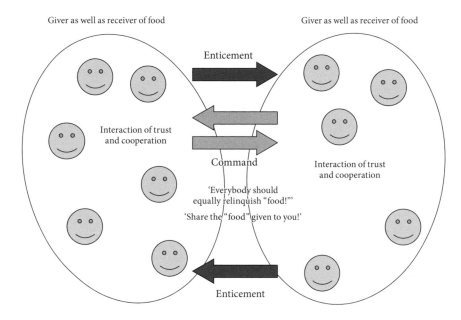

Analysis of the subsistence mechanisms of the Inuit reveals that to create groups of equal individuals who trust one another through the sharing of food it is vital that members of the group create and share a symbolic representation of "commander" who is imagined to be external to the group and to command all of them to equally share food. This applies to reciprocal gift giving, in which two or more groups command each other to relinquish their own resources and simultaneously share within their own group the resources received from the other group/s. The structure of reciprocity can be understood as being derived from the sharing system in the Inuit subsistence framework, which offers a glimpse into the primordial mechanism of human group formation.

Introduction: A unified world as a starting point

To achieve a universal understanding of human social groups, we should not automatically take for granted either the biological or the cultural meanings ascribed to the category "human", but rather start with the following question.[1] What is the mechanism by which certain forms of human social groups are differentiated and realized from the midst of a sea of various exchanges that unfold between the varieties of entities in the universe, of which humans are one? In other words, what are the kinds of communication on which the various states of human sociality are based, and what are the mechanisms by which they are constructed so as to emerge as "human" social groups from the midst of a variety of other exchanges? And, if there are particular characteristics of human sociality, what might they be?

The solution to these questions that is provided by the Inuit as they go about their subsistence activities[2] is instructive in terms of addressing the above issues. As I have explained in detail elsewhere (Omura 2007a), the Inuit unfailingly practice their subsistence activities on the basis of principles which guide them in their interaction with various environmental factors, of which they see themselves as a constituent; and they weave and maintain their "human" (*Inuk*) social group through this web of circumstances that is comprised of undifferentiated categories or, in other words, is a unified world. In the sense that the way that this sociality is understood is, at the very least, not along the lines of a bias towards a "human/nature" dualism, we can see it as being based on a considerably more universal perspective than that of the modern West. The means of identifying a "human" social group from amongst a sea of interactions between various entities as the Inuit carry out their subsistence activities without regarding the "human" category as obviously distinct provides us with clear guidelines for considering the issues of human sociality and groups from a universal perspective.

What, then, is the solution that Inuit subsistence provides to this problem? Also, what can we learn from this Inuit way of group formation? In this chapter, that solution will be made clear by scrutinizing the means through which the Inuit create their own social groups as they tune their interactions to a variety of environmental factors in the course of practicing their subsistence activities. In addition, this chapter will consider the principles provided by Inuit subsistence mechanisms in relation to the historical evolutionary basis of human society.

Subsistence activities in Inuit society: Systems of social interaction

Since the 1980s, an ongoing revision has been conducted regarding anthropological subsistence studies of the Inuit of the Arctic region, which had tended to
overemphasize adaptation to ecological circumstances. The new focus has been on
the social function and symbolic meaning of Inuit occupations (see, for example,
Bodenhorn 1989, 1990; Ellanna 1991; Fienup-Riordan 1983, 1990; Kishigami 1996,
1998, 2003, 2007; Nuttall 1992; Omura 2007a, 2007b, 2008; Stairs and Wenzel 1992;
Stewart 1990, 1991, 1992, 1995, 1996; and Wenzel 1991). These studies have shown
that even today—after having undergone rapid lifestyle changes since the 1960s—in
Inuit society subsistence activities not only operate to maintain social relations
via the distribution of resources including meat, they also function to preserve
relationships with nature, not least of all with wildlife.

Following the period from the 1950s to the 1960s, when, under the Indigenous
Peoples Policy of the Canadian federal government, the Inuit people were forced
to transition from their migratory lifestyle governed by seasonal cycles to a settled
existence, a wave bringing radical change swept over every aspect of their lives.
Through sales of items such as skins and carvings and also as a result of engaging in
wage labor, they have come to be dependent on the world of the industrial capitalist
economy; through the permeation of systems such as those of formal education,
healthcare, welfare, law and the monetary economy, they have been integrated
into the Canadian nation state; and they have been exposed to the mainstream
culture of mass consumption that has come flooding in through the mass media.
As a consequence of all this, today's Inuit are living in a high consumption society
that is no different from our own.

However, even in these conditions, the subsistence activities of the Inuit people
have not lost their importance as the foundation that supports both their lives and
their identity. The forms that these activities take today have certainly changed
considerably; many hunters now engage in wage labor and participate in subsistence
activities on a part-time basis. This is the result of the fact that activities such as
hunting have become highly mechanized through the use of equipment—including,
for example, high powered rifles, four-wheel drive buggies and metal boats with
outboard motors—and also as a result of the need for money to meet the costs of
procuring and maintaining this equipment and for buying gasoline and ammuni-
tion. And yet, despite all this, subsistence activities are being vigorously pursued,
and it is even said that "Inuit who do not engage in subsistence activities are not
Inuit" (Omura 2002a; Stewart 1995, 1996). In addition, even though using cash in-

come to buy processed foods has become general practice, the meat from wildlife hunted through subsistence activities is venerated as the indispensible "real food" (*niqinmarik* or *niqituinnaq*) that maintains ethnic identity, and the distribution of this meat continues to function as one of the cornerstones of the preservation of social relations (Kishigami 1996, 1998, 2007; Stewart 1992; Wenzel 1991).

For example, in the case of the Inuit from the village of Kugaaruk in the Canadian central arctic, before the move to permanent settlement in the 1960s there were five types of meat distribution: sharing meat with several set exchange partners (*niqaiturvigiit*); distributing meat to other tents and dwellings (*payuktuq*); giving meat to visitors (*minnak*); inviting particular people to a meal (*akpaaqtauyuk*); and the distribution of meat by hunters (*ningiq*). This system of meat distribution performed the function of preserving and regulating social relations within extended family groups, which were the basic units of social life, as well as social relations between extended families (Damas 1972; Kishigami 2007). Although some aspects of distribution, such as *niqaiturvigiit*, ceased following the permanent settlements of the 1960s—as lifestyles and the manner in which subsistence activities were carried out changed as a result of the introduction of high-powered rifles, for example—the distribution of meat continued to be practiced on a daily basis, along with its function of preserving and regulating social relations (Kishigami 1995, 1996, 2007). Relatives, who form the most basic unit of Inuit social relations, are "related people who may go away but come back and then share food, help each other, and stay together" (Balikci 1989: 112). This sharing of "real food"—that is, meat that has been obtained by hunting—is at the heart of the construction of social groups.[3]

Moreover, it has been shown that subsistence is also a means of preserving their bonds with "land" (*nuna*), establishing reciprocal relationships between them and wildlife (Bodenhorn 1989, 1990; Fienup-Riordan 1983, 1990; Kishigami 1996; Omura 2007a, 2007b, 2008; Stairs and Wenzel 1992; Stewart 1991; Wenzel 1991). The Inuit do not see wildlife as mere resources, and nor is the relationship between humans and wildlife a one-sided affair along the lines of "kill and consume/be killed and consumed". It is thought that all wild creatures possess a "spirit" (*tagniq*) with a human form, and that they are "non-human persons" with an autonomous will. In addition, it is thought that the "spirit" of each wild creature is immortal, and that in order for this "spirit" to be reborn in a new body its flesh must be consumed by people possessing both the appropriate intentions and attitudes.

Hence, from the viewpoint of wildlife, the capture and consumption of their bodies for the purpose of sustaining human life are acts that lead to the rebirth of their own "spirits" in new bodies; and, in this sense, through the medium

Photo 5.1 Fishing with a stone weir and capturing arctic charrs with a fishing spear
At Kugaaruk, Nunavut, Canada, in 1991.
Source: Courtesy of Henry Stewart.

of subsistence, humans and wildlife are bound in reciprocal relationships. The following is an example of the reciprocity that exists between seals and humans. Seals, as "non-human persons", voluntarily give the body in which they reside to humans, thus helping the humans to survive. Meanwhile, through sharing and consuming the meat that they have been given by the seals in a manner that is profoundly respectful to the latter, humans are helping the seals' "spirits" to separate from their bodies and to assume a new form.

The important point here is the belief that in order for these types of reciprocal relationships to be established between humans and wildlife, hunters must carry out subsistence activities with the appropriate attitudes and intentions. What is meant by this is: firstly, that hunters show profound respect for wildlife; secondly, that they intend to use the body of their prey for food; and, finally, that they do not intend to monopolize this food but will share it with others (Kishigami 2007; Stairs and Wenzel 1992; Wenzel 1991). Consequently, practicing subsistence properly is not simply a matter of hunters having a well-developed knowledge of the environment and wildlife or of excelling at hunting and fishing, they must show respect for wildlife, and they must not monopolize but rather share their meat.

This way of thinking remains extremely widespread in contemporary Inuit society. For example, I was in the village of Kugaaruk in the Nunavut Territory of Canada in December 2004 to learn the Inuit language, and when I carelessly let slip the remark "*tuktu ihumakittuq*" (caribou are stupid), the hunters around me scowled. This was because it is thought that all wildlife have a "spirit" and are constantly paying attention to the words and thoughts of humans, with the consequence that wildlife will not approach those who say bad things about them. Hunters have also been repeatedly warned by elders that wildlife are not to be excessively injured or tormented when they are captured, and they have been instructed, time and again, in the moral lesson that hunters who injure polar bears frivolously and who do not administer a fatal blow will be visited by misfortune and poor yields. There was also a case, in the summer of 1989, when a group of young hunters returned home leaving behind a caribou that they had shot, only to be severely scolded and sent back to retrieve its body. This was because it is considered insufficient to be certain of having killed a wild animal; one must ensure to bring it home and make good use of it by eating it. Furthermore, one of the Kugaaruk elders recounted the following story during my interviews.

When I caught a fish with my *kakivak* (fishing spear) the very first time, at Lake Inirjuaq, right here [indicates location on a map], I cried because I could not keep it. My uncle who was fishing nearby came to me and filleted the fish as soon as I had pulled it out of the fishing hole. Then, the people from the camp ate it. I did not want to give them my fish, because I was a child. I cried a lot. It was always like that in those days. Our ancestors did the same thing. Then, my uncle did the same thing again in the spring of that same year. When I pulled a big trout out of a fishing hole over there at Lake Tahirjuaq, he came and filleted it again. As soon as I pulled it out, the people at the camp started to eat it. When they started to eat, they told me to eat with them. However, I cried again because I wanted to keep it for food. I did not want to share it. I still remember those fish. We believe that if one shares and eats one's first game as soon as one takes it, one will have considerable luck at hunting, and become skilled at hunting animals. This is why my people always tried to consume the first of any animal I took, right away. One could be smart later in one's life, could be smart at hunting animals, if the very first of any animal he took were eaten right away, whether that be seal, caribou or fish. Because my grandfather wanted me to be smart, whenever I made my first catch the people always tried to consume it right away.[4]

As this story clearly shows, one of the conditions for being a great hunter is generously sharing and eating one's catch with others. By the same token, it is held that no matter how good a hunter's knowledge is or how excellent his skills are, if he does not show respect to wildlife and share their bodies with others in a way that is not wasteful, then wildlife will not approach him.

In this way, the subsistence that is practiced in Inuit society operates not only as a means of securing the resources needed for survival, but also to continuously build and reproduce the relationships between people and between people and wildlife. Inuit subsistence is not just a technical process for acquiring the necessary resources for survival; it is also a social process that concludes the relationships between the Inuit themselves and between them and wildlife. Moreover, it is through this social process that these relationships rely on one another in a mutual and cyclical manner. In order to regulate and preserve the social relations between humans through the sharing of "real food", this food must be given by wildlife, and in order for this to occur it is necessary for humans to share this food so as to help the "spirits" of wildlife to be reborn in new bodies. The construction of Inuit social relations is simultaneously the creation of reciprocal ties between the Inuit and wildlife, and the cyclical process of practicing subsistence is the device that continuously forges human social groups while binding the Inuit in relationships with wildlife. It is precisely because of this that even today, when as a result of immense lifestyle changes people can access resources for their survival by means other than subsistence, it continues to be a popular practice.

What, then, are the concrete mechanisms by which the practice of subsistence creates relationships between humans and forges ties with wildlife? Simply saying that the relationship of sharing "real food", which gives rise to human social groups through the practice of subsistence, simultaneously works with and creates relationships in which humans and wildlife cooperate, does not clarify the concrete mechanisms by which subsistence functions to create these two relationships. Indeed, it is a matter of course that the voluntary act of wildlife giving their own bodies to humans results in aiding their survival. How, then, is it that the sharing of food amongst humans assists the "spirits" of wildlife to be reborn in new bodies? If these questions are not resolved, then the process by which the creation of human social groups through the sharing of food is intertwined with animal-human mutual aid will remain a mystery.

In the following section, as we examine this question, we will specifically trace the process by which the practice of subsistence connects human relationships with those with wildlife, and we will also consider the mechanism by which Inuit society is created via the latter.

"Trust" and "enticement": "Humans" and "wildlife" in Inuit subsistence

As discussed above, in Inuit subsistence, human social relationships are regulated and preserved by the sharing of "real food", and humans and wildlife form reciprocal relationships in which they cooperate. Seen from the Inuit's standpoint, relationships amongst humans and those between humans and wildlife are of two distinct types: "food sharing relationships" and "reciprocal relationships of mutual assistance". Consequently, through the practice of subsistence, Inuit social groups are formed by the differentiation and simultaneous coupling of human "food sharing relationships" and "reciprocal relationships of mutual help" regarding wildlife. How exactly, then, are these two relationships coupled, while also being differentiated, in the practice of subsistence?

The first thing that needs to be made clear when considering this issue is that it is not humans but wildlife who hold the initiative in Inuit subsistence. In order for wildlife to give their bodies to humans, it is essential that the humans who receive these bodies always and without fail share and consume them as "real food". If this does not happen, then humans will cease to receive these bodies from wildlife. However, it is up to the wildlife as to whether or not they relinquish their bodies to humans. Even though wildlife can only be reborn once humans share and eat their bodies, whether or not a wild animal wishes to be reborn is a matter entirely for the will of that animal. Although humans generally need the bodies wildlife give them as food, the wild animals' need for help from humans only ever arises when they wish to be reborn, and even though the term "cooperating" is used, this is an asymmetrical form of cooperation in which it is always humans who are dependent on the will of wildlife.

Consequently, the voluntary giving of the wild animals' bodies to humans is, in effect, a matter of humans sharing and consuming those bodies, and an edict to humans to help the animals' "spirits" to separate and be reborn in a new body. But, no matter how much humans keep sharing and consuming these bodies, this does not amount to a directive from humans to the wildlife that they relinquish their own bodies. Even so, it is not the case that humans have no effect whatsoever on wildlife. Through the subsistence process of sharing and consumption, humans enable the rebirth of the wild animal's "spirit" in a new body, and this becomes the incentive for the wild animal to voluntarily give its body to humans. As a result, while in Inuit subsistence wildlife order humans to share and consume the bodies that they have given up, humans also provide the enticement for wild animals to voluntarily relinquish their own bodies.

The difference between these directives and enticements is important. This is because whilst both entail a positive appeal for the other party, the relationship between oneself and others is completely reversed. In contrast to directives, which appeal to the other party from a strong supportive position, enticements appeal from a weak position that subjects one to that other party. This is due to the fact that enticements "are a positive appeal to others, but, nonetheless, one stands in a weak position which is dependent on the other party" (Tachikawa 1991: 39). Consequently, in Inuit subsistence, although humans and wildlife certainly cooperate in reciprocal relationships, humans appeal to wildlife on the basis of enticement, the act of a dependent, weak person, while wild animals appeal to humans through directives, the act of an in control, strong person; and, in reality, humans are under the control of wildlife. The reciprocal relationships that exist in subsistence are not cooperative relationships between equal parties; they are relationships between dominant wildlife and dependent humans.[5]

The important points here are that in subsistence it is only the wildlife that issue orders from a dominant position, and that the Inuit do not issue directives to either humans or wildlife; the Inuit are not in a dominant position with regard to any one or thing. It is wildlife that give the order to "share", while the Inuit, from their position as the dependent, weak party, merely entice the wildlife. Moreover, what emerges amongst the Inuit as a result of obeying the wild animals' directive are relationships forged by sharing and consuming their bodies; in other words, relationships in which the same act of "eating" one particular object is performed cooperatively. What this achieves is trusting relationships in which the same act of "eating" takes place, as mutually autonomous and equal individuals, none of whom order each other about from a position of dominance, cooperate and share the expectation that betrayal through such acts as seizing or plundering the food of others will not occur.

Accordingly, there is considerable meaning in Inuit subsistence regarding the construction of asymmetrical reciprocal relationships, in which Inuit entice wildlife in order to live, while wildlife command humans so that they will be reborn. This is because it is via the cycle of these asymmetrical reciprocal relationships that the Inuit, through the mediation of directives from wildlife, are able to generate equal relationships of trust in which none in their midst holds a dominant position. By entering into asymmetrical relationships with wildlife and entrusting them to issue orders from a dominant position, the Inuit completely forbid anyone from issuing orders, and they have also banished relationships of dominance and dependence from their midst. In this sense, Inuit subsistence does not simply create Inuit social groups while building relationships with wildlife; it is also a device for ridding these human groups of the troublesome relationships of dominance and dependence.

However, the price of freedom from relationships of dominance and dependence involves confronting a major issue in their relationships with wildlife. This is the question of what they should do in order to ensure that wildlife will give them their bodies. As far as the Inuit are concerned, they must get "real food" from wildlife both for their continued existence and for the construction of egalitarian, cooperative social relationships. However, no matter how much the Inuit continue to share and consume this food or how avidly they help the "spirits" of wildlife to be reborn, this amounts to no more than enticement, and it is not the case that wildlife will always willingly give up their own bodies to humans. In the first place, rebirth in a new body is not as essential to the daily lives of wildlife as are the food requirements of humans. At the very least, it is certain that the frequency with which wildlife desire to be reborn is not greater than the rate at which the Inuit require food, and it is a matter dependent entirely on the will of the wild animals themselves as to when they might give their own bodies to humans. This forms an extremely uncertain dynamic.

Consequently, because the Inuit cannot expect with any certainty that after having shared and eaten the "real food" that was given to them by the wildlife, and thus having helped their "spirits" to be reborn in new bodies the wildlife will necessarily give them their bodies in future, they are faced with the need to find some way of appealing to the wildlife so that they feel inclined to be reborn in a new body. If a carnivorous animal such as a polar bear counterattacks or charges, the charge must not be understood as part of the reciprocal relationship, but as the bear declaring its intention, "I am not inclined to be reborn yet", and the hunters must flee. Unsurprisingly, having shared and consumed the body given up by a wild animal, it is not simply a case of waiting for new wildlife to give up their bodies.

However, all that is permitted to the Inuit to use in making this appeal is enticement, the technique of the weak. As we have seen in our examination thus far, this is because by virtue of being the weak party with regard to wildlife, the Inuit have been able to banish relationships of dependency and dominance from their own society; if they were to order or use forms of compulsion to make wildlife give up their own bodies, then the result would be that the relationships of dependency and dominance would come back into their society via the wildlife. If a hunter who encountered a wild animal were to order or forcibly compel it to surrender its own body, then would this not amount to that hunter, via the body of the wild animal, compelling other Inuit and ordering them to share? In order to ensure that they do not stand in a dominant position in relation to their fellow Inuit, Inuit hunters have no alternative but to assume the stance of the weak party and use enticement in their dealings with wildlife.

In reality, the fact that they appeal to wildlife through enticement is clear from the nature of the Inuit hunters' knowledge of wildlife. As investigated in detail in other works (Omura 2002b, 2005a, 2005b, 2007a, 2007b, 2008), this is because Inuit knowledge of wildlife is based on a "tactic" that was characterized as a technique of the weak by Michel de Certeau.[6]

According to de Certeau (1987), a tactic is a maneuver in which one seizes an opportunity that appears for an instant in relations with another, when one is under the dominance of the other who is stronger than oneself, and makes use of that opportunity to achieve one's own purpose. Just as in jujitsu where "softness overcomes hardness", one can, while being reconciled to the dominance of the other person who possesses overwhelming strength, bring about a reversal in the power relationship by using the other's movements to guide them and intervening in the movements in the space of an instant. This tactic is not a case of infringing on the other person's freedom or of coercing the other to do anything; for the person concerned, it is nothing more than the art of enticement which guides the other's will in precisely the manner that that person freely wills. In order to make skillful use of the other party's movements, in which they are freely engaged, during the execution of the tactic the other party must be allowed to act freely and independently right to the end.

It is the art of this tactic that constitutes the main Inuit knowledge of wildlife, and this is the means by which they entice them. When, where, in what condition and what kind of wildlife were encountered; by what kind of exchange of actions were the wildlife enticed; and how was the relationship with them turned into one of "giver/receiver of food?" These types of practical methods in the form of memories, that include the starting point of one's own personal experience along with the experience of contemporary hunters and one's ancestors, are comprehensively accumulated in Inuit knowledge of wildlife. Guiding the other party—whatever action they may take—by making skillful use of their behavior, and, ultimately, mastering the art of tactical enticement that will lead the wild animal to willingly become the giver of food are achieved by means of this knowledge.

Once we understand this way in which the Inuit appeal to wildlife through the art of tactical enticement, both what is meant by the expression "spirit" and why the sharing of "real food" helps the "spirit" of wildlife to be reborn in a new body become clear.

As examined earlier, the sharing and eating of "real food" is the Inuit method for helping the rebirth of the "spirit" of wildlife in new bodies; and via this sharing, the Inuit construct social groups in which they act cooperatively in an establishment of trust. Once it thus becomes possible to act cooperatively on the basis of mutual trust, then, in the same way as food given by wildlife is eaten in common, it becomes possible for the sharing of methods for tactically enticing wildlife to be held in common. Then, as a matter of course, once these means for enticing wildlife become common knowledge amongst the Inuit the stock of methods is enriched, and when a new individual wild animal is encountered, the possibility of entering into a relationship of "giver/receiver of food" with that wild animal is increased. In short, the sharing and eating of the food given to the Inuit by wildlife is what helps the process of rebirth into a new individual wild animal that will itself enter into this "giver/receiver of food" relationship.

Consequently, the "spirit" of wildlife is the relationship between the Inuit and wild animals, and the rebirth of the "spirit" in a new body is the rebirth of the relationship of "giver/receiver of food" between the Inuit and the new individual wild animal. The reason, then, for the Inuit having to share and consume the food given by wildlife is so as to hold in common and enrich the stock of techniques for tactically enticing wildlife and to prepare for future subsistence activities by building relationships of mutual trust and cooperation amongst themselves.

By this stage, the concrete mechanism for the creation of Inuit social groups, via the relationships with wildlife through the practice of subsistence, should have become clear. Through the following cyclical mechanism, Inuit social groups are created via connections with wildlife (see Figure 5.1).

Firstly, the Inuit entice wild animals to give their own bodies and the enticed animals, by giving their bodies to the Inuit, command the Inuit to "share". By sharing and consuming the food given to them by the wildlife in accordance with this command, the Inuit create social relationships amongst themselves in which there is mutual cooperation and trust, and in the midst of this cooperation and trust they share and richly cultivate the tactical techniques for enticing wildlife. Through the cultivation of these techniques of tactical enticement, the Inuit again

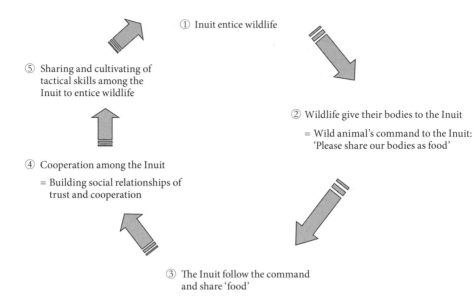

Figure 5.1 Recursive system of Inuit subsistence

entice new individual wild animals, and once again create the relationship of "giver/receiver of food" between themselves and these individuals. Thus, through the cyclical process of subsistence, the mutual acts of "enticement/command" between the Inuit and wildlife are developed as they become intertwined with those of "trust and cooperation" between the Inuit themselves, and the differentiation of "Inuit (human)" as "those who ought to trust one another and cooperate on an equal basis" and "wildlife" as "those whom we entice and whose orders we ought to obey" emerges (see Figure 5.2).[7] The consequence of this is that Inuit social groups, in which equal individuals cooperate in mutual trust, are continuously created through the intermediary of asymmetrical reciprocal relationships with wildlife.

The ontology of sociality: "Sharing" as the historical evolutionary basis of human societies

This kind of Inuit subsistence mechanism is instructive regarding a number of aspects of the historical evolutionary basis of human societies.

To begin with, it is not inevitable that humans will share and eat food equally amongst themselves. As we have seen in our investigation thus far, the Inuit do

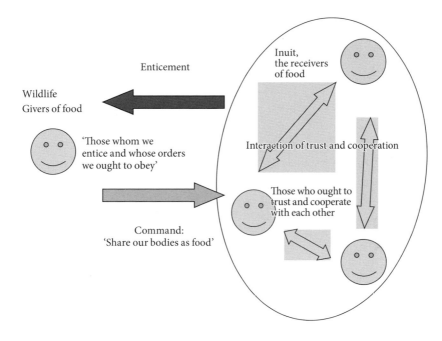

Figure 5.2 Relationship between the Inuit and wildlife

not share food as a matter of course. They share on the basis of a stern command issued by wildlife that would otherwise refuse to provide this food. Thus, the existence of at least one society that needs a directive of this kind to share food speaks eloquently to the fact that humans do not feel that sharing is an obvious necessity, and that some sort of model/criterion for doing so is required. If the sharing of food were a universal biological characteristic of humans—that is, if it were a necessity for the human species—there would be no need for them to be ordered to do so by wildlife. They would most likely simply share of their own accord. In short, the act of sharing food is unlike the universal act of eating food; it is an unnatural and artificial act. In other words, it is not a natural biological state for humans.

Put the other way around, humans are living beings who tend to monopolize food resources. Moreover, given that humans are also one type of living being, their daily survival is not merely a case of them reproducing themselves self-sufficiently on their own—they need to be supplied with food by others apart from themselves.

Photo 5.2 Drying arctic charrs
At Kugaaruk, Nunavut, Canada, in 1992.
Source: Courtesy of Henry Stewart.

In short, humans find themselves in a state of regular want, and in order to check this want they must procure food, from others, that will satisfy it. However, even if they try to obtain this food, wildlife, as living beings, also feel the imperative to go on living, and are not likely to voluntarily die and become food for others. Nor is it likely that other humans, having managed to acquire food by some means or other, and given that it is usual to monopolize rather than share, will part voluntarily with this food once obtained. For this reason, humans—and wildlife—must generally steal food from those around them. Naturally, since this applies to all humans, they are in relationships of mutually stealing food from one another.

Consequently, the natural state of all human individuals that provides the logical premise when humans form social groups is one of mutual distrust in which it must be assumed that when someone has managed to get hold of food it will be stolen by those around them. It cannot be expected that people will share food with one another, because the opposite is usual: stealing is the natural state. Of course, when there is a surplus of food, the need to steal disappears, and as a result we even see the emergence of conditions in which people share with one another. However, even in these circumstances, because it is usual for humans not to share, and because they are usually concerned with satisfying their own wants, they continue to view the others around them as potential enemies who could at any moment steal their food.

This is extremely important when considering the historical evolutionary basis of human society. Humans are always in a state of mutual distrust with all others around them concerning food—the minimum necessary resource needed for survival—and because they always see others as potential enemies, at the very least this state of mutual distrust must be dealt with in order to form any type of human group. Others who could possibly steal or monopolize the food that is the minimum requirement for existence can pose a threat to one's own survival, and it is far too dangerous to eat together or cooperate with them. When eating together at the same place and time, the other person might steal one's own food, or even if one were to cooperate in bringing down some prey, if the other were not to share that prey it would then become necessary to fight over it. The mechanism of Inuit subsistence teaches us that some form of artificial arrangement is needed to eliminate this state of mutual distrust concerning food in order for humans to be able to create any kind of group and act cooperatively.

Secondly, the Inuit mechanism of subsistence teaches us that in order to form groups in which humans are linked by equal relationships, something external that is at a distance from these relationships is needed. Undoubtedly, the sharing of food brings to an end the state of mutual distrust in which people see each other as potential enemies and brings into existence a state in which people cooperate on the basis of faith in the fact that others will not steal their food. In brief, equal relationships in which autonomous individuals rely on one another are created. However, if sharing food is not the true nature of humans, and stealing food is, then some form of compulsion becomes necessary in order to realize this sharing. Moreover, if it is someone from within a human social group who carries out this compulsion, then the groups that emerge as a result of this compulsion become not groups sharing equal relationships but ones dominated by the person who compels people to share. As a result, in order to create groups of individuals who act on the basis of equal cooperation and mutual trust, as exists in the Inuit subsistence mechanism, it is essential for that compulsion to be located outside the group.

For that reason, groups of equal individuals created through the sharing of food always seek an external entity that does not share a particular group's food but commands them to share in order to distribute food amongst themselves. The Inuit look to wildlife, with whom they do not share food, to act as this external entity, and by entrusting these wild animals to command them to share, the Inuit realize the creation of relationships of equality and trust amongst themselves. Of course, it does not specifically need to be wildlife; anyone who did not share their food could act as the external entity compelling them to share. For example, a transcendental being, such as a god, could also fulfill this function. Alternatively, even a human who had been ejected to a position outside of the group, and lived this same kind of transcendental existence, could serve the same role. Furthermore, in instances where there are several adjoining groups in search of an order to share, all they have to do is to ensure that the groups always obtain food from each other. In all groups, in order to obtain food from other groups, everyone in one's own group must equally renounce food, and although this is an inside-out process, the sharing of the act of renouncing food comes as a directive from outside the group.

What is important here is the fact that the final method mentioned above is the same as the practice of exchange called gift giving (see Diagram on page 124). As identified by Lévi-Strauss (2001), typically seen in the practice of giving away women, this exchange takes the form of a mutual act in which two or more groups prohibit the personal consumption of their own resources amongst themselves, and instead give these resources to each other. The mutual ordering of the other group to ensure that all members equally forgo food is a prohibition of the personal consumption of their own resources; and this prohibition acts as a command to exchange resources between groups. This prohibition of personal consumption is demanded so as to command the equal renunciation of resources by all members, and it is the precondition needed for sharing the resources obtained from the other group—the concomitant of this equal renunciation of resources. The "obligation to give", the "obligation to receive" and the "obligation to reciprocate", which Morse (1923–4, 1973) identified as essential elements of gift giving, are phenomena that emerge secondarily as requirements for the mutual command of groups to share equally in the renouncing and sharing of resources within the group.[8]

Consequently, we must appreciate moves to locate the source of the command to share outside one's own group, in order to construct groups of individuals who share their food and experience mutual trust, as giving rise to the form of reciprocal gift giving. The way that this reciprocity works is not by bringing about a situation of want in closed, autonomous groups through the prohibition of personal consumption of their own resources, thereby giving rise to communication between isolated groups. Groups of individuals who trust one another equally, produced through the sharing of food, are most certainly not isolated, closed groups, but the groups that begin to seek the other groups commanding them to "share" beyond the borders of their own group at the time of the group's first crisis of survival; they always seek communication with the outside. The Inuit mechanism of subsistence shows us that the gift giving-type of communication between groups is preconditioned, from the outset, by the need created by groups of equal individuals through the sharing of food to seek the other group's directives to one another, and that it is precisely the sharing of food itself that is the driving force behind communication between the groups.[9]

Finally, the Inuit mechanism of subsistence shows us that for groups of equal individuals who trust one another to be created through the sharing of food, and for gift giving-type communication to occur between these groups, it is essential to create the symbolic representation which is imagined to be external to one's own group and to command them to share food. As has been discussed in this chapter, the Inuit are able to entrust to wildlife the order that commands them to share food because wildlife symbolize "dominant living beings that they need to entice"; not because wildlife actually dominate and issue commands to the Inuit. It is this use of wildlife to symbolically represent the orderer which is imagined to command them to share food from outside their own group that provides the starting point for the establishment of the Inuit mechanism of subsistence, the possibility of the unnatural act of sharing food and the creation of social groups in which the Inuit cooperate on the basis of equality and mutual trust. This is the same in both the case of a directive to share coming from a being that transcends one's own group and in the case of groups issuing orders to one another to give food and to share equally in the relinquishment of food. Unless people imagine the being that transcends their own group in symbolic terms, and symbolize other groups as beings with whom one exchanges, then orders will not function effectively.

The important point about this is that the imagined symbolic order from outside of the group will have no efficacy whatsoever if sharing of the symbolic order does not take place within the group. If someone in the group were to assume that all the members of the group must share food in complying with the order issued by the imagined symbolic orderer, such as wild animals, then should the rest of the group not share the assumption, they would just end up having their food taken by those around them for whom it was not usual to attempt to share. Moreover, it is not to be expected that people who are not in the habit of sharing food in the first place would attempt to share the symbolic orderer that commands them to share. Accordingly, this is simply a case where the sharing of food and the sharing of the symbol have come into existence at the one time in an intertwined fashion. The subsistence mechanism of the Inuit shows us that in order to create groups of equal individuals who mutually trust one another and cooperate, it is necessary to establish the sharing of food and symbols at the same time.

Thus, the subsistence mechanism of the Inuit teaches us that in order to create groups of individuals who trust one another and cooperate on an equal basis, it is essential to transcend, at a stroke, the major obstacles of sharing food and symbols. There is no doubt that sharing food was a major ordeal in the case of humans for whom exchange of this nature was not a part of their true character as a species. We do not know with any certainty why groups appeared amongst humans or just what abilities were required in order to transcend the obstacles to sharing food and symbols. However, the most likely ontological support for the establishment of human groups of individuals who trust one another and cooperate on an equal basis is to be found in the acts of sharing food and symbols; and it is fairly safe to say that the realization, at a stroke, of the sharing of food and symbols provided the historical evolutionary basis for human societies.

6 Violence and the Autopoiesis of Groups: From the Ethnography of Pirates and Feuds

Ikuya Tokoro

The afferent/efferent nature of violence and groups

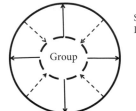

Solid lines: efference
Dotted lines: afference

Violence and group formation

(A) Conventional conceptual view of groups: the group as
protagonist (state, village etc.) exists first and carries out violence

(B) Action-centred view of groups: the boundaries of the group
carrying out the violence are recreated each time it engages in violence

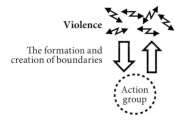

In, for example, cases where a large village splits into two sub-groups as a result of conflict, as far as the original village is concerned, violence has an efferent nature which urges division, while for the newly created sub-group, it encourages unity amongst its members (see Figure A). Thus, the efferent/afferent nature of violence is relative depending on one's point of view, and "enemies"/"allies" are duly created through conflict. Common wisdom is apt to see the group as the protagonist carrying out violence but, actually, groups and the boundaries between them are created every time that there is violence (Figure B).

Violence and groups

In the main, when discussing human life, people find violence an unpleasant and, at the same time, unavoidable topic. As we will see below, alongside its relationship with human evolution, the subject of violence has been repeatedly discussed in anthropology and other related social science disciplines. This chapter, unlike existing research to date, does not discuss violence in terms of it being directly related to human evolution. Rather, the aim here is to attempt to analyze the violence (pirating and feud disputes) that occurs in the Sulu Archipelago in the Southern Philippines, the site of the author's fieldwork, in terms of the relationship between the formation and production of groups. However, in the course of this task, the hope is to re-examine the groups that are found in human society—in more general and abstract contexts—using, as it were, an action-centered paradigm.

Our general understanding, when considering group violence in an everyday context, is most likely formed along the lines of there being pre-existing and fixed groups or categories of groups, and of these groups—as existing entities—having some form of mutual relationship (whether peaceful relations, such as night visitation marriage and trade, or violent relations, such as those found in disputes and wars).[1] However, in a deep consideration of what we call groups, it is instead possible to form the totally opposite view that "groups (categories) do not exist first, *a priori*, but rather that groups (categories) are formed (produced) through mutual acts". There are, in fact, ample cases, when one examines actual social phenomena, that hint at precisely this opposite viewpoint (let us provisionally refer to it as "the action-centered view of groups"). This chapter is an attempt to discuss the formation/production of groups and group categories by focusing particularly on the "violent" acts that make up part of the variety of human actions. While sharing a common consideration and problem consciousness with that found in Kawai's chapter of this book—regarding raiding—the discussion here is an attempt to examine the relationships between groups and the violence that exists in the societies of the maritime peoples of Southeast Asia, who have considerably different ecological and social environments as well as livelihoods from the peoples of the cattle breeding societies of Africa, as presented by Kawai.

This issue of violence and the creation of groups (and also the related topic of social order) is a subject that has been vigorously discussed by René Girard (1972), Hitoshi Imamura (1982, 1989b, 2005a) and others, as the basic motif of their social philosophy, but—as will be discussed below—it has, in recent years, also been referred to in more specific contexts such as anthropology and primatology. The first part of this chapter refers to the indications and arguments concerning violence made in the leading anthropology and related social sciences research. The second part presents a consideration of this issue on the basis of an examination of specific cases taken from ethnographic surveys conducted in the field.

Why violence? Disputes concerning human evolution

There are significantly divergent points of view regarding the question of how to understand the violence that occurs in human societies; and, in the past, this issue has often led to fierce conflicts of opinion and disputes. Foremost amongst these are the questions of whether we regard human nature as being fundamentally violent or whether the phenomenon of violence is seen as having been learnt *a posteriori*, with human society having maintained a long-term trend of peaceful existence. The conflict between these contrasting views of humankind is deep-seated. Eisei Kurimoto, in his book that synthesized anthropological research on war, summarizes his conclusion that this is a conflict between the "Hobbesian view of man" and the "Rousseauian view of man" in the following terms:

Since the nineteenth century, the so-called issue of nature versus nurture—originally, the issue of whether human aggression was an innate feature based on instinct or an acquired feature based on learning—has been the subject of dispute. This could be restated as a dispute between the Hobbesian and the Rousseauian views of man (Kurimoto 1999: 15).[2]

In this way, disputes regarding the violence that is found in human societies readily tend towards a bipolar split along the lines indicated by Kurimoto. Kurimoto has already provided a detailed and clear analysis of both of these positions, and as there is not the scope within this chapter to enter directly into any of the nuances of these arguments, I will only touch on them in the most cursory manner.

There are frequent references to the so-called "hunting hypothesis" as one setting in which we first see *man* as living an originally violent existence. This began with the interpretation of Australopithecus fossils by Raymond Dart, a scholar of prehistory, and then gained widespread recognition as a result of the efforts of the author Robert Ardrey (1961, 1976). Broadly speaking, this view has as its foundation the idea that "Australopithecus hunted other animals and in addition to his existence as a carnivore, he was a killer ape who also murdered his companions".[3] The basis for this argument was, in part, Dart's interpretation that the hole in the fossil of the Australopithecus skull had resulted from an attack by fellow members of the same species. Alongside hypothesizing that early humans took the form of "Man the Hunter", Dart argued that the origins of man's violent nature are to be found in hunting activities.

However, there have been reinterpretations of this Australopithecus fossil, upon which Dart had based his hypotheses: subsequent brain research has supported the plausibility of the argument that the hole in this skull resulted not from violence on the part of Australopithecus, but rather from a canine tooth when the individual became the prey of a leopard. In addition, there has also been considerable criticism in recent years of views, including the hunting hypothesis, that assert the original violent nature of man. In one recent example, in their book, *Man the Hunted: Primates, Predators and Human Evolution* (2005), Hart and Sussman advocate the view that rather than being a killer ape, early man was prey, and that it was not hunting activities but specifically the actions that he took to defend himself against predators that were likely to have contributed to human evolution: they put forward the "Man the Hunted" hypothesis, thus turning the "Man the Hunter" position on its head.

Things are not quite as simple, however, as being able to claim that the view that regards *man (homo sapiens)* as having lived an originally peaceful existence is the established theory in anthropology and related disciplines. It is certainly still the case that the idea that "war is a characteristic feature of societies that have developed centralized states and civilization, while in primitive societies people lead fundamentally peaceful lives"—Rousseau's image, as it were, of the "noble savage"—continues to be deeply held. Also, while there are researchers who, on the basis of studies of modern-day hunter gatherer societies, agree with the portrait of "peaceful hunter gatherer societies", there are, in opposition to this, a considerable number of advocates of the proposal that "non-westernized primitive societies were beset by group violence, including war". Jared Diamond (1992, 1997), for example, whilst calling attention to the fact that war and mass slaughter had also existed in the past in non-western societies, such as that of the Maori, expresses skepticism regarding the depiction of non-western primitive societies in terms of the "noble savage".[4] Perhaps we can understand the controversy in the field of cultural anthropology—between Lévi-Strauss, who saw peaceful exchange as the basis of human society, and those such as Pierre Clastres (1974), who were critical of this, pointing out instead that war was inherent in primitive societies—as variations on the controversy between two contrasting views of violence in human societies (the Hobbesian and Rousseauian views of mankind referred to by Kurimoto).

Beyond "Hobbes versus Rousseau"

Thus, at present, it is still the case that there are two contrasting views of how to regard violence in human societies, and it cannot be said that there has, as yet, been any decisive resolution of the dispute between the two sides.[5] In addition, aside from the preeminent point of controversy on the issue of "nature versus nurture" within the dichotomy of the Hobbesian and Rousseauian views of man, one is also likely to encounter a point of controversy on another perspective: whether violent tendencies are prevalent, as a matter of course, in interactions in human societies or whether, conversely, the existence of these societies is strongly characterized by peaceful tendencies. I would like to confirm that the foremost point of controversy—"nature versus nurture"—is ultimately a problem of logical possibilities; and the other points of controversy—whether the phenomenon of an inherent tendency towards violence exists as a practical and real problem in human society, or whether humans have coexisted peacefully—are strictly likely to be those with differing logical phases. One could, for example, mount the following argument (regardless of its merits or demerits): "the human being is discontinuous from other animals, and—in contrast to other animals—humans rely to a lesser extent on their genetic and innate behavior mechanisms, relying instead to a considerable extent on acquired mechanisms such as culture, language and social systems. Nevertheless, and for this very reason, man can be even more violent than other animals".[6] To reveal my own position—regarding the point of controversy as to whether violence is a matter of nature or nurture—I sympathize with the view that this is a matter of nurture: in other words, the position that holds that violence is primarily a matter of culture and learning rather than the result of heredity and "instinct". However, this does not necessarily mean that I unilaterally extol the Rousseauian view of society—that "non-western primitive (and pre-literate) societies, in which man originally lived a peaceful existence, are fundamentally peaceful". I acknowledge the existence of numerous cases, including my own field of Sulu, to be discussed below, that provide proof that contradicts this Rousseauian perspective, and would go so far as to consider that the violence found in the creation of order in human social groups is an important subject that demands attention. My tentative conclusion, at this point in time, is that it is more productive to ask "whether, in human societies, violent interactions can be brought about under any circumstances whatsoever and whether, under different circumstances, these could, conversely, converge towards peaceful interactions" rather than inquiring, in an essentialist way, into the two absolute alternatives—"Is man, in his original state, fundamentally (essentially) violent or peaceful?"[7]

The issue of group formation and violence

In this chapter, I do not propose to delve any more deeply than I already have into the arguments that have been made about violence to date. I would, instead, like to take up an issue that has not necessarily been dealt with directly by either side in the dispute. That is the question of violence in the formation and production of groups. There may well be readers who feel that this reference to "violence in the formation of groups" is bizarre. Even if violence has tended to be seen as the cause of disruption, division and the dismantling of order in society and groups, this is because it seems natural to think that it is the very antithesis of the formation and maintenance of order in such collectivities. In fact, Hobbes, who placed tremendous stress on the origins of violence in human society, saw this "violence" as meaning practically the same thing as a lack of social order. Group violence such as individual crimes—and, needless to say, wars and civil insurrections—are, without doubt, fundamentally exceptional pathologies as far as society is concerned; and, the recognition that violence results in destructive and negative effects on social order may be sensible and not necessarily a mistaken recognition.

However, counter to this sensible viewpoint, there also exists an argument pointing out that occasionally violence contributes significantly to the formation of social and group order, or, that there may be cases when violence directly provides an opportunity for the formation of societies and groups.[8] René Girard and Hitoshi Imamura's theory of "scapegoat" and the exclusion of the third element, and Pierre Clastres' *Society Against the State* are well-known representative arguments of this type (see H. Imamura 1982, 1989b; Girard 1982; and Clastres 1974, 2003).

The arguments of all three appear to be in agreement regarding their conception that "it is not the case that groups exist *a priori* with actions taking place within them, but rather that actions precede groups". However, since Tanaka's article in Chapter Twelve of this book deals with all three in some detail, I will limit myself to some brief comments regarding Pierre Clastres. Clastres is an anthropologist who is well known for having proffered an original view of society, based on his research amongst indigenous societies in South America. Clastres' discussion of war is unique in having taken as its basis the perspective of theories on the formation of social order while at the same time criticizing Lévi-Strauss' argument that war is a resolution of the failure of exchange. Clastres points out the particular importance of war in primitive society, and opposes "naturalistic interpretations of war" of the sort that depict hunting as having developed into war (the hunter hypothesis type of interpretation). At the same time as doing all this, Clastres offers up the comment that "human violence lies in the domain of sociology, and not that of zoology" (2003).

Clastres argues the following about what enables a centrifugal effect and autonomy vis-à-vis the tendency that, through war, leads primitive society towards a centralized state and the class divisions that come with it.

> To the extent that primitive societies are made up of a variety of equal, free and independent socio-political units, war constitutes a privileged form of existence. (Clastres 2003: 105)

> What are primitive societies? They are all communities that are not endlessly divided and that follow the same centrifugal logic. What kind of mechanism expresses and at the same time guarantees the permanence of this logic? It is war: war as the truth of mutual communal relations; and war as the major sociological means of resisting the centripetal forces which bring about integration and promoting the centrifugal forces which lead to decentralization. [...] The existence of primitive societies is entirely based on war; without war, primitive societies cannot continue to exist. The more war that there is, the weaker the degree of integration; the best enemy of the state is war. To the extent that primitive societies are "warlike societies", they are societies that resist the state. (Clastres 2003: 110–111)

In this way, primitive societies manage, through war, to avoid being swallowed up by a centralized state, and they are seen as being able to exist as "societies that resist the state" (Clastres 1974). If Girard and Imamura's arguments about the mechanism of the creation of social order through violence appear to be quite speculative theories containing a highly abstract dimension, then Clastres' arguments are of considerable interest in that they pointed out in a more practical way—using actual ethnograpic examples from South America—that individual groups in primitive societies are able, via war, to achieve autonomy and decentralization from integration into a large state. In my view, however, Clastres' descriptions pay scant attention to the question of how the individual groups in primitive societies, who are the protagonists that carry out war so as to avoid being swallowed up by the state, were formed in the first place. Accordingly, what I hope to do in this chapter is to carry out a more concrete examination of examples of the connections between the formation of groups and violence (violent actions) by focusing on actual cases of piracy and feuds among the maritime peoples of Sulu. By way of preparation for this concrete examination, however, I would like to add just one more point about violence, from a somewhat abstract perspective.

"Violence I" and "Violence II"

The second half of this chapter will take up both acts of piracy and feud disputes from the perspective of group violence, but first I would like to point out that there are considerable differences in the typology of violence in both of these cases. To put it another way, acts of piracy and feud disputes can, strictly speaking, be said to belong to different stages and types of logic.

The present day pirate activities that are introduced in this chapter are made up of three basic components: the pirate group, the victims who are the targets of the raids and the property possessed by said victims. If we were to express this as a threefold relationship consisting of self, other and "things", then the overriding objective of acts of piracy—while allowing for some exceptions—would be, most importantly, the acquisition of "things" in the form of resources. Expressed another way, if it were possible to acquire "things", then there would be no necessity for the pirates to use violence against others. In fact, as is clear from the life history of a pirate that will be introduced later, the act of piracy involved in *kulukulu* (the act of plundering trade vessels at sea), is tied along a continuum to trade, and there is no absolute necessity to use violence against others.[9] Violence of this kind employed by pirates is violence as a means of acquiring "things"—piracy's overriding aim. Let us tentatively call this the "Violence I" level.

The important point here is that the violence that occurs in feud disputes, although still violence, is on a different level from "Violence I". What is characteristic of feud disputes is that, even in cases when they start off as disputes to acquire some sort of "thing", at some point during the dispute, achieving the reciprocity of the threat of "fighting back to repay what was done to them"—what one group has sustained and simultaneously the injuries that they have inflicted on the other group—becomes its own goal. In other words, it is in feud disputes that the very act of using violence against the other group readily undergoes metamorphosis (becomes its own goal). In feud disputes, violence is not for the sake of competition for objects in the form of resources—"things"; it is the assessment itself of the very nature of the relationship between oneself (one's group) and the other (the other's group) and the nature of the abstract relationship between "honor" and "shame" that is the aim of disputes and violence. In this respect, feud disputes occupy a different position—in terms of logical level and type—from the level of

"Violence I", which, as in acts of piracy, basically target resources and objects, that is "things". We can tentatively say that the violence in feud disputes falls under the level of "Violence II".

To comment briefly on the relevance of this to human evolution: if we regard examples of animals attacking other individuals of the same species as the earliest cases of the phenomenon of violence—or as original violence—then original violence that is concerned with the acquisition of items such as food and territory could probably be said to be able to be detected amongst quite an extensive number of animal species. To put this in terms of the language used in this chapter, this corresponds to the "Violence I" level. In contrast to this, the phenomenon of using violence which has become its own goal for purposes such as feuds between groups, and which is directed at other individuals in the same group as the "enemy" group—in other words, the "Violence II" level—can be thought of as the sort of phenomenon that makes its first appearance in species that are evolutionarily relatively close to man, rather than amongst living creatures in general. However, in this chapter, I will avoid any further adherence to this point, proceeding instead to a solid investigation of the connections between violence and group formation that hereafter will be based on more tangible evidence.[10]

The ethnography of violence in Sulu (Part 1): Pirates

Group violence in Sulu

The maritime people of Sulu, amongst whom I conducted my fieldwork, are located in the southernmost part of the Philippines. The world of these sea people spans the islands that extend to the border zone between the northern part of Sabah (Borneo) in Eastern Malaysia and Indonesia as well as to the maritime waters around these islands. In terms of the populations that make up this society, the Tausug and Sama language groups ("Sama Dea", the land Sama, and "Sama Dilaut", the sea Sama) form the main ethnic groups, and consequently these are the languages that are predominantly spoken.[11] Furthermore, nowadays, all Tausug and Sama language groups are over ninety percent Muslim and, if we exclude the agricultural portion, they have also come to be known as maritime peoples whose livelihoods are traditionally based on engaging in activities such as maritime trade and fishing. Located as it is on the periphery of the Republic of the Philippines—often described as a "weak state"—Sulu finds itself in a situation where the order and effective administrative rule of the central state does not extend in any effective manner to its territory. Even if we leave aside matters such as crimes by individuals, the

Photo 6.1 A Sulu maritime village
Typical residences of the Sama and Tausug fishermen.

local area is known, as we shall see below, for the frequent eruption of all manners of group violence.

The largest scale of violence appears in the armed conflict between the Philippine National Army and a variety of Muslim secessionist forces. The armed conflict that broke out in the 1970s between the neighboring island of Mindanao and all parts of the Sulu Archipelago is known as the "Mindanao conflict". I would like to leave this somewhat macro-political violence to previous research and proceed in this chapter to an investigation, based on ethnographic data, of micro-group violence, which does not have any political ideological background to it.[12] The two main focuses of investigation here will be what are known as "piracy" and "feuds".[13]

"Piracy" in Sulu: Its historical antecedents

Acts of "piracy" in Sulu were at their height in the eighteenth and nineteenth centuries, but are still a regular occurrence today. Nowadays, when "piracy" is mentioned as a legal concept, "pirate" generally refers not to a state institution but to someone acting privately to carry out violent acts which threaten the safety of other ships on the seas (and also those carrying out these acts).[14] In this chapter, however, the term "pirate" is not only used in the narrow sense of international law, but rather in its usual meaning of "thieves who roam the seas, attacking any passing ships and coastal areas and plundering property".[15]

Let us begin with a discussion of the history of piracy in Sulu. Pirate groups who used Sulu as their base in the earlier colonial period, particularly from the eighteenth to nineteenth centuries, are famous for having assembled a fleet and mounted expeditions that stretched from the island of Sumatra in the west to Papua New Guinea in the east, covering the whole of the islands of Southeast Asia. At that time, Sulu's pirate groups used ships that were close to thirty meters in overall length and weighed about six tons, and furnishing themselves with weapons such as eight to twenty-four pound guns, spears and kris, they amassed fleets of around 100 ships for their expeditions. According to Warren (1981), who has conducted detailed historical research into the Sulu during that period, the objective of pirate expeditions was the acquisition of slaves. Some of these slaves were in high demand as a labor force for gathering and processing products such as the sea cucumbers and sharks' fins that made up the export goods in the long haul maritime trade, which had the then prosperous Sulu Sultanate at its centre.

Piracy and ethnogenesis: The birth of the "Bangingi"

The case of the group known as the "Bagigi" (Bangingi) is of considerable interest in connection with the Sulu pirates of the pre-colonial period.[16] The Bagigi are seen largely as a particular ethnic group within the Sama language group—in its broad sense—in the area of the Sulu Archipelago but, historically, their origins are as a pirate group having as its parent body the Sama Dea (land Sama) of Balangingi Island in the northern part of the Sulu Archipelago. The category "Bagigi" appears from the late eighteenth century onwards, and there is no notable mention of it in the historical records before then. According to Warren (1986), the appearance of this Bagigi category is a consequence of slave raiding by the Sulu Sultanate: the result of the creation (ethnogenesis) of a new ethnicity by people who were forcibly transported from various parts of Southeast Asia and incorporated into pirate groups.

The origins of the Bagigi was in mixed groups made up of the slaves who had been brought from various places and the original inhabitants, with these slaves then going on to find partners in the local area and to assimilate; it even became possible for them to rise to higher levels of society, including to the position of chief. The group that was formed in this way created a new category of "group"—the Bagigi. Thus, confirming that either in conjunction with or as a result of the local pirate activities (slave raiding), in the pre-colonial period in Sulu a new ethnic category was created (Warren 1986).

Present day pirates

The words used for "pirate" in the Sulu region these days include *mundu*, a general term for pirates and "outlaws", but there are also terms which categorize according to style—as in the pirates called the *salusu* and *pagayau* who were engaged in the traditional slave raiding of the Sulu Sultanate era, and the pirates known as the *kulukulu* who carry out ambushes and give chase on the open seas. In addition, it is also possible to categorize pirates, roughly speaking, in terms of those that attack on land, those that carry out ambushes on the open seas and those who hijack, and so on. The numbers of people whose sole occupation is that of piracy are actually small; they are practically all people engaged in some other occupation such as trade or fishing. In order to clarify these various actual types of pirates, let us consider the life history of an ex-pirate, recorded during my fieldwork.

Life history of a pirate

Sakib (pseudonym) is an ex-pirate from P village on the island of Tawi Tawi in the southern part of the Sulu Archipelago (a Sama Dea), and at the time of my research (1993) was (approximately) sixty-three years old. At present, he is no longer active as a pirate, and lives with his family on S island (pseudonym) near the national border with Sabah.

From the 1950s through to the 1960s, along with a group from the same village, Sakib was a leader of a pirate group that was active over a wide sphere that included the southern Philippines, Sabah and all parts of Indonesia. In the 1950s, he also repeatedly took part in acts of piracy known as *kulukulu*.

In *kulukulu*, anywhere from around one, two or three ships would sail the Celebes Sea and wait for days on end for an opportunity. If they were to spot any of the Bugis people, a prosperous group at that time, or the Sama traders, who hailed from between Sulawesi in Indonesia and Sabah, they would give chase at full speed. When they caught up with them, they would threaten them with small arms and steal their cargo of copra or whatever else they might be carrying. This was a time when both the assailants and, of course, the trade vessel under attack did not have engines; it was the age of sailing ships.

It also happened on occasion that small boats would set off with a load of say, rice or sugar, and "barter" with another trade vessel. During this exchange, if the other ship was armed and both parties were fairly evenly matched in the power relationship, then the exchange of goods would proceed in a peaceful manner. If, however, one vessel detected that it had superior power over the over, then it would either plunder the other vessel's cargo or enforce an unequal exchange of goods akin to plunder. In these plundering expeditions, gangs were frequently assembled of people from a number of different islands.

Sakib also took part in various other pirate activities aside from *kulukulu*, participating in expeditions that attacked coastal villages in Sabah. At the time of an attack on K village (pseudonym), near Sandakan on Sabah, he rode out with ten other pirates on a boat with an engine. The pirates targeted the shops of prosperous Chinese traders in K village, and fled after having stolen their property. Apparently, although the expedition was, as one would expect, armed with rifles and other weapons, there were no killings because their prey did not resist.

When raiding the coastal villages of Sabah, a plan would be worked out at least one month in advance; scouts would, as a matter of course, be dispatched to reconnoiter factors such as the condition of the island, targets for plunder, how well it is guarded and possible landing areas. Preparations would even extend as far as secretly clearing bamboo groves so as to construct a ready approach into the village from the landing area. Apart from Sandakan, they also often raided the whole of the east coast of Sabah including places such as Lahad Datu. When setting off on these types of raids, a variety of practices associated with mystical beliefs were held to be indispensible, such as wearing various types of talismans called *anting-anting* and *hajjimatt* and reciting secret incantations.

Sakib had many friends in the Philippine military who knew of his participation in raids and *kulukulu* activities on Sabah but, since a portion of the booty obtained from plundering was paid to the soldiers, the latter never challenged these activities. The participants split the booty from their raids equally, and used some of this money to furnish themselves with new rifles and engines for their next expedition. For reasons including the strengthening of border security by Indonesia from the late 1970s onwards, Sakib ended his pirate activities.

It is possible to discern a number of characteristics common to pirates from the Sulu Archipelago from Sakib's life history. From the episode in which bartering readily transformed into the pirate activity of *kulukulu* once the nature of the relative power relationship between those involved in the trade became clear, and also from the fact that Sakib had a long experience of both long distance trade and piracy, we can ascertain that there is a certain type of continuity between piracy and trade.

Photo 6.2 An armed inhabitant guarding against pirates
Located in the southern part of the Sulu Archipelago.

In recent years, what has become conspicuous is a "modernized" style of raiding in which pirates set off in speed boats equipped with high-powered American or Japanese built engines, armed with automatic weapons such as M14 and M16 rifles and communicating with each other on Japanese ICOM or mobile phones. It is important to note, however, that even amongst these modern day pirates there persists, just as before, a strong faith in a variety of talismans and secret ritual knowledge.

The formation of pirate groups
Included in the examples of practices that pirates engage in as ritual knowledge, known as *ilmu*, and related rituals, are: things that give the pirate a body that is impenetrable to bullets, making it immortal, and things that can give pirates forewarning of various dangers. An example of the latter would be one form of the divination ritual known amongst pirates as "the ritual of preserving *inyawa*". What is of particular interest to the discussion in this chapter is the fact that this ritual is a mechanism that is employed each and every time for the self-organization of the pirates into their action groups—over and over again, as it were. Below is an attempted simple explanation of how this works.

Firstly, *inyawa* means "spirit" or "breath" in the Sama language. This divination is carried out before the pirate expedition embarks; applicants and hopefuls for participating in the pirate expedition will have been called out by the leader and lined up around him beforehand. Then, the leader draws in a breath, gulps down saliva and nimbly pats the back of his head. It is held that those who have been able to hear the actual sounds made during these actions will not meet with any danger on the expedition, and they are selected as the members of the expedition. If, however, there is anyone who was not able to hear these sounds, then that person is not allowed to accompany the pirate expedition. The reason for this is that it is believed that if he were to be allowed to go along, then his fate would undoubtedly be to lose his life en route and, in addition to this, he would also risk endangering his companions. In fact, around one or two in ten of the men originally called out will definitely be left out for this kind of reason.

According to Sakib, mentioned above, the area where the leader calls out to assemble the pirate expedition centers on where his relatives and in-laws live; there is no predetermined rule that this must extend also to friends and acquaintances who may be applicants and hopefuls. There are also occasions when the membership of the pirate's action group exceeds the bounds of the pirate group. In fact, in Sakib's pirate expeditions there were cases of members of the Ubian and the Tausug participating in the Sama Dea pirate group and similarly occasionally cases of Sama Dea, such as Sakib, taking part in Ubian and Tausug pirate groups. The important point to note here is that although the pirate action group can assume the potential participation of these candidates, any extension of the actual group and its borders will always be decided and organized on an ad hoc basis for each expedition; this will occur on the basis of relationships with the particular leader and the ritual knowledge that was discussed previously. Contrary to what might have been expected, what emerges is that a characteristic of the formation of pirate action groups is the absence of a necessarily predetermined membership or ethnic category that serves as the basis for inclusion.

This fact applies equally in the relationship between pirates and the objects of their raids (the victims). Pirates like Sakib tend to stress the following sorts of statements regarding pirate raids that they have organized themselves: "We will not attack fellow Muslims. Only infidels are attacked". In fact, there is definitely an observable tendency to make the shops of the relatively wealthy Chinese (most of whom are Christians or Buddhists) the targets during raiding expeditions to the east coast of the island of Borneo. When looking at present day cases of piracy as a whole, however, we see that, in fact, attacks by Muslim pirates on fellow

Muslims are actually quite frequent; and Sakib himself also repeatedly attacked the trading ships of the Muslim Bugis. According to the author's fieldwork survey data, the damage inflicted by cases of piracy fell overwhelming not on Chinese or Christians, but on fellow Muslim fishermen and contraband traders. In concrete terms, of all the cases of piracy between 1992 and 1994, more than ninety-five percent were on fellow Muslims and more than thirteen percent of all cases were against the same ethnic group; in other words, these were cases in which the victims and the perpetrators belonged to the same ethnic group.

In general, what can be identified as the striking characteristics of acts of piracy in Sulu today are that: (1) the pirate action group is organized (its membership is decided) on an ad hoc basis for every particular expedition via divination; and (2) the relationship between the victims and the perpetrators in piracy is not predetermined on the basis of religious affiliation or ethnic category.

The ethnography of violence in Sulu (Part 2): Feuds

Feuds and their idiom in Sulu

In present day Sulu, feuds and feuding disputes are known by names such as *mamauli*, *magbalos* and *magkontra*. In the Philippines, alongside the northern mountain region of the island of Luzon, Sulu is also known as a place of repeatedly occurring feud disputes, and these have continued to be frequent in the area from the 1990s. Although there is considerable diversity in the origins and progress of these disputes, generally, the pattern is as follows. They start as quarrels over goods, land and political resources (for example, political posts), or troubles concerning "honor" and other issues to do with women; then, once these quarrels develop into cases with casualties, they continue as prolonged violent disputes on the basis of the principle of negative reciprocity, as it were, which is at work within both the perpetrator and victim group and which expresses itself in the thinking—"If they get you, get them back" (Kamilian 2007).

There are a several characteristic cultural idioms surrounding the feuds found in Sulu, such as "honor" and "shame". For an adult male, for example, not to take action when he is treated unjustly by another person brings "shame" both to him and to his family's honor—it is regarded as incompatible with his value as a man. Thus, acts such as challenging others to a fight and feuding in order to restore honor, without any concern for danger, are admired as acts of "bravery".

Enemies in feud disputes—particularly the main personages from the opposing side who have inflicted harm on oneself (one's side)—are called "enemies" (*banta*), while one's allies and members of one's alliance in a dispute are called "friends" (*bāgai*). There is a leader (*nakura*) in the feud dispute group, and this leader is regarded as being in a very close dual relationship with those below him and his followers (*tendog*). The leader and those in his alliance as well as those below him are linked by "obligation" (*buddi*); that is, an obligation of moral reciprocity.

Being someone who is in a relationship of "friend" (member of the alliance), for example, indicates that one is in a reciprocal relationship in which there can be mutual expectations of the alliance (allies) during armed disputes and conflicts (particularly in the case of the Tausug (Kiefer 1986)). It is also possible to expand the network of alliance relationships by making it possible for people who do not necessarily live in the neighborhood and those beyond the sphere of relatives to also become "friends". By making use of this kind of network, it becomes possible for one to build an alliance group that goes beyond the bonds of relations and territory.

Phases/various aspects of violence in Sulu

The following is an introduction to several cases of feud disputes that actually broke out in Sulu, taken from my field notes.

Case 1: An example of piracy amongst the Tausug that developed into a feud dispute

The areas around the island of Sitangkai see a large number of pirate and feud battles. In 1989, a band of five pirates from the maritime village of Sapasapa, along with their Tausug leader Omar (pseudonym), attacked a passenger boat despite it being owned by other Tausug. This ship was en route from Semporna (a town in Malaysian Sabah, in the northeastern part of the island of Borneo) heading towards Sitangkai (an island in the southernmost part of the Sulu Archipelago: Philippine territory). The pirates then killed a total of seven passengers and crew after having robbed them of money and goods. This had been done to silence one of the passengers—a fellow Tausug known to the offenders by sight. However, one of the children amongst the passengers survived by having pretended to be dead. The offenders' identities came to be known as a result of information provided by this child. Immediately after this, the bereaved family group belonging to the murdered Tausug went to Sapasapa to feud, and shot Omar's father dead. Omar and the others fled away to Sabah and other locations.

Case 2: Conflicts between ex-relations by marriage (amongst the Tausug)

This is a case in which troubles concerning a feud between former relatives by marriage that followed the divorce of Amboi and Usman (both pseudonyms) from Tongbankao escalated into an incident between both families involving casualties. Amboi's son sought reconciliation with the ex-wife (Usman's family), but was rejected resulting in a gun battle between the two families that resulted in Usman being killed. As the result of a large scale battle between a unit of around 100 Philippine police, who were dispatched to deal with the incident, and eight members of Amboi's family, the maritime village of Tongbankao was completely destroyed. Subsequently, Amboi surrendered to the governor of Tawi Tawi province. The governor pardoned Amboi, and he is now engaged in cultivating seaweed in one of Tongbankao's villages.

Case 3: Conflicts and disputes concerning marriage (failed attempts amongst the Sama Dea: Simnul Island)

Mansur (pseudonym) from Simnul carried out a "kidnap" marriage with his now wife Aisya (pseudonym) that lasted until just before the battle with his wife's parents, who were subsequently reconciled (details follow).

Mansur was Aisya's third cousin, but they were in love with each other and wanted to get married. However, his parents were not keen on any talk of marriage. Learning of this, Aisya's parents also opposed any marriage with Mansur. Mansur went to Aisya's house with his father's forty-five-caliber pistol in hand. Aisya's parents were out; only Aisya and a female cousin were at home. Mansur took Aisya by the arm and attempted to lead her away. "What are you going to do?" Aisya asked. Mansur replied "If I cannot get permission to marry you, then I will marry you by kidnapping you". Having seen this, the cousin went and found Aisya's parents to let them know. Aisya's father rushed home, and took up a harpoon. In response to this, Mansur picked up the forty-five-bore and declared "I'll shoot if you come any nearer", and then fled with Aisya. Mansur then rushed into the home of a male first cousin. Mansur's family and relations assembled in this house. In preparation for the beginning of a battle with Aisya's family, Mansur's family had armed themselves before coming to the meeting. Since they had been shamed, Aisya's side also, understandably, began preparations for battle. Thereupon, Aisya's mother went to Mansur's home alone and as a result of her negotiations, she managed to conclude a provisional settlement saying: "We will allow an official marriage soon, but for now I am taking Aisya back home". Along with the fact that Aisya's mother was related to Mansur's family and the inappropriateness of any violent behavior against a woman who had come alone, a compromise was concluded and matters did not develop into an armed conflict between the two families. Mansur and Aisya subsequently married peacefully.

Case 4: A feud dispute between relatives in S village (amongst the Sama Dea)
March 1993: Dato Farok (pseudonym, hereafter DF: Sama Dea), known as an ex-MNLF[17] commander from S city and also a pirate leader, was shot dead by a hit man hired by relatives of the mayor of the same town, Liguan (pseudonym, hereafter L). The origin of this matter is to be found in a feud resulting from fact that seven years earlier DF had killed L's nephew over troubles concerning a woman. DF's father had also been killed as part of a feud with L's side, several years before DF's own killing. Gamul (pseudonym, hereafter G)—my key informant in this case—is both a first cousin on DF's mother's side and also L's nephew. Furthermore, L and DF were also related to each other by marriage. Following this incident, DF's siblings initially targeted L and his siblings, but due to the latter's intense vigilance there were no opportunities. G, who was related to both sides, found himself in a difficult position. Then there was the added fact that at the start of this case, G lived in the same village as DF's siblings and was their ally. After a while, however, as DF's siblings began to have doubts and suspected G of secretly betraying them to L's side, G—fearing for his safety—fled to Bongao (See Figure 6.1: note, however, that this is not a comprehensive genealogy).

As can be seen from the various cases introduced above, the causes, background and then the process that unfolds for each actual feud are varied, but if we were to state only one characteristic relevant to the topic of this chapter, then this would be the point that of the above cases, there is not one incident of disputes "amongst" different ethnic groups, rather, all cases are of disputes "within" the same ethnic group (namely, disputes between the Tausug and disputes between the Sama Dea). This does not just hold for the four cases given here; it is a fairly general trend. For example, Table 6.1 shows the totals for the separate ethnic group affiliations of the people involved in the seventeen feud disputes that I was able to confirm during my stay in the southern part of the Sulu Archipelago from 1992 to 1994.

The interesting point that emerges from the above table is that while feud disputes do break out both "amongst" ethnic groups (for example, Tausug versus Sama Dea) and "within" ethnic groups (for example Sama Dea versus Sama Dea), a higher frequency of disputes occur "within" the same ethnic group—in fact, more than eighty percent of all cases break out between people belonging to the same ethnic group. In addition, I have not been able to identify, within the scope of my research and including the four cases outlined above, any cases of feud disputes between direct blood relatives such as parents and children. We can see, however,

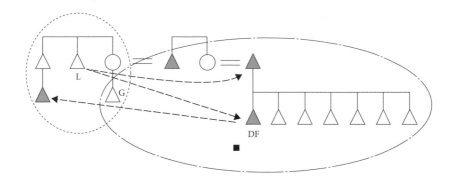

Figure 6.1 Relationships between relatives in feud disputes in S City (Case 4)

Table 6.1   Types of feud disputes seen between ethnic groups

Relationship between ethnic groups of people participating in feud disputes	Number	%
Sama Dea vs Sama Dea	9	52.9
Tausug vs Tausug	5	29.4
Tausug vs Sama Dea	2	11.8
Sama Dea vs Tausug	1	5.9
Total	17	100

Source: Based on a survey conducted by the author in the sate of Tawi Tawi in the southern part of the Sulu Archipelago, 1992–1994. However, these figures also include feud disputes that broke out before 1992.

that there is an extremely high level of changeability in the boundaries and confines of who is an ally ("friend"). Put another way, the borders of enemy group/ally group cannot be predetermined according to principles of ethnic category (as we have seen), residence (village) or kinship. On the one hand, it is possible for complete strangers to be mobilized and accepted into the ally group via "friends" and networks of alliance relationships and formal kinship relationships. As we have seen in cases two, three and four above, it is also a fact, however, that people belonging to the same ethnic category and those who are related to each other, including in-laws, and who live in close proximity to one another can experience a polarization—a redrawing of borders—to become "self/others" or "enemies/allies" via a dispute that arises as the result of some motive or other.

Violence and the autopoiesis of groups

The self-production of "enemy" group and "ally" group via action

I would now like to undertake a somewhat abstract and general consideration of what can be said on the basis of the cases of group violence in the form of the piracy and feuds in Sulu that have been introduced above. Although there are a large number of cases involving changing borders and also cases of belonging to two categories and of changing categories of belonging, it is undeniable that at the present time, the ethnic groupings (and categories) of Tausug, Sama Dea and Sama Dilaut in Sulu, in general, continue to exist as relatively stable standard frameworks. However, as has been discussed in this chapter, the subjects and relationships in acts of violence are not necessarily decided according to these ethnic group categories; rather, we can observe a dynamic process in which it is via the very act of violence that the action group—the protagonist of this violence—is formed/produced (autopoiesis).

I would now like to distinguish two distinct modes for establishing action groups through violence on the basis of the ethnographic cases introduced in this chapter. One is cases of the group formation that accompany the emergence of the action group category: this corresponds to cases of the formation of ethnic group categories known as "Bagigi", linked to the pirate expeditions (slave raiding) in Sulu in the pre-colonial period. The other is cases which could rightly be depicted as "the formation of groups lacking categories", and it is this mode that is most widely at work in present day ethnographic cases. Present day cases such as those of the self-production of piracy action groups and "enemy/ally groups" in feud disputes can concretely be included in this mode. What is characteristic of this latter type of case is that, as we have seen in the analysis of cases in this chapter, it is conspicuous in present day piracy. Action groups are organized, however, at the time of every single pirate expedition and enemy/ally groups in feud disputes also do not have a predetermined membership that has been decided on the basis of local ties, blood ties or other principles. The most remarkable characteristic then is that the self-production (autopoiesis) of the range and the boundaries of groups duly occurs, recursively as it were, as a consequence of the very links in and continuation of these acts (feud disputes). The term "autopoiesis" is used here because this process, formally expressed by Hideo Kawamoto (2000), can be used to describe the situation that we have been depicting: "the formation (production) of boundaries as a result of links between actions".[18]

To date, group violence such as warfare has, generally, been assumed to involve pre-existing groups made up of protagonists and the parties concerned, in the form of the state and the people, village communities, kinship groups and the like. From the cases of group violence in Sulu that we have examined in this chapter, a more accurate understanding of this situation would instead be that on the outbreak of violence, the groups that will duly become the protagonists will arise as groups on each individual occasion; precisely, as it were, through an action centered view of the group.

In this context, the theories of Imamura, Girard and Clastres, referred to earlier, can be seen as providing a uniform standard suggestion when considering these sorts of concrete ethnographic facts. However, there is a propensity in Imamura's theory—because of the treatment of the formation of societies and communities as abstracted general theory—for the diversity and nuances of circumstances that emerge from concrete ethnographic cases to be treated as abstractions. Consequently, both Imamura and Girard's hypotheses, in particular, display an emphasis on the establishment of a single, unified community and, in its way, a stable social order through the exclusion of particular members (the third element). However, I would like to point out that circumstances during the actual formation of groups can be far more fluid than this. Similarly, Clastres' cases assume from the outset the *a priori* existence of small-scale, autonomous communities—the protagonists of war—and the aspect of the formation of communities and groups cannot necessarily be said to receive adequate consideration.

As has been shown with regard to this in the ethnographic cases introduced in this chapter, the groups and group boundaries temporarily established at any given time—whether piracy action groups or the boundaries of enemy/friendship groups in feuds—are characterized by a *dynamism* in which re-establishing boundaries and reorganizing are made possible by the constant dismantling and restructuring that accompanies changes in the actual situation and in mutual action. It is also important to bear in mind that at the same time as a group is established within a certain sphere, another group may be dissolved at some other level. The following section discusses this latter point in terms of the question of the efferent and afferent nature of violence.

The efferent/afferent nature of violence

As has already been mentioned, it is generally accepted that violence operates as one of the elements that disrupt social order and groups. Indeed, it is, generally, precisely the effect of destroying and demolishing/disturbing the order of society and groups that is the most widespread understanding of the prominent characteristic of violence. If we were to provisionally refer to this as the centrifugal effect of violence on group order, then we could see the cases in this chapter as the, as it were, centripetal effect of group violence. What we refer to as centripetal here is, put simply, a situation in which the group is able to heighten its internal afference by stirring up conflicts and wars against external "enemies".

The point that should be carefully noted here is that the afferent/efferent nature of conflict is relative depending on one's point of view. The same violence could simultaneously be afferent or efferent as far as a group is concerned, depending on one's viewpoint. Let us assume, for example, a situation in which a group such as a large village community splits into two sub-groups as the result of a dispute. As far as the original large village is concerned, the group violence that prompted the split in the village is centrifugal, but for the sub-group that was born through this split, the effect of this same violence would have to be regarded as operating centripetally.[19] The establishment of "enemy"/"ally" groups via feud disputes that has been examined in this chapter is a representative case in which groups that are hostile to one another, and the borders between these groups are duly created through the feud disputes; and, accordingly, these disputes can be held to operate centripetally within each hostile group.[20]

A deep examination of the issue of "violence" ought to simultaneously include a deep probing of the set of issues that are the other side of the coin (in a broad sense) of violence: "peace", "reconciliation" and "dealing with disputes", but we have come to the end of our discussion and there is not the space to do so in the present chapter. I will return once again, at the next opportunity, to an examination of the issues of "peace" and "reconciliation" in the cases from Sulu.

7 Forming a Gang: Raiding among Pastoralists and the "Practice of Cooperativity"

Kaori Kawai

The mentality of the raiding situation

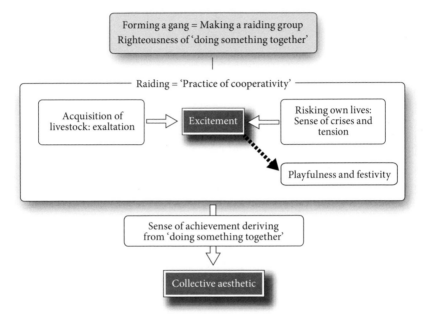

The Dodoth, one of the East African pastoralists, is known to maintain an autonomous livelihood, but when it comes to doing something together, they show a remarkable level of social cooperativity. Raiding is an act with the clear purpose of looting livestock and is always conducted in a group. This "cooperative practice" creates the mentality of "excitement", entailing a mix of exaltation from trying to acquire valuable livestock and a sense of crises and tension for risking their own lives (and this excitement often accompanies an essence of play and festivity). In the Dodoth, "doing something together" is a socially righteous act and the achievement leads people to the height of what may be called a "collective aesthetic".

Introduction

In an attempt to discuss the phenomenon of "grouping" in human society, this chapter will focus on "raiding", an act aimed at looting livestock, among neighboring East African pastoralists. From a diachronic perspective, alternating periods of antagonistic relationships, during which raiding frequently takes place, and peaceful non-antagonistic relationships have been repeated in these neighboring groups (K. Kawai 2004). Here, I will examine the inter-ethnic relationship, with particular attention paid to antagonistic periods. Specifically, I will focus on the actual interactions taking place at the scene to reveal what cannot be observed by simply looking at the inter-group relationship from a macro perspective. In other words, I will examine the antagonistic mutual negotiations which may occur between a group formed for the purpose of raiding (hereafter "raiding group") and one formed by members of a settlement or livestock camp who became the target of raiding (hereafter "raided group"). The "antagonistic relationship" between the two opposing ethnic groups does not resemble all-out war but in fact represents the sum of the antagonistic mutual negotiations that take place between the raiding and the raided.[1]

In my discussion below, I mainly focus on two aspects of raiding. The first is related to the fact that antagonistic mutual negotiations accompany violence. Up until now, raiding has been regarded as an act of vandalism, in which "belligerent pastoralists" loot each other's livestock using force. This view has been supported by the accounts of colonial administrative officers, missionaries and anthropologists. By the late twentieth century, the region's pastoralists became strongly influenced by the modernization campaign in the area, while firearms, such as automatic rifles, became increasingly available. Given such changes, it was pointed out that raiding was no longer a phenomenon that voluntarily occurred within closed pastoralist societies, but had been influenced by the society's political and economic relationship with the wider regional groups, the nation and even internationally. However, it should be noted that the act of raiding itself has never been closely examined. Below, I first critically review previous anthropological studies on raiding. Specifically, I:

- sort the terminology pointing to inter-group relationships in this region
- point out the risks in the way raiding has been positioned as a phenomenon that may lead us to understand the origin of human violence and war, simply because it is closely associated with these events

- introduce the hypothesis that supports the possibility of labeling raiding as "a part of livelihood", in the sense that it is a method, albeit rather extreme, of acquiring livestock for pasturage which forms the foundation of subsistence. Here, raiding is regarded as a form of established "trade". In other words, strictly speaking, raiding is not a form of "war" but a form of trade. War and trade are clearly two completely different concepts, and we can contend that raiding is not even a failed form of trade (Clastres 2003)[2]—yet it could be called war trade.[3]

The second aspect is related to the temporary and contingent nature of the raiding group formed by the participants. We will begin by simply asking why they need to raid in a group and investigate the purpose and benefits of doing so. A raiding group is a practical working group with the single clear-cut goal of intending to "acquire livestock". However, I believe that what the group actually gains is not just livestock, but something related to sociality and the participants' mentality. The key notions here would be a state of excitement,[4] such as tension and exaltation, festivity and play, as well as a sense of aesthetics and identity.

Image of "belligerent pastoralists"

Repetition of antagonistic/non-antagonistic relationships

In examining inter-ethnic group relationships, the East African pastoralists, including the Dodoth of Uganda and the Turkana of Kenya to be discussed in this chapter, have long been branded as "belligerent pastoralists" in the ethnographic arena due to their frequent exchange of violent raiding. What is unique about the inter-ethnic group relationships in this region is that one group does not view another group as "enemy" in a static way, and antagonistic relationships are not meant to last forever.[5] On the contrary, relationships keep shifting between antagonistic and non-antagonistic phases, with neighboring groups raiding each other at times and living peacefully at others (K. Kawai 2004). From a diachronic perspective, it would be fair to say that each ethnic group has experienced antagonistic relationships with each of its neighboring groups at one point in time (Itani 2009: 244–246).

Let us now briefly sort the terminology related to the unique inter-group relationships in the region. As above, we have seen the repetition of antagonistic and non-antagonistic phases in the East African pastoralist groups' relationships. Until now, the terms "antagonism" and "friendship" or "alliance" have often been used to explain these phases. In this chapter, however, I will avoid using such terms that attach a prescribed meaning to the counterpart "relationship" of "antagonistic relationship", and instead use the more neutral term, "non-antagonism", that seems

Photo 7.1 Herd-boys and herdsmen that start grazing at seven am
At the time this photo was taken, a large-scale camp was set up to accommodate some 100 groups of cattle. Marking the beginning of the pastoralist tribe, the Dodoth, the overwhelming scene of thousands of cattle about to be grazed was combined with the sounds of the footsteps of crowding cattle, bells ringing around their necks, and the clouds of dust they produce.

to reflect the reality in a more appropriate and accurate manner. The relationship that has been referred to by the term "friendship" is not a metaphor that represents their ties at an ethnic group level. If such relationships are to be recognized, they are *relationships between individuals* who belong to different ethnic groups at a strictly personal level, and may involve visiting each other or offering or exchanging livestock. "Alliance" is also an inappropriate term in that, in this region, alliances formed at an *ethnic group level* to achieve a common purpose are rarely seen. It is not uncommon for members of different ethnic groups to get together to form a single raiding group to raid another group. However, here, members participate at an individual level or as an aggregate of individuals. Raiding is never a matter that involves a whole ethnic group.

Let us now look at the Dodoth, a group of pastoralists who live in the northeast end of Uganda, and examine their relationship with neighboring pastoralists the

Turkana of Kenya, who live to the west of the Dodoth at the bottom of a steep escarpment (created with the formation of the Great Rift Valley) with an altitude difference of 1000m, signifying the border between the two countries.

The risk associated with the question "why do humans fight?"

Raiding, as a form of antagonistic mutual negotiation, accompanies force and has thus often been confused with or at times treated as a form of war, i.e., the primitive form of war.[6] However, raiding in the region discussed in this chapter is aimed at "acquiring livestock" more than anything else, or at least, that is what the people claim its purpose is. Indeed, raiding is an act that involves force and on the surface it looks very similar to war or ethnic conflict, thus it has often been categorized as a form of "war (in uncivilized societies)" (Clastres 2003). Among such studies, the Bodi of the southwest regions of Ethiopia, in particular, have until very recently often been used and discussed as a perfect example when considering the origin of violence, based on a description published in a colonial government report in the 1950s that claims "they easily kill people" (Fukui et al. 2004).

However, the actual studies of pastoralists in Southern Sudan by researchers including Katsuyoshi Fukui (1984, etc.), David Turton (1997, etc.), Sharon Hutchinson (1996, etc.) and Eisei Kurimoto (1996, etc.) began in full swing from the 1970s. The period coincided with the outbreak of Sudan's civil war when the purpose of raiding shifted from the acquisition of livestock to violence including massacre, extermination, arson and looting of livestock, assets, women and children. Countless lives of non-combatants were lost during this time. During this period of war, researchers reported the confusion taking place and the influence of the civil and ethnic war. While interpreting the situation by looking at the political and economic relationships between the pastoral society and the outer world and at how the former responded to the situation, the researchers pursued their studies exploring the origins of war and the evolution of violence as innate to human beings (Fukui 1996, 1999; Fukui et al. 2004).

Considering the political and economic situations within and outside the region as well as international relations, it is evident that the antagonistic mutual negotiations observed among ethnic groups in the East to North East African countries such as Kenya, Uganda, Sudan and Ethiopia cannot simply be labeled as attacks intended "purely" for the looting of livestock. Clearly, we cannot ignore the influence of the ethnic and civil war that was intensifying year by year. However, it seems that the raiding carried out among the Dodoth, where such influence has been relatively limited, is fundamentally different to "war". As far as the participants' awareness is concerned, raiding is purely intended for the "acquisition of livestock" and from a realistic perspective, such an act is clearly different to the acts of violence as seen in the above civil and ethnic war in Southern Sudan. This interpretation of raiding may sound rather idyllic, but one thing for sure is that attacks on other groups not connected with the looting of livestock are nonexistent in this region. Casualties from raiding are undoubtedly on the rise, due to the increasing inflow of inexpensive weapons such as AK47 automatic rifles and bullets in recent years. However, this is simply due to the modernization of weapons, and at least at this stage, raiding does not fall into the category of violent acts of "war" which aim at killing with the intention of ruling, conquering and destroying the antagonistic group—in other words, a primordial form of war.[7]

Raiding as a form of primitive trade

In the case of the Dodoth, when focusing on the people's subjective intention and the outcome alone, one may realize that the act of raiding, despite its violent nature, is not solely related to violence but may in fact be a method of transferring livestock between neighboring groups to ensure people's survival (K. Kawai 2004). While it is, without doubt, an act of antagonistic mutual negotiation, it is not the antagonistic relationship itself that prompts violent behavior in an autotelic manner. Raiding is simply a way to acquire livestock when this becomes necessary. Though not identical to what we are about to discuss, let us now introduce another view on raiding that supports our claim that in the case of the Dodoth "raiding is not a form of war". This theory recognizes raiding as a form of trade, but we need to take into account that it also looks at raiding from a rather abstract point of view. This is a perspective that distances itself from all the agonies, struggles and bargaining which may be experienced by human beings through the reality of raiding, and rather looks down on raiding and the various behaviors surrounding it from way above, almost from an evolutionary perspective.

Based on my field data (K. Kawai 2004, 2006), social philosopher Hitoshi Imamura likened raiding between the Dodoth and the Turkana to "(long-distance) trade" (H. Imamura 2000: 59). He claimed that their acts were homologous to those of long-distance trade merchants (H. Imamura 1994b: 66–68).

In Imamura's view (private communication, September 26, 2004, etc.), the Dodoth are "traders" and thus, by definition, their acts can be regarded as "trade". He claims that the primordial form of trade, at any point in time and at any location on earth, can without exception be termed "looting". In the case of the pastoralists, their possessions (movable assets) are livestock and thus livestock, as the mutual "commodity" of the ethnic groups, become the target of trade, whether it be through aggressive looting or peaceful negotiation. Long-distance trade merchants are adventurers who take risks, so when they set forth on expeditions (in this case, raiding), it is "essential" that they conduct thorough "research" (market research, in modern terms) on the areas they intend to visit. They will first "collect information" through geographical observation, check on the adversary's movements and the status of the counterpart's movable assets etc. and then evaluate the situation in the context of "trade/raiding" in detail. This is Imamura's perspective (1994b: 67). As with the case of Ethiopian pastoralists called the Meen, introduced by Jon Abbink (1993), the Dodoth read the intestines of livestock after they slaughter them (hereafter "intestine reading") (K. Kawai 2006, 2007). This is a kind of divination where the intestines symbolize Dodothland and its livestock camps and settlements, and by combining these sets of information the Dodoth can read where and when certain events are going to take place. Imamura interprets the intestines to be a sort of map conceived for the observations of the raiding target. He claims that, in modern terms, intestine reading would be equivalent to a quantitative regression analysis. It is an absolute requirement to make an "action plan" before taking any sort of action, and thus it is necessary to make such a plan for trading, i.e., raiding. The Dodoth call the intestine rite "prophecy", but this is without doubt an action plan. In the way of the Dodoth, the planning does not extend further than the prophet's indications and suggestions, and all the members discuss the actual details in a meeting. Imamura claims that this would be equivalent to a tabletop discussion using intestines as a map, which modern people like to call a "blueprint".

Imamura continues as follows. The people's movable assets (livestock) are "not infinite" and are indeed limited. In the trading of limited assets, the level of movable assets diachronically ends up being about equal, despite the presence of winners and losers. This is because if the disparity becomes large enough to force one party to fold, the social relationship itself will disintegrate. Even if there are winners and losers, the asset level in the regional groups as a whole has to remain at a certain level.[8] This is equivalent to posting an absolute assumption that "the society will continue to exist", and this assumption must be adhered to. Trade is not a war of extinction, and that is why we are claiming that raiding is a form of trade rather than "war". War and trade are completely separate concepts. The act of raiding may seem similar to war, but it is actually something quite different.

In this respect, the act of modern piracy described by Tokoro in the previous chapter of this book may also be regarded as a form of trade. According to Aristotle (1961: 49), piracy was a legitimate livelihood in Ancient Greek society.

Some, who could not fully provide for themselves, pursued both pastoralism and looting to cover the shortfall (Aristotle 2001: 27). Imamura argues that pirates are "armed trade groups", and that the Dodoth perfectly fit into Aristotle's definition. For example, according to Tokoro's thesis mentioned above, piracy was often conducted as a side job, and pirates had other subsistence activities in their daily lives such as fishing and trading. Dodoth raiding groups are also organized by those who are usually engaged in pastoral activities. Imamura further points out that, in ancient times, trading equaled looting and looting was equivalent to trading. Thus, though it caused much pain in people's lives, there was also "an essence of play" in it at the same time. Looting as a form of trade does not intend to destroy the target. In that sense it is trade rather than conquest. From the communication perspective, raiding is a mutual act, so if the target refuses peaceful negotiation and responds negatively with an attack, the group would have no choice but to fight back. Indeed, Imamura concludes that the historic reality would have been that there was no other choice but to trade by looting. From a logical perspective, we cannot conclude that raiding is trading simply because it is not war. However, there are a number of examples in the discussion to follow that supplement Imamura's view.

The Turkana become the Dodoth's outright "enemy" only through the event of raiding (both when raiding and being raided) and, at other times, they are by no means the Dodoth's "enemy". Dutch primatologist, Frans de Waal, known for his work *Chimpanzee Politics: Power and Sex Among Apes* (1982), expressed the following view on how primates, including human beings, lead a peaceful life in his book, *Peacemaking among Primates* (1990: 259, 303): "Conflicts between people are inevitable, and logically speaking, with the long evolutionary history of aggression, there should be a powerful mechanism for problem solving in place", and "in the case of competitors who depend on each other, it is very rare to see an absolute victory". This view is in line with Imamura's perspective on ethnic group relationships that entail raiding. Both the Dodoth and the Turkana, as with any other pastoral group in a potentially antagonistic relationship, would be in trouble if their counterpart were to be completely wiped out.

The raiding program

The formation of raiding groups and the path to raiding

The size of raiding groups varies, but one common characteristic is that raiding always occurs "in a group". The reason behind this is simply given as follows:

"because it is meant to be done so". Let us just say here that the rationale is the legitimacy and fitness of the act of raiding itself. The target of raiding may be a large-scale livestock camp with multiple herds or a relatively small herd out grazing, at a watering place or in a pen in a settlement. Besides raiding, there is another way of acquiring livestock known as *akoko* (stealing). This is a method of robbing livestock, in a relatively small group, usually in threes or fours but sometimes in groups of up to ten, on the premise that members of the counterpart group will not be injured. This is done by quietly taking stray animals from the pasture or sneaking into settlements and small livestock camps in the middle of the night and stealing the animals by breaking the fence. Sometimes, robbers find young shepherd boys who are grazing their animals on their own or in a small group and chase them away by threatening them, tie them up or, on rare occasions, even kill them to abduct the animals under their care. Conversely, what they regard as true raiding is referred to by the word *ajore*, a term used for assault and combat, where they organize an armed raiding group to loot livestock without hesitating to use force. A group from about a dozen to sometimes tens and hundreds of members is organized to raid a herd of animals out grazing or drinking water or large herds with tens and hundreds of animals in a large livestock camp.

The raiding group does not necessarily have to be systematically organized by the whole Dodoth community, regional group or settlement, or by existing so-cial units such as clan or lineage. Given certain conditions, a raiding group with dozens of participants may be swiftly organized by a single protagonist.[9] The ini-tiator often has a good reason for raiding and acquiring resources (livestock), for example, having lost a significant amount of livestock through a raid or needing a large number of animals for the payment of bridewealth. He will ask or request his friends and relatives if they would come raiding with him and they usually agree on the spot, as if it were the most natural thing to do. Indeed, it *is* a natural thing to do and as we shall see later, participating in a raiding group has a signif-icant social aspect. However, it should also be noted that there is no problem with refusing simply by saying "not this time". Having a friend in the target settlement or livestock camp often seems to be grounds for refusal, but it could be that one may simply not be feeling up to it. In any case, the reason for refusal is usually not questioned. The Dodoth, as well as the Turkana, are very individualistic. There is nothing wrong with refusing to cooperate (participate) and there is no need to explain one's actions (Itani 2009: 405).

Those who live with the possibility of being raided, conversely, display a sense of crises and anxiety, not only in their words but also in their everyday lives. The

Dodoth create a sturdy settlement fully surrounded by high fencing and young, armed men sleep next to the livestock pen. When going out grazing, shepherd boys always remain alert, constantly carry guns, frequently change the feeding and watering places to avoid ambush and patrol the water place and pasture before the arrival of the animals.

We can often recognize a "process in which raiding becomes increasingly real" in their daily lives. Raiding gradually becomes realistic as people who sense the danger of raiding take various countermeasures. For example, if people hear of the news that a certain area of Dodothland was raided, it will prompt them to take precautions and reinforce their lines of defense, even if the news itself was groundless. In addition to practical measures such as reinforcing the fencing of their livestock pens, doubling the fencing around the settlement, stocking up on bullets at the borders of Sudan and intensifying patrols, various ritual practices such as establishing something as a fetish, conducting rituals to repulse the enemy and painting their bodies with a certain color as a charm known to protect "lives" come into play. Intestine readings are also enthusiastically conducted at these times.

As such, the Dodoth's sense of crises and tension rise by the day through the practice of various anti-raiding measures. In addition to anecdotal information, people begin reporting sightings of footprints and traces of fire and gradually, the neighboring group (such as the Turkana), who has always been regarded as a potential threat, emerges in an unmistakable "enemy" act. Whether to prepare to fight the "enemy", to take the most certain measure of avoiding raiding by moving the livestock or to first conduct rituals to repulse the enemy is discussed on a daily basis, and these measures are put into action one by one (K. Kawai 2002a).

The Dodoth are constantly conscious of their "enemies" in their daily lives and do not hesitate to risk their lives in order to maintain or acquire livestock. However, these practices cannot simply be dismissed by attributing them to the Dodoth's "belligerent" and "volatile" nature. What should be noted here is the fact that a mechanism that prepares people for antagonistic mutual negotiation is inherent in Dodoth society. People are prepared to face raiding through a "process of substantializing the enemy" by taking practical action in response to the sense of crises and tension and the various emotions that accompany such actions such as fear, nervousness and exaltation.

The difference between raiding and "war"
Raiding is an act of collective looting with the goal of acquiring livestock. How, then, does it differ from "war"? We would like to discuss this point below.

First of all, Dodoth raiding lacks certain aspects of "war" in general, such as retaliation and revenge based on hatred and hostility. While the act of retaliation motivated by a deep rooted hatred, as portrayed in "Violence II" by Tokoro in the previous chapter, is an act of violence whose sole purpose is retaliation itself, Dodoth raiding is a fundamentally different phenomenon, as explained in the previous section, despite the superficial similarity of both being "armed" acts. In this regard, Dodoth raiding remains at the level of what Tokoro defines as "Violence I"—a measure of acquiring resources. In fact, that is why the Dodoth, after the looting of their livestock by the Turkana, then raid their southern neighbor, the Jie, rather than take revenge on the Turkana. Raiding groups are formed through the powerful desire for cattle; in other words, the motive for raiding does not originate from external factors but more from individual or group initiative. It is definitely not an act conducted in reciprocity.

However, it should be noted here that though the raided group shows the above reactions when their livestock are completely driven into the enemy's territory, they display a strong sense of cooperativity in attempts to recapture looted animals while still on their way to the enemy's territory or in pursuit of a failed raiding group. Men rush outside without any hesitation, with a gun on their shoulder, the moment they hear a high-pitched voice or whistle emitted as an alarm call to announce that raiding is in progress. They literally put themselves at risk without a second thought. They believe that stealing Dodoth livestock is an unacceptable act and it is in their sense of justice to protect their animals from the raid, even if it means killing the members of the raiding group. Once the alarm call is emitted, whether the raided individual is a friend or a stranger does not make a difference. Rushing to support a group member is almost regarded as an absolute order in their unconscious minds.

Secondly, I would like to mention the Dodoth's rather surprising, sometimes even bizarre, comments after being raided. Being raided is an event that would deprive them of the basis of their pasturage livelihood. Besides this economic aspect, losing one's animals is a psychologically traumatizing event for the victim. However, curiously, they never seem to denounce or deny the act of raiding itself on ethical or moral grounds. For example, they say that they feel angry at the Turkana for planning and conducting raids against the Dodoth, but they also show understanding by saying that there must be an inevitable reason for them doing so. Obviously, this does not mean that they submissively accept raiding by other groups. However, we may posit that this is a way of thinking that is pre-prepared in their society and consciousnesses in order to allow them to accept and overcome the painful experience of being a raiding victim. At the same time, the Dodoth and the Turkana's mutual recognition may also underline that they are "equal" partners, in that they are both pastoralists who heavily rely on livestock, and may experience or may have already experienced raiding from the opposite side at different times.

Interestingly, the raided Dodoth may even go on to show compassion for the members of the group who have looted their animals. This is what one of them had to say: "They (Turkana) live in a harsher environment than us as they have no means of farming"[10] and because of that "everyone, including women, children and elders need to live in camps, moving from place to place all the time".[11] After sympathizing with their circumstances, he continued by saying that "they must deeply love the animals". I am not sure if this could be a convincing explanation for the logic behind his comments, but one thing for certain is that I cannot help but feel that this example reveals how the Dodoth and the Turkana respect each other as fellow "pastoralists who rely on cattle". In other words, it seems that the Dodoth, in comparing their fellow pastoralists with others, see the Turkana as "peers" who belong to the same broader grouping. The Dodoth are constantly concerned about what is happening in other pastoralists' groups, not just the Turkana, and discuss events such as who raided who on a daily basis. The Dodoth's neighbors include the Ik (hunter-gathers) and the Acholi (cultivators) of Uganda as well as pastoralists the Jie of Uganda, the Turkana of Kenya and the Toposa of Sudan, but they do not really seem to have any interest in or feelings towards the non-pastoralists. The Ik and the Acholi hardly possess any livestock and are thus not considered candidates for friendship or marital relationships, which require the offering or exchange of animals. They do not even become targets of raiding, as they do not own any animals.

Thirdly, the Dodoth do not embark on political actions such as "invasion" or "ruling". Raiding is merely a scramble for livestock, and though the casualties that accompany it is an issue that cannot be ignored, it never progresses in a political context to determine ruler and ruled or invader and invaded. To begin with, the concept of invasion itself is difficult to implement in this region, as the boundaries between the neighboring groups are not clear. The concept of territory is also rather vague. For example, during non-antagonistic periods the Dodoth tolerate the Turkana climbing up the escarpment and grazing and watering their animals in Dodothland, or transiting in Dodothland to sell their livestock in the township of Kaabong that forms the center of that region. The Dodoth-Turkana border range is commonly used, but as far as access to resources needed for pasturage such as sites for livestock camps, grazing fields and watering places is concerned, the general trend is that the Turkana who live in a more arid environment migrate into Dodothland with its watering holes and grazing fields that do not run out in the dry season and use their resources. The Turkana only enter Dodothland during the dry season for grazing purposes and return to Turkanaland in Kenya during the wet season. The Dodoth tolerate all this until the Turkana embark on acts of raiding.

"Cooperative practice" in raiding

Emotions, collective aesthetic consciousness and formation of identity in raiding

It is not difficult to imagine the emotions participants experience when they go raiding. In addition to the exaltation of acquiring "beloved cattle", there is the fear and tension of the possible death they may face. According to a raiding anecdote, participants are filled with extreme exaltation such that they "keep pushing forward to the enemy land for three days and three nights without eating". This does indeed sound quite extraordinary. Once they acquire the animals and return home, they repeatedly talk about the event proudly and with an exaltation that they just cannot contain. Raiding trips from the past are also often repeated. The younger generation who listen to their tales also get excited by the stimulating stories and long to participate in the raids as soon as possible, promising to themselves, "I am going on the next one". As such, these gatherings offer an opportunity for the raiding reserves to foster their identity.

One thing to note here is the fact that these heroic tales of raiding are not intended to praise a single hero. Indeed, every single member of the raiding group is considered a hero. On January 14, 2000, when a group of men visited a site that they had once raided for ritual purposes, they showed intense excitement and were filled with heightened exaltation. It was the site where the men had raided the Turkana in a large-scale attack in the dry season of 1998, resulting in the successful acquisition of many animals. The Dodoth's assault squad ambushed the Turkana, who were expanding their grazing area into Dodothland, at the slope of the two hills and proceeded to attack them. The young men in turn explained the details passionately; who was stationed where, the direction the Turkana cattle arrived from, who attacked the arriving Turkana from where and which direction the Turkana fled towards, referring to the geographical circumstances, pointing out details of the terrain such as little peaks, streams, rocks and Euphorbia trees in an elaborate manner using their whole bodies (K. Kawai 2002b: 75–76).

Photo 7.2 The Dodoth men gathered for the enemy-repelling ritual

The men place a black, young ox as a sacrificial offering and bury its head together with three kinds of plants into the ground on the mountain path taken by the Turkana when coming to Dodothland. The men believe that the rite will make the enemy Turkana become frightened or fall out amongst themselves and eventually return to Turkanaland.

When departing on a raid, the men are filled with mixed emotions including nervousness for entering the enemy land *together in a group*, fear of possible death, expectation, anxiety and thoughts as to the successful outcome of a raid and how many animals they could potentially acquire. These emotions eventually burst through and are elevated into a state of what may be called a "collective aesthetic consciousness". During the raiding, participants are filled with various emotions that lead to such aesthetic consciousness, which encompasses exaltation, tension and fear, and these emotions will be talked about with the same level of emotional intensity for years to come.

Raiding groups, in most cases, are organized by acquaintances but, as mentioned above, they are not dependent on particular social units (categories). Raiding groups are always organized by collective participation. In fact, these groups may even extend beyond the ethnic category of the Dodoth and can be organized by men from varying ethnic groups. What has been described as an "alliance" of two or more voluntary ethnic groups in previous studies on East African pastoralists was, in fact, a single raiding group organized by individuals from two or more non-antagonistic groups raiding a third group, and was in fact not an "alliance" formed at the *ethnic group level*. It is natural to assume that a cross-boundary raiding group organized by two or more ethnic groups would contain some sort of sub-group based on ethnicity, possibly in the form of units organized by their ethnic groups. In reality, however, this is not always the case. The emergence of transboundary units and leader groups have proven that borders at the ethnic group level, which we believed to exist, are often easily transgressed. In a raiding group organized by members from several ethnic groups, ethnic identities take a step back as the raiding group itself is strongly highlighted, and a cross-ethnic identity is temporarily fostered. At this stage, there is no meaning in distinguishing a group organized purely by Dodoth members and one organized by those from several ethnic groups. The key factor is the formation of a raiding group by collective participation.

We can expect the men to be feeling intense excitement, not only during the raiding itself, but also in the formative process of the raiding group. The various characteristics associated with the raiding group and the act of raiding itself, such as exaltation, unity, synchronicity, equality and extraordinary experiences give us the impression that raiding is accompanied by a sort of "festivity". As Imamura indicated in the second section of this chapter, "looting is trading" and "thus, it had caused much pain in people's lives, but there was also 'an essence of play' in it at the same time". This comment seems to apply to Dodoth raiding as well. The members are about to go on a mission to acquire cattle, the only asset in which they invest the utmost economic, social and cultural importance. How could they not feel excited? They even literally put their lives on the line for the cattle. Raiding is an antagonistic mutual negotiation that occurs between those who try to snatch livestock and those who endeavor to prevent it from happening (Kitamura 2004: 488). What Imamura refers to as an "essence of play" is the act of trying to score (loot the animals/protect the animals) in a game of "raiding" that is extremely dangerous with the possible consequence of death and is at the same time filled with excitement such as tension and exaltation.

I wish to reiterate an important point here, which is the fact that the "game of raiding" is not a solo race but a team game played within a *peer group*. In this sense, raiding is a "cooperative practice" in which the participants get to experience the feeling of "being united" with each other. The sense of unity further supports the effort of trying to become one, and the effort then inverts to create a new sense of group cohesion. The loop of these two senses and action reinforces itself and generates further excitement. This, then, allows them to experience legitimacy, convincing them that they are doing the "right" thing, which would generate a further escalation of exaltation.

Social cooperativity in raiding

Despite its temporary and contingent nature, a raiding group is formed by those who share the intention of "looting livestock together". We have already discussed the possibility, from a social perspective, that the act of looting "together" may have as much importance as the obvious substantial gain through the acquisition of animals. From an economic point of view, the Dodoth family units, which form their livelihood (production and consumption) units, are thoroughly autonomous and individualistic. However, when it comes to "doing something together", whether it be for ritual practice or raiding, they show an incredible level of cooperativity. Having said that, while they place high value on it, the nature of the Dodoth's cooperativity does not restrict individuals from doing what they want to do, as is evident from the example of how the Dodoth are free to turn down requests to join raiding parties.

If I would add something new here, the way the acquired animals are distributed among the members of the raiding group is also quite interesting, as they never seem to "divide equally" or distribute the same number of animals to each member. Then again, it does not seem to be that they are allocating the distribution of raided animals based on individual members' contribution to the actual conflict. I am sure that there must be a long negotiation before members reach a decision, but no one has ever complained about the distribution method after the number has been finalized, or at least, I have never heard of such protests. I once passed by a raiding group comprised of seven men who had just returned from a successful raid in which they acquired twenty or so cattle from the Turkana. They were ebullient, somewhat proud and victorious, and all looking thoroughly content. However, when I confirmed later, I found out that in the case of this raiding group one member (the proposer) received half of the cattle, the other three gained one to three animals each and the remaining three did not receive any at all.

As in this case, participants in raids may end up not gaining any animals. In fact, this case is not a rare exception. Why do members participate in a raid, risking their own lives, in order to loot livestock that may not even become their own? This is why we cannot assert that the ultimate purpose of raiding, if such thing does exist, is solely the "acquisition of livestock", no matter how strongly the Dodoth insist that it is. We are assuming that the social cooperativity of "doing something together", especially in the social context that we touched on earlier, is demonstrated here as an equally important element. The fact that it is quite normal for them to "agree on the spot" when requested to join a raiding group also supports this view. The way they always go raiding in a group and the fact that the raided party is also extremely quick in creating the interception/pursuit group both demonstrate the Dodoth's consistent attitude of accepting the task of "doing something together with one's friends" on the spot as an entirely natural thing to do.

So far we have explained that raiding is an antagonistic mutual negotiation between a raiding group and a raided group and not one that takes place between the Dodoth and the Turkana at the ethnic group level. Having said that, it should be noted that once the looting begins, the categories at the ethnic group level, i.e., the categories of "the Dodoth" or "the Turkana", suddenly come to the fore. The act of raiding becomes the trigger to manifest a framework that complies with the ethnic group category. It is indeed a process that instantly activates the dormant ethnic group categories. Responding instantaneously and rushing to support on hearing the news that a livestock camp has been "attacked by the Turkana" is without doubt an act that complies with the ethnic group category. We have seen that members of these groups have the mentality and attitude capable of not unconditionally conforming to the ethnic group category in the way they make friends and form marital relationships with members of groups that could potentially be "enemies", and in the way they organize cross-ethnic raiding groups. However, as soon as the attack begins, they are also capable of complying with the ethnic group category of the Dodoth. In other words, Dodoth society contains both aspects of not being bound by and yet making use of their group category. From the evolutionary perspective, this is a unique ability acquired only by humans.

Conclusion

The Dodoth assert that the purpose of raiding is to "acquire livestock". However, as discussed above, it is quite normal for individuals to end up not gaining any animals, despite a successful acquisition by the raiding group. The reason why this is acceptable is possibly because forming a raiding group is a justification for the collective cooperative practice of "doing something together". Here, by venturing into the risky task of raiding "together in a group" with a sort of "pride"—which could well be *that* "aesthetic consciousness" formed through raiding—the members gain the "experience that justifies doing something together", as well as a sense of achievement by "doing the right thing" by the group. We believe that this opportunity for cooperative practice forms the situation in which their unique form of excitement, entailing a blend of various emotions, arises. Needless to say, they do not embark on raiding missions and risk their lives simply *for* the experience of excitement. What motivates them to form a raiding group is the "attempt to acquire livestock" as well as "doing something together" with their friends who are in dire need of livestock. Here, we can see the way of the Dodoth society that assumes such acts as natural and righteous.

Finally, though it may seem rather abrupt, I wish to compare the state of excitement of the raiding group we have seen so far with that of non-human primates. How far does the cooperative practice of raiding connect with the excitement groups of higher primates, such as great apes, share as a community?[12]

In this chapter, we have looked at raiding as a form of inter-group antagonistic mutual negotiation accompanying violence. Despite some similarities, in that they both accompany violence, we found raiding to be completely different to the inter-unit group killing of chimpanzees that Kuroda introduces later in this volume (Chapter Eleven). Raiding is not at all equal to the intense antagonistic mutual negotiation among chimpanzees, coincidentally named "primitive war" by Jane Goodall (1994: 154–173), which sometimes ends up completely wiping out the other party.[13] The fact that the intention is not to rule, invade or destroy the other party also tells us that raiding is not an event that can be referred to when considering the origins of war. Instead, raiding is an exchange of movable assets (livestock) between several groups, on the premise of each other's existence, in what may be called a form of trade by attack and looting.

Contrary to these differences at a macro level, when we looked at the violent behavior at the micro level, we noticed that the excitement expressed by the individuals participating in raiding is homologous to that of chimpanzees during their collective violent acts. The above phenomenon of chimpanzees' killing at the unit group level only occurs on very rare occasions. However, when we compared the mentality of people during raiding with the excitement, accompanying tension and fear, seen in the more frequent chimpanzees' collective hunting behaviors (T. Nishida 1981: 221–229, 1994: 52–55; Hosaka 2002: 231–232) while hunting for animals such as red colobus, the two were strikingly similar. When chimpanzees go hunting in a group, become extremely excited and cause a big commotion, they apparently show not only exaltation but also tension and unmistakable signs of fear. Conversely, the various acts surrounding raiding allow both the raiding and the raided groups to share, as a cooperative state, the psychological excitement entailing elements such as exaltation, tension and fear. In other words, raiding provides them with an opportunity to sympathize with each other in a sort of "emotional buzz". In that respect, I wish to suggest the possibility that raiding may share the emotional foundation of "collective excitement" with what has been observed in the collective cooperative practice of higher primates such as great apes, especially chimpanzees and bonobo (see Kuroda's theses in Chapter 11 of this book).

Article 2

Yesterday's Friend is Today's Enemy: The Huli Society of Papua New Guinea

Masahiro Umezaki

Papua New Guinea Highlands

In Papua New Guinea, with a population of six million, more than 800 languages are spoken. In contrast to the low-lying coastal areas, which house a large number of language groups with fewer than 5000 speakers, in the area known as the "Highlands" language groups with large numbers of speakers exist, such as the Enga, Merupa, Kuman, Huli and Kamano (Ray 2006). The ecological characteristics that these language groups share are the progress of improved land productivity and the intensification of pig rearing as well as rapid population increases, all sparked by the introduction of the sweet potato in the early eighteenth century. As a consequence of this process, dubbed the "Sweet Potato Revolution", the scale and structure of highland society populations have taken on considerably different characteristics from those of the lowland areas.

My survey involved people belonging to the language group known as Huli, who live in the Tari Basin in the Highlands. During my research there, the phrase, "Yesterday's enemy is today's friend" kept coming to mind. My survey focused on a group called the Wenani, comprised of three sub-groups: the Haroba, Madija and Hirua. At the time of my first study in 1993, the Haroba and Madija were united as enemies of the Hirua. It was virtually impossible for me, as a friend of the Madija, to enter Hirua territory. When I conducted my 1994 survey, the Haroba and Hirua were united as enemies of the Madija. As I had decided to live in a hut built for me by the Haroba at this time, it was very difficult for me to associate with the Madija. Incidentally, when I set out again in 1998, there were no group alliances among the three and all were enemies. In 2007, the Haroba and Hirua were again hostile towards the Madija (see Figure A2.1 and Photo A2.1).

The Huli "population"

Over 200 groups, called *hameigini* (*hame* = siblings, *igini* = children) live in the Tari Basin, of which the Wenani are one. *Hameigini* is a name indicating a group

Figure A2.1 Map of the Wenani

The Wenani are the *hameigini* located to the west of the Tari Airport. Constructed using aerial photographs, this map shows the fields for which the three sub-populations of the Madija, Haroba and Hirua respectively, hold cultivation rights. According to an analysis conducted by the authors, thirty-three of the 103 active members of the Wenani in 1980 had moved out to other *hameigini* by 1995, and fifty-one of the 154 active members of the Wenani in 1995, had migrated in from other *hameigini* over the fifteen year period from 1980 (Umezaki and Ohtsuka 2002).

that asserts usage rights over a certain area of land, and it is also used as the place name for that land. The names of *hameigini* often have their origins in the names of patrilineal founders dating back five to ten generations.

Hameigini are organized on the basis of the following two criteria:

1. each *hameigini* designates all of the founder's descendants—irrespective of whether they are matrilineal or patrilineal—as potential members
2. *hameigini* are exogamous units, and members do not marry those of the opposite sex from the same *hameigini*.

Photo A2.1 A man carrying a homemade shotgun
There has been a trend towards increased numbers of victims in recent years as a result of the use of homemade shotguns in 'disputes.'

Theoretically speaking, the fact that *hameigini* are exogamous units means that all of the ancestors of a given individual are bound to hail from separate *hameigini*. However, the number of a given person's ancestors from the previous n generations is two to the nth power. As is clear from the fact that in the Tari Basin today there are no more than about 200 *hameigini*, the criterion that *hameigini* are exogamous units operates on the level of people's memory or consciousness rather than the biological.

In order for potential members to open up new fields or build new houses on a *hameigini*'s land—after having explained the genealogical link to the founder of the *hameigini*—they need to show evidence of their ancestors having actually lived on that *hameigini*'s land; examples of this might include displaying knowledge of an ancestor having dug a particular trench in a field, planted trees around a field, or of the existence of an ancestor's gravesite.

Moreover, one is usually required to make a substantial contribution to the continuance of that *hameigini*, by, for example, giving support during a dispute or contributing bride price. Usually, any one individual is a potential member of five to ten *hameigini*, and possesses houses and fields in several at once (Photo A2.2).

Photo A2.2 Trenches dug between hameigini *or* emene

Large trenches of this type are dug at the boundaries between *hameigini* or *emene*. Fields are spaces into which no one other than those working them and their supporters may enter, while, by contrast, these trenches are public spaces in which anyone may walk. Most trenches, however, are extremely muddy and difficult to walk in.

Hameigini have sub-groups known as *"emene"*. Taking the Wenani as an example, the three sub-groups, the Haroba, Madija and Hirua, are *emene*. In most cases, *emene* are named after the children of the patrilineal founder of the *hameigini*, and this name is also used as the place name for the lands within its territory. Members of the *hameigini* who can trace their membership back to the founder entirely along the patrilineal line (that is, their grandfather's grandfather's grandfather's … grandfather) are called *tene*, and those who interpose descent through the matrilineal line somewhere along the way (that is, their grandfather's grandmother's grandfather's … grandfather) are known as *jyamuni*. *Tene* do not necessarily have any more influence or decision-making power than do *jyamuni*.

Figure A2.2 is the genealogical chart of all the *tene* (except for children) who comprise the Wenani. The entries within boxes are members who were alive at the time of the survey, and those found inside boxes outlined in bold are those with houses in Wenani. According to this figure, there are five sons in the Wenani: Madija (MD) is the eldest, and Pera (PR)—Haroba and Hirua's father—is the third son. At the time of the survey, there were twenty-nine *tene*, with eight of these having a house in the Wenani. Of the thirty-eight households in the Wenani, ten were headed by a *tene* and twenty-eight by a *jyamuni*.

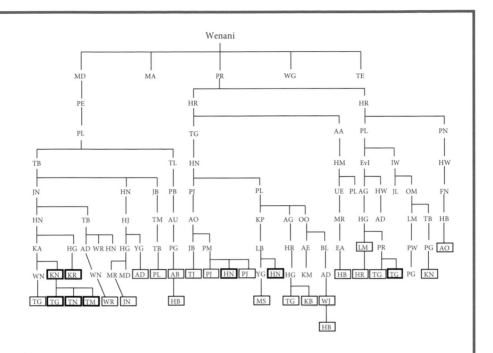

Figure A2.2 Genealogical chart of the Wenani

The members enclosed by boxes, are people who were alive at the time of the survey, and those enclosed by boxes with a darker outline are those who had homes amongst the Wenani. At the time of the survey, there were thirty-eight households living amongst the Wenani, and twenty-eight of these were descendants with a matrilineal link (*jyamuni*). This family tree was put together on the basis of what was told to me by members of the Madija; it is a systematization of their combined understanding following repeated discussions in various locales. The results are fluid, and probably not biological. Although the Haroba and Hirua broadly share the same outline of this genealogy, they stress the legitimacy of a different version regarding those parts with which there have been problems in past disputes.

"Dispute" units

The genealogical chart that is acknowledged by each of these respective *hameigini* or *emene* has an important meaning in terms of the practical response of individuals to the "disputes" that occur in the Tari Basin. The reason for this is that the units of "dispute" in that context are not individuals but groups—namely, *hameigini* or *emene*.

Matters that readily become the cause of disputes include conflicts over land use and suspicions of adultery or pig theft, but even more trivial matters such as a squabble while having a drink, for example, can lead to a dispute. The catalyst for the 1994 "dispute" between the Madija and the Haroba/Hirua was an argument over cultivation rights for a single field. This field had lain fallow for about ten years, and until that time, it was believed to have been in common use amongst the Madija and the Haroba. The chronological passage of the "dispute" went as follows.

1. When a Haroba household mowed the grass of the field and attempted to plant sweet potatoes, after an interval of ten years, a Madija household asserted that the field actually belonged to them.

2. The "dispute" that could have been avoided had one party offered the suggestion "Why don't we share it, as in the past?" escalated into a full-scale conflict when the Madija assumed an unyielding attitude, assembled potential Madija members from throughout the whole area of the Tari Basin and established a house for them.

3. The Hirua, who had been in dispute against the Madija/Haroba alliance the previous year, decided to support the Haroba and fight against the Madija on this occasion. The Haroba/Hirua also called together the potential members of both of their *emene*, and set up houses for these males.

Given the situation, both the Madija and Haroba/Hirua camps made preparations for a fight, and since they did not know when the other party might attack them, they could no longer venture outside their own territory. At the time of this "dispute", both parties stewed over the issue for about two months as they each remained alert within their own territory, with the "dispute" ending on an indefinite note. The field concerned was abandoned uncultivated (Photo A2.3).

The important element so as not to lose out in a "dispute" is gathering about the same number of supporters as the opposing group. Usually, by paying bride price and supporting friends in their disputes, one can expect a lot of support in one's own disputes. Put another way, people who do not usually pay bride price or lend their support in disputes cannot expect to gain support for their own disputes, and end up losing out.

These sorts of conflicts have become everyday occurrences in the Tari Basin. When a dispute breaks out, all Huli men take into consideration the "connections" between themselves and the two groups concerned—that is, their links in the genealogical chart and the actual results of support thus far as well as payments of bride price. In the event that they judge there to be no "connections" whatsoever, they are permitted to take a neutral stance for the time being. If they have strong ties with one of the parties concerned in the matter, then they are not allowed to remain neutral. They hasten to the scene of the "dispute" in order to support the group with whom they have "connections". Hurrying to the scene is also a necessity

Photo A2.3 A pig feeding
Giving a pig to one's adversary's group is important in bringing about the end of a 'dispute.' An individual pig passes through the hands of many owners in the course of a single lifetime.

in terms of personal safety. The reason for this is that when a particular individual has strong "connections" with one of the groups, he is automatically seen as an enemy of the other, and there is thus the possibility that he may be attacked. There are also many cases in which hastening to the scene of a "dispute" has become the occasion for men to be invited to build a house and take up residence within a *hameigini*.

When there are "connections" with both of the parties concerned in a "dispute", it becomes a matter of weighing up one's options while taking into account the relative importance of what has happened in the past as well as future developments, and then deciding whether to support one side or to take refuge in a *hameigini* that is far removed from the scene. Thus, whenever various "disputes" occur, depending on who the parties concerned are, yesterday's enemy may become today's friend and later yesterday's friend may become today's enemy; population movements occur between *hameigini* and there is a considerable changing of places by members. Friends who usually enjoy a pleasant chat at the market may end up attacking each other with bows and arrows because of a dispute that has occurred in a place that has nothing to do with either of them.

When this type of "dispute" logic is applied to those between adjacent language groups, the settlement of conflict becomes a difficult matter. In an area to the north of the Tari Basin, separated from it by mountains, is a mountain gold mine called Pogera, inhabited by the Ipiri people who speak Ipiri—a different language from that of the Huli. Below is an outline of the circumstances that led to the dispute that broke out between the Huli and the Ipili in 1995.

1. A Huli, who was working at the Pogera gold mine, was killed by an Ipili.
2. Since the murdered Huli belonged to the Agana (a Huli *hameigini*), the men of the Agana killed an Ipili man who was in the Tari Basin. People say that this murdered Ipili man who had come to the market in the Tari Basin unaware of the troubles back at the gold mine in Pogera, was enjoying a chat with a friend when he was attacked. The man had his head cut off with a bush knife, from behind.
3. A Huli man (of the Hau *hameigini*), who happened by chance to be sitting next to the murdered man, was showered in blood in the process, and died of shock.
4. Following this incident, the Ipili back at Pogera killed a Huli man (of the Phi-Nagia *hameigini*).

In the course of these events, the Ipili people regarded all Huli as enemies, and killed Huli who were in their territory. Meanwhile, the Huli man (of the Hau *hameigini*) who died of shock in the Tari Basin and the Huli man (of the Phi-Nagia *hameigini*) who was then killed in Pogera were victims of the troubles that the Agana men stirred up with the Ipili. As a result, the Hau and Phi-Nagia sought compensation from the Agana, and became hostile towards them. The conflict between the Huli and the Ipili, groups with different languages, gave rise to a "dispute" between *hameigini* within the same Huli language group—the Agana and the Hau/Phi-Nagia. This turned into a drawn-out conflict that was ongoing at the time of my 2005 survey.

Models, along the lines of those used to describe groups in the lowlands of Papua New Guinea—villages that are subordinate populations of language groups, and function as living units—do not apply to Highlands society, the Huli included. *Hameigini* are extremely fluid "groups", altering their very membership depending on the distance of individuals from the endlessly occurring social incidents; and their membership can be seen as an ephemeral assembly that is an expression of the decisions made by individuals regarding each "dispute". Meanwhile, the fluidity of the *hameigini* leads to an active flow of information between people in the language group, and the 70,000-strong Huli do often behave as one unit against the other language groups. The increased ease of movement around the Tari Basin, which has resulted from the infrastructure improvements that began in the 1950s, and the increase in daily movements and opportunities for contact between people have also spurred this trend (Photo A2.4).

Photo A2.4 Children relaxing at home

When children reach approximately the age of the child on the left, they leave their mothers, and move into either their father's house or one where young men are living. It is said that in this way, they turn into men who are able to participate actively in battles.

Is yesterday's friend not today's friend?

As far as people who have gone to live in cities such as Port Moresby are concerned, the "disputes" that occur in the Tari Basin are long-standing, serious problems. They pay constant attention to those that have occurred in the Tari Basin, far removed from the space in which they live, because they determine whether the friend who has come to visit from the village is an "enemy" or not, and also whether they ought to be afraid of friends who also live in the city and who may or may not have suddenly become "enemies". It appears that there was actually a case where a Huli, who had moved to Port Moresby to live, was killed by a friend who—before he had time to realise it—had been transformed into an "enemy" because of a "dispute" back in the village, of whose existence the Huli man had been unaware. With a sense of danger about this situation as the backdrop, around 1995, Huli living in the city held a discussion and made a pact with each other not to follow the logic of "disputes" in the villages, and—on the basis that yesterday's friend is today's friend—vowed to stop the practice of changing social relationships as a result of "disputes" that occur in places that are far removed from their own living spaces. They are attempting to lead lives in which they have made a distinction between the Huli "group" that lives in the Tari Basin and behaves as a coherent unit and their own city-dwelling "group".

Meanwhile, there is also a discernible movement amongst the Huli themselves to differentiate hospital staff from the logic of Huli disputes. The impetus for this was supplied by an incident in which the sole doctor in the sole hospital in the Tari Basin was killed by an "enemy" who came to the hospital pretending to be a patient, and who was in the doctor's care without the latter being aware that a "dispute" had broken out. Because majority opinion amongst the Huli did not support this action and, moreover, because as a consequence of this incident, no doctor was deployed to the Tari Basin hospital for a considerable period of time, it is now common to hear the majority view that people working in public institutions should be treated as distinct from "disputes" and from the logic of "groups". Police authorities, church officials and companies engaged in gold and copper mining are also bodies that repudiate the logic of Huli disputes and groups, repeatedly asserting that the responsibility for disputes should be legally judged on an individual level. The issue of turning the group that is the recipient of the royalties that arise in the course of mining—given by the company that carries out the mining—into members of a limited *hameigini*, rather than potential members of a *hameigini*, is one of the challenges facing mine management.

On the basis of significant circumstantial evidence, it is assumed that the Huli of 300 years ago, when sweet potatoes were introduced, constituted a mere 2000–3000 people. If the logic of "disputes" and the logic of "groups" evident today was created as a result of the sweet potato revolution that followed, then these could also be seen as a flexible policy response to the rapid population increases and the ecological changes in living spaces that accompanied this shift. People's flexibility in having been able to move their place of residence was, in fact, significant as a micro response to problems that we see in the Tari Basin today, such as continuing soil degradation and a declining wild animal population. There is no doubt that in future, as a result of the circumstances confronting Papua New Guinea, the logic of "disputes" and "groups" will continue to arise in an altered form in a variety of situations. An examination of the resulting influence that this will exert on the health of the lives of the Huli people is a vital task for future research.

Part III
The Formation and the Development of "We" Consciousness

8 From the "Here and Now Group" to the "Distant Group": Hunter-gatherer Bands

Hideaki Terashima

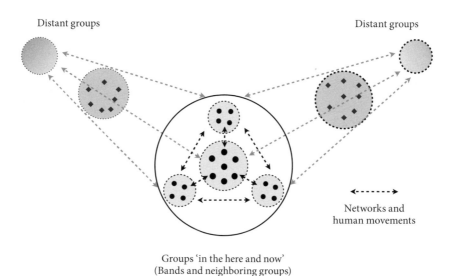

Distant groups   Distant groups

Networks and
human movements

Groups 'in the here and now'
(Bands and neighboring groups)

It is networks of people, built on of a variety of bonds, and the bands that emerge at these particular nodes that mold hunter-gatherer societies. People collect in bands, which are "groups in the here and now", and lead cooperative lives with neighboring groups. However, these are not closed spaces; there are always routes leading to "distant groups". In response to the needs of the moment, people move in search of different ties, and reaching a "distant group" they switch to make this their "group in the here and now". Based on a network of bonds and the free movement of people, these two groups overlap with one another and support people's identities.

Introduction

The statement that hunting and gathering is the oldest way of living for humans is not likely to generate any controversy. "Band" is the term that has come to be used to depict these groups of hunter-gatherer people. However, there have long been various opinions, and there is certainly no general agreement, regarding the following fundamental questions: how are we to comprehend the structure and function of the groups that we call bands; and, what position can we assign to them as human social groups in the evolutionary process? Such is the state of affairs that there are even those who argue that the very term "band" is not appropriate to refer to the groups actually formed by hunter-gatherers (J. Tanaka 1971; Lee 1979). There do not, however, appear to be any doubts regarding the actual existence of concrete living groups of people who engaged in the unique form of living known as hunting and gathering. The issue is what to call these, and it may be that the term band has "reached the end of its useful life in anthropology" (Kelly 1995), but no other appropriate word seems to have been found to replace it. This is because, as described above, there was not an accurate understanding of the actual word "band", and this remains the case. An understanding of the original human groups that we call bands, and which led the oldest form of collective lifestyle, is also extremely important to evolutionary considerations of human society. The current attempt to focus on the evolution of human groups and to inquire into the nature of bands is thus a significant undertaking.

Hunter-gatherer peoples live communal lives in groups called bands as they procure the resources that they need to live, either cooperatively or individually, and engage in mutual sharing. Bands are independent economic units and they are also politically independent. In hunter-gatherer societies there are no distinct social organizations that exceed bands. It is for this reason that these are also known as "band societies". That is, "hunter-gatherer societies = band societies" (Service 1979). This equivalence is, however, beset by two problems. One arises from the fact that hunter-gatherer societies display an extraordinarily large number of variations (Kelly 1995). In places such as Africa, for example, there are nomadic hunter-gatherers societies that have an "immediate return system" (Woodburn 1982). On the other hand, the native peoples in the area of the Northwest Coast of North America formed class societies made up of nobles, commoners and slaves, and these settlements were permanent with visible signs of the accumulation of wealth through prosperous trade (Prentice and Kuijt 2004). The vast majority of hunter-gatherer societies, however, can quite reasonably be seen as being mainly composed of groups of small numbers of people who move around in keeping with the availability of food resources each season, and who live within egalitarian social systems (Lee and Daly 1999). It is this type of hunter-gatherer society that will be considered in this chapter.

The other problem is that while it is certainly the case that in terms of visible, concrete groups there are no social organizations larger than bands or indeed any mechanisms to unite these, there are invisible networks of people that provide the backdrop to actual groups and perform important functions. It is not possible to understand the special characteristics of a band society by simply looking at one band. Bands need to be understood as existing alongside other band groups and, furthermore, alongside far off groups spreading out beyond space and time.

Bands are social groups that are formed on the basis of groupings of individuals, and in order to understand them we need to seek the fundamental relationships between individuals and groups. Considerations of the relationships between individuals and groups proceed from two positions. One is the stance that sees the group as central, and could be called group centrism or society centrism. Accordingly, the individual is treated as one of the factors constituting groups and society. The other position is that focusing on the element of the individuals who make up groups and viewing the group from the individual's perspective, which could be called individual centrism. From a biological point of view, the merits or demerits of groups are argued firstly from the viewpoint of the individual—in terms of factors such as the acquisition of food, opportunities for breeding and safety (see Nakagawa's essay in Chapter Four of this volume). Meanwhile, from a sociological point of view, the structure and functions of groups are discussed from the perspective of the points that go beyond the individual, such as organizations, systems and ideologies that influence the whole of society.

Very soon after the Second World War, advocating the need for animal sociology, Kinji Imanishi opened up the field of primate sociology. Imanishi promptly turned his attention to the importance of the relationships between the groups that occur amongst animals that are higher in the evolutionary hierarchy and the individuals who form these groups, attempting to understand groups by fixing the individual as his pivot. A representative study of this type was "identification theory" (Imanishi 1957) with which he attempts to explain the development of group-centered behavior by focusing on individuals in Japanese macaque society. The explanation that is given is that group-centered behavior is a cultural element that is acquired via male individuals identifying themselves with the alpha male during infancy. This identification theory has been criticized for being impossible to actually prove (Itani

1991b). However, Imanishi's thoughts on groups themselves, his central interest, are extremely interesting. Namely, Imanishi argued that groups were gatherings of individuals who had the fundamental ability to live individually; "Are groups not something that birds and animals, who having reached a higher stage of evolution and becoming aware for the first time, came up with themselves *in order to satisfy their sociability*?" (Imanishi 1976, emphasis added). In writings from his later years, we also find statements such as "[t]he goodness of living in groups… has as a major trait in that members of the group *enjoy group life*" (Imanishi 1987). Statements such as "sociability" or "enjoying the lifestyle" sound extremely anthropomorphic, but when we consider the evolution of groups leading to human group formation, there is certainly also a need to take into consideration the value of some things that go beyond the biological needs of the individual.

Imanishi held that as animals attained a certain level of higher development, an irresistible desire for groups was born in individuals, and he expressed this using the word "sociability". His argument applies largely to groups of birds and beasts and primates other than humans, but it could also be seen as applicable to humans. It is extremely interesting to consider the use of the word "sociability" in this way to indicate the root cause for the existence of groups alongside Hitoshi Imamura's idea of the "social", which is discussed below. Imamura questions the relationship between society and social, and concludes that social is not simply the adjective for society. "'Social' is the force of social formation which enables all other social relationships; alternatively, it is 'social bonds'" (H. Imamura 2000). It is the social that comes first, and society can only begin to exist as a result of the social. "Social bonds" are "the nature of human relationships imbued with emotional meaning", "fellowship", "moral obligation" and "mutual aid". Imamura argues that they are the most fundamental relations between people: generous hospitality and assisting others. In this chapter, the aim will also be to show that the groups found amongst hunter-gatherers are characterized by the social, in this sense of the term, and by related networks.

The approach that Imanishi employs is one of trying to understand groups from the position of the individual; and while he rejects biological individual centrism—that is, individual reductionism—he similarly does not regard the group as absolute. The stance upon which the current chapter is based also resembles this approach. That is to say, it is not an approach that looks at the structure and function of bands from the viewpoint of bands as the central subject; it is not an attitude that is "band-based". Nor is it an individual reductionist approach that thinks about bands by disaggregating them into individuals. It is a stance that explores the nature of bands as groups by taking a wide-ranging view of the state of bands as they relate to the self, from the perspective of the individual who always exists in connection with the group. The scene that is visible from this standpoint is one of networks of individual self-reliance and those that connect people to one another, and also the figure of the band that is created at these nodes, as ever-changing in appearance. These are human groups with links to the world of primates that preceded humans, and they are also forms of groups that have acquired a human originality.

The band model

Patrilineal/patrilocal bands

In twentieth century anthropology, hunter-gatherer bands have been discussed on the basis of a variety of factors related to equality and power, kinship, marriage systems, rules governing residence, territory, use of resources and the sexual division of labor. Let us quickly review these. The two representative models that can be mentioned are the sociological model and the ecological model.

The sociological model has its origin in Radcliffe-Brown's study of Australian indigenous society. Radcliffe-Brown saw the "nuclear family" and the "horde" as basic components of Australian indigenous society. A horde (=band) is a small group that possesses and has exclusive use of a specified territory or hunting grounds; they are the groups that form the units of the hunter-gatherer way of life. Succession within hordes is patrilineal, with membership being life-long. Hordes were also held to be self-reliant groups that are mutually independent and unique entities as political groups (Radcliffe-Brown 1930/31).

Subsequently, research with an interest in social evolutionary theory continued this line of inquiry. Julian Steward's (1955/1979) "patrilineal bands" and Elman Service's (1971) "patrilocal bands" can be seen as an extension of Radcliffe-Brown's theory. Patrilineal bands are made up of males and their families who are related to one another in the male line. Steward gives ecological advantage for the hunting activities of males in sparsely populated areas as the reason for the formation of these bands. Conversely, Service, rather than identifying hunting motives, saw male unity as important for preventing attacks by other hostile bands, and stressed the importance of the solidarity of close male relatives such as siblings and cousins. Service explained that bands inevitably became agnatic to ensure that these males would remain within the band and choose patrilocal marriage.

Steward and Service clearly dealt with the structure and function of bands from a group-centered perspective. Whilst there is a major difference in that one saw ecological factors as fundamental and the other put military concerns first, both considered that a state in which groups were based on a social system of patrilineal/ patrilocal rules and band exogamy was the basic form of hunter-gatherer bands. The sociological model is characterized by its plain and simple nature, and was emulated by many anthropologists, becoming the standard model for hunter-gatherer bands. Actual bands, however, were never that simple or elegant.

Until the mid twentieth century, the classic image of patrilineal/patrilocal bands presented by anthropologists was that of the Australian indigenous bands, but in subsequent surveys, it became clear that close patrilineal group composition and the monopolization of territory by patrilineal groups was rarely observed. Patrilineal clans certainly did exist amongst indigenous Australians, and they held land rights that had been inherited from their ancestors. However, it is not the case that these kinds of patrilineal clans form actual bands. In reality, bands arc gatherings of people who are joined by a variety of blood relationships and marriage ties (Keen 2006). Furthermore, the people of a particular patrilineal clan do not live forever on their inherited land, and nor is it the case that the use of land resources is limited to patrilineal relatives. People are able to use lands outside of their own patrilineal group via a variety of relationships. In addition to patrilineal descent, maternal kinship is also used. Links to land can be forged as a result of the single event of it being the place where one's mother conceived (Myers 1986; Morton 1999).

The problem with patrilineal/patrilocal bands is to be found in thinking that is band-centered; thinking that sees a band as a container that entails rules and a system and prescribes membership. In this type of model, individuals have but a passive existence determined by the particular rules and system of the band. When this type of model encounters cases that do not fit, the only way of dealing with these is to designate them as "exceptions". The "composite bands" category that can be found in the patrilineal/patrilocal bands model (Steward 1955/1979 and Service 1971) is one such example.

Flexible bands

Following the end of the Second World War, fieldwork on hunter-gatherers was energetically carried out all over the world, and new empirical data was amassed. At the 1965 international conference held in Chicago titled "Man the Hunter", up-

and-coming researchers who had hastened from their field sites all over the globe presented materials that urged a major revision of the image of hunter-gatherers to date, unleashing heated discussions (Lee and DeVore 1968). They also disputed the existing patrilineal/patrilocal bands model as the form of bands. The major point at issue was the fluid nature of bands. This group of researchers confirmed that in actual bands there was an incessant coming and going of members, and that these were not at all groups with clearly defined borders. It also became clear that territory either did not exist or, even if it did, it was not exclusive as access was granted to a variety of other people in addition to patrilineal members. As a consequence of this, the "band = patrilineal group" equivalence became unfeasible and the concept of band exogamy also lost its import.

It was the Bushmen who enjoyed the limelight as the most typical exponents of hunter-gatherer peoples living in these sorts of bands with flexible borders. Jirō Tanaka, who carried out surveys of the Bushmen of the Central Kalahari, held that it was inappropriate to call the Bushmen's residential groups "bands" because these were open and had a composition that was rich in fluidity (J. Tanaka 1971). He pointed out that the family was the only permanent social unit in Bushman society, and that the unit of living was actually a temporary residential group that ought simply to be called a "camp". Richard Lee, who carried out surveys of the !Kung Bushmen, also advocated the use of the term "camp" and not "band" for the same reasons. A camp is "a noncorporate, bilaterally organized group of people who live in a single settlement and who move together for at least part of the year" (Lee 1979).

What about this idea, however, that hunter-gatherer groups are nothing but camps, in the sense of temporary gatherings? Thinking of the band as a container, when its contents are people, the fluid nature of these contents does not necessarily negate the existence of the container. The problem lies in the thinking that says that the container prescribes the contents. What about, conversely, thinking of this in terms of the changing contents inventing the container that suits them on each specific occasion? There is no need to think of containers that come into being in response to the movements of people as fixed entities. It should be better if only we get some critical distance from the stereotype of the patrilineal/patrilocal image. Even amongst the Bushmen, family and close relations mutually connect people from the inside, and spontaneous cooperation and distribution of food that is not constrained by external rules and systems develop people's cooperative natures. Bands are not at all solidly organized, but nor are the people who assemble within them an odd miscellany; it is clear that they are actively engaged in cooperative living. It is for this reason that in this chapter the word "band" is used to include

Photo 8.1 View of the Ituri Forest

The forest is located in the northeastern region of the Democratic Republic of the Congo. Stretching as far as the eye can see, the forest is covered with trees between thirty and fifty meters high. Tens of thousands of pygmy hunters live in the forest that spreads some one hundred thousand square kilometres. They organize into bands, each constituting a few dozen members and move from one place to another in the forest.

the Bushmen's groups. The problem is not the outward appearance of bands, but their contents.

The true state of fluidity

Following "Man the Hunter", "fluidity" became the buzzword of the day. From the perspective of ecological adaptation, flexible bands that are able to change freely in response to time and place have an advantage over patrilineal/patrilocal bands born out of firm membership and the retention of territory. However, fluidity is a result, not a cause. There are a variety of reasons that cause individuals and families to change bands. Thought needs to be given regarding what it is that makes this movement of people in and out possible—that is to say, what enables the band's fluidity?

The following is an examination of this issue drawing on cases from my own surveys (Terashima 1985). The Mbuti Pygmies who live in the Ituri Forest (Photo 8.1), in the rainforests of the Democratic Republic of the Congo, formed groups of a type that at a glance appeared to fit the patrilineal/patrilocal model. In their society, there were groups that should be called patrilineal descent groups, and most bands included at least one of these and were known by the name of this group. Marriage was generally patrilocal, leading to the impression that this is the embodiment of the patrilineal/patrilocal band model (Harako 1976; Tanno 1976; Ichikawa 1978). At the time of my initial survey work, I also thought that the patrilocal band was precisely the right principle for describing these residential groups, and even while I noted that there were changes in membership, that was how I reported on the groups (Terashima 1984). Five years later, I conducted my second round of fieldwork, and the structure of the five bands that were the subjects of my investigations varied widely in their composition. If we assume that bands come into existence because of genealogical relationships and residence rules, then we would clearly not expect the normal state of affairs to be fluidity of membership. However, I confirmed that the fluidity of members was understood to be a very common occurrence.

During my first round of surveys, I attempted to understand bands in terms of post-marriage residence rules and the residential system. Then, during the second round, as a result of attempting to pursue the status of the individuals in each band from the perspective of the individual, a very simple state of affairs became clear. The relationship between an individual member of a band and other members fell into three categories. One was people who lived together with other patrilineal relations. The second was those residing with their matrilineal relations. The third was people living with parents-in-law or brothers and sisters-in-law, either in the husband or wife's band (Terashima 1985). Thus, everyone was related in some way, sometimes through marriage, to other members in the band. The vast majority of people gathered together on the basis of patrilineal relations, but no one considered residence within the band by people other than this group as being in any way an unusual state of affairs. They might be residing with the wife's or the mother's relatives, but if this is what people wanted to do, not only was it possible, but this in no way gave rise to any particular difficulties. This situation confirmed that living with one's patrilineal relations was not obligatory but rather up to the choice of individuals.

Why, then, were the Pygmy bands of the Ituri Forest judged to have been patrilineal/patrilocal? This was the result of the fact that there were key members within each band, and that the overwhelming majority of people in the band were in patrilineal relations with those people. It may be inevitable that when a decision is made on the basis of the external composition of the Pygmy bands they will be seen as patrilineal/patrilocal. If, however, they are viewed from the perspectives of individual members gathered within the bands, then they are not gathered there because of the prescriptions of a patrilineal/patrilocal system. The fact that the main body of the band is made up of people in patrilineal relations with one another is not the same thing as the band being composed along lines prescribed by patrilineal/patrilocal rules. We must see the preponderance in bands of people who are in patrilineal relations with one another not as a prescription, but as being based on other factors. Without going into any detail, my conjecture is that the origin of the external appearance of patrilineal/patrilocal band composition may be the symbiotic relationship that the Pygmies have built up over many years with the neighboring agricultural peoples (Terashima 1985).

Leaving aside questions as to whether this conjecture is right or wrong, the prevalence of patrilineal relationships in the residence choices made by individuals is not founded in "patrilineal ideology". On the basis of the three kinship bonds mentioned above, Pygmies are able to reside with various people apart from their own patrilineal relations. There were families that moved alternately between the husband's and the wife's band and also families that resided in bands on the basis of completely different connections. It is probably the case that within the wide range of options, close patrilineal relations such as one's own parents, siblings and cousins, were the easiest relations to consider for historical and social reasons.

Fred Myers, who has researched the bands of the Pintupi people who live in Australia's Western Desert region, also argues that bands are not "givens" constructed on the basis of prescriptions, but that they are socially achieved through individual choices and decisions. The composition of real bands is to be explained solely through the history and processes of individual affiliation (Myers 1986). This, together with the above description of the situation of the Pygmies, should clarify the importance of individual decision making in band societies. Patrilineal/patrilocal bands were thought of in terms of the band deciding who ought to become a part of it, and this was the product of band-centered thinking. The reality is the opposite of this: it is precisely the active choices of individuals that bring about the creation of the actual band.

The characteristic features of hunter-gatherer peoples that became clear following "Man the Hunter" were the autonomy of individuals and families, the freedom of action and an emphasis on the equality of individuals (Lee and DeVore

1968; Woodburn 1982). There is no "compulsion" in hunter-gatherer society (Clastres 1974). People are basically able to act without being controlled by the will of others. They dislike being ordered about by others and, similarly, avoid issuing orders to others. This is one manifestation of egalitarianism, the representative value of hunter-gatherer peoples; and the fluidity of bands also has its source in this. Belonging to a particular band and leaving that band are both the choice and the decision of the individual. The fluidity of bands is born as a consequence of choice, which is supported by this kind of individuality and independence on the part of individuals.

It is also important, however, to exercise caution regarding an emphasis on the terms "individual" and "individuality" in this context. This is because these terms are both concepts that were originally spawned in modern times, and because of this we cannot simply introduce these concepts, as they are, into hunter-gatherer society. Reluctantly, no detailed examination is possible here, but even when "individual" and "individuality" are used in reference to hunter-gatherer society, what is intended is not a subject of the sort that constructs its self, makes decisions and acts while in opposition to others and the environment; nor is it a subject that includes modern individualistic thinking. Instead, this is an ego, a subject, that is constructed while cooperating with and adapting to others and nature.

If we assume that bands are created in this way, on the basis of free decision-making by individuals, our next challenge is to ask what lies at the foundations of the choices and decision-making exercised by individuals, and at the basis of the reasons for residing with other people? Although people are said to be able to wander freely between bands, they do not gather and separate purely as a result of chance. In the background to fission-fusion, we would expect there to be things that draw people, and things that bind individual-to-individual and family-to-family. These things create the group we call a band, become the basis for the identity of the people living in the band and, simultaneously, give rise to its fluidity. These are the bonds and networks between people that, while including the band as an entity, transcend and extend beyond it.

The true state of bands: From "here and now group" to "distant group"

The "here and now group"
In our investigation thus far, it has become clear that bands are not characterized by ideology or territory. Let us now look at the figure of the band as seen from

Photo 8.2 A Pygmy band and young girls in the forest
This photo shows a pygmy band in a small forest clearing. Each family builds their own dome-shaped hut, from which they go out every day to hunt and collect plants in a nearby forest. Visitors from other bands often drop in.

the perspective of the individual. First of all, the band has an extremely concrete nature—it exists as a residential group (Photo 8.2). The coordinates along which the residential group is positioned are in a time that is "now" and a place that is "here". The people who assemble in a band are linked to each other, first and foremost, by directness and contemporaneousness in a format that has co-residence as its basis. Every day, people see each other's faces, talk to one another, give and take things—they repeat mutual acts that put them in physical contact with one another. From their outward appearance, they are collectives that might resemble primate groups.

Meanwhile, it is shared lives at a point in time that directly supports bands as residential groups, with indirect support coming in the form of a shared past leading up to the present which includes a host of experiences. The most important mutual act that makes viable the cooperation of bands that share lives in the present is the cooperation that occurs in a variety of activities and the sharing of food. The people who assemble in groups have a "we" consciousness as a result of the very act of having gathered together. This is also a "we are the same" consciousness, and it is because they are "the same fellow people" that cooperation and sharing occur naturally. As numerous others have already stressed the importance of cooperation and sharing (Kishigami 2003, Ohmura's essay in Chapter Five of this work and others), there is no need to go over this again here. Clearly, this surpasses the earlier primate level, and is the world that is characteristic of humans.

In addition, "sameness" is identical to being "equal". As pointed out earlier, hunter-gatherers are people who strongly seek sameness and unity, enjoy cooperating and sharing with others and are intensely sensitive to either discriminating against others or being discriminated against. They take pleasure in jointly planning the same activity, carrying it out and then sharing its fruits. Cooperation and sharing that is based on these feelings of equality bring about a strong sense of unity within the band, reinforce the links between members and, moreover, lead to the creation of even more stable egalitarian relations. The opposite situation—namely, the failure of cooperation and sharing, and then the loss of a sense of equality—will even lead to the collapse of the band. Cooperation and sharing, in both a material and spiritual sense, form and preserve the band as a "visible group".

Meanwhile, the band, as a group, is sustained on the emotional front by familial and kinship bonds that connect members such as parents and children, siblings and brothers and sisters-in-law. Even in the bands of the Bushmen, where fission-fusion occurs repeatedly, there is a dynamic unity of the family which centers on links between, for example, parents and children and brothers and sisters, and that constitutes the nucleus of the camp (J. Tanaka 1971). Amongst the !Kung Bushmen, families and relatives as well as relatives by marriage reside together (Lee 1979). As mentioned above, the Pygmies will also gather in a band based on relations between parents and children, siblings and cousins. In Pygmy society, there is a preference for sister exchange marriages, and the couples from the two groups conducting the marriage often live in the same band (Ichikawa 1978; Terashima 1984). In the Pintupi groups of Australia, bands are also composed of people linked together by kinship bonds in the broadest sense (Myers 1986). However, the point that particularly needs to be borne in mind is that these types of kinship bonds—far more than being based on the ideology of the patrilineal clan or lineage—are always bonds that centre on oneself or one's family, and extend to both the patrilineal and matrilineal lines.

As one might expect, living in the same place and sharing the same environment as the locus of hunting and gathering activities (=land and resources) are also factors that form a part of shared lives. An individual band will consider a specific piece of territory to be their own place and will hold the rights to its resources. Living together in the same place forms people's identity, making it what it is. Links to the land make people's reciprocal ties visible. The nature of belonging that is mediated by "the same land", the nature of belonging to a living space that the !Kung Bushmen call *n!ore* (Lee 1979) and the nature of the relationship that Pintupi call "one countrymen" (Myers 1986) are all examples of this. Ties to the earth, with its permanence, confer on the group an identity that transcends time.

The ownership of land, in the form of "territory", was held to be of particular importance in the sociological band model. However, territory that is regularly defended is practically non-existent. The ownership of land in hunter-gatherer societies is not of the kind assumed by the sociological model, in which there are exclusive borders and its use by other groups is prohibited, but rather it is a type of ownership that emphasizes the building of bonds between people by approving access rights to the same land—that is, its shared use.

As discussed above, there is an assertion that "we are the same" amongst people who gather together in a band. The word "same" can be replaced by the word "family", to give: "we are all family". The "same people" are "family", and "family" are the "same people". The shared experience of living together makes people family. Of course, this is not in the narrow sense of the meaning of "family" and "patrilineal group". Also, this assertion includes a continuous temporal view, from the past to the future, regarding the bonds between people. This is because the family has a history and a future. People who have become a family remain family even if they separate, and these relationships are passed on down the generations.

Although bands are residential groups, their members frequently act independently, and it is often the case that a portion may be absent, having travelled far away on a hunting trip. They will also often call in and stay with another band. There are even cases of these visits lasting as long as several months or over a year. Thus, even in the band as a residential group that is visible to us, there are actually always absentees. The band exists as something that includes these absent members as well. We also need consider these people when thinking of the band's overall image and unity. Consequently, it will be clear that in the expression "the band in the here and now", notwithstanding, we are not only talking about directness and contemporaneousness. Connected to this is the dimension of a world that is cut off from immediate time, or from immediate time and space.

Bands as "distant groups"

Bands in the form of "groups in the here and now" are the most familiar and concrete groups, but they do not, of course, exist in isolation. There are neighboring groups that encompass multiple bands in a number of friendly relationships, and each respective band exists within these interactions.

There are frequent mutual visits and the exchanging of information between neighboring bands. They also exchange food items and other subsistence goods. In cases where the distance between bands is particularly small, one also sees considerable cooperation in subsistence activities such as hunting, fishing and gathering. Major reorganizations of bands occur following mutual fission-fusion. As is the case with bands, neighboring groups also lack clear borders, organization or territory as a whole. Neighboring groups form one part of the intermarriage pool for the respective groups, and people are linked together by a variety of kinship and marriage ties. Neighboring groups still have a strong aspect of directness and contemporaneousness.

"Groups in the here and now" and neighboring groups appear on the surface of normal life, but occasionally there are comings and goings by people from beyond neighboring societies. This is the considerably wider world that extends beyond the sphere that is visible on a daily basis. These are distant lands that require more than a few days travel to reach, and occasionally people come from and go to as yet unvisited lands. Alternatively, they may be groups of people who are beyond recollection. I would like to call these "distant groups". They are temporally and spatially remote, but there is a cognizance that these people are "same people". The dimensions of directness and immediacy have been lost between these types of people. In their place exist ties that are recognized through the mediation of factors that transcend time: links of belonging to a certain shared piece of land and also of being related. Distant groups are people who live in memories of the past and in images of the future; they are people held in one's heart.

One may feel that these sorts of groups are very generally the same as those that are called "patrilineal groups" or "kinship groups". However, "distant groups" are identified as those using a variety of bonds, including patrilineal and matrilineal ties. Distant groups are also self-centered ones that always expand outwards from the starting point of the individual, and although called "same people", they are comprised of people who are different according to each person. Hunter-gatherers' kinship identity is not an identity with ideological patrilineal groups even in a society that has a patrilineal kinship system; it is, in a practical sense, an identity with "people who are linked to me".

The people in distant groups usually do not have any direct contact with one another. These people exist only in their hearts. However, inter-marriage groups usually extend as far as these remote people. There are always paths that connect these groups, and people may actually undertake a trip to visit a distant group when they want to do so. Trips intended solely to fulfill the requirements of "sociability" are also not rare. Sometimes, people head off to meet the people in their hearts without any concrete objective. And when people arrive at the distant group in their midst, it would immediately, from that point onwards, become the "group in the here and now". In this way, the far off, distant group represents the span of the world to people who live in bands, enlarges their network of bonds and supports identities that go beyond time and space. Hunter-gatherer people are not people who seclude themselves within their territory; they are people who will travel great distances and who seek connections with others.

People who live in bands: Bonds and independence

As described above, while hunter-gatherer bands are concrete entities, they also possess a dimension that goes beyond the concrete. In hunter-gatherer societies, there exists a network of individuals based on a variety of bonds and, in addition, the "we here and now" who are divided according to directness and immediacy. At the same time, there are also the "distant we" who transcend the time and space of the "here and now" and extend far beyond. The totality of the lives of people who live in hunting and gathering societies is shaped via an overlap of both of these dimensions. Consequently, it is not possible to simply reduce bands to the level of ecological adaptation. At the same time, nor can we simply leap to the level of relatedness, such as clan and lineage. These are but one aspect of the attributes of bands, and it is hard to say that these are the forms that bands take as a whole.

Bands do not have set boundaries in the same way as objects; we cannot define them on the basis of systems and rules. They appear when needed, in places where they are required and in the form in which they are needed, enveloping the focal points of bonds between people. They persist as entities, but their form has many manifestations depending on the accumulated history and aspirations for the future. The choices for these changes lie in the hands of individuals, and the future is always filled with considerable fluidity. Even with regard to bands "in the here and now", there is variety in the manner in which individuals participate in bands and in the state of these ties; each member has their own image of the band and configures the band into which they will place themselves. That is, it could be said that the coalescence of the overlapping images of a band that are held by all of its members is the form of the band that is visible to us.

It is the vast array of bonds that link people to one another that provide the cohesive power for those who live in these kinds of bands. Bonds form networks, and bands are to be found in the focal points of these networks. Finally, let us take a closer look at these bonds. The importance of bonds is not restricted to African hunter-gatherers. Myers (1986) also notes that Pintupi society emphasizes "relatedness" above all else. That is what connects, unites and maintains bonds. It is precisely people's relatedness to one another that forms the basis for regulating the whole of social life. People seek relatedness; they lend their ears to other people's requests, show sympathy and empathy and negotiate with them. The independence of the individual is important in Pintupi society, however it is always linked to others; that is, it is understood that it must be expressed through the maintenance of relatedness. Willful and selfish types of demands are not evidence of the independence and liberty of the individual; they are merely "things of which one ought to be ashamed". In this sense, equality is also closely linked to relatedness. Positive links with others are nothing less than egalitarian relationships.[1]

Incidentally, the concept of "bonds" is not one that simply reveals the external relationships between people; it also embodies an internal sense. This results from the fact that the word "bonds" is perceived at the deep level of the "acceptance" of connections between you and me, and between one person and another (Ikegami 2004). Yoshimasa Ikegami says that unconditional "acceptance" lies at the heart of religious salvation. Through unconditional acceptance, people can have a personal experience of having been saved. Acceptance is, however, replete with the dangers of constraints. Strong bonds forcefully constrain both parties. If one hates constraints, then acceptance is not possible. However, without being accepted, people cannot go on living. This is clearly a paradox, but Ikegami suggests that it is in this that we can find the significance of belief in the powers of religion and the supernatural.

In hunter-gatherer society, people gather in bands because they are seeking acceptance, but in these cases also there is a considerable possibility of the paradox mentioned above coming into existence. However, in hunter-gatherer society, it is possible to get out of this acceptance quite easily should it become difficult. It is the network of bonds between individuals that makes this possible. There is no doubt that what lies at the root of the independence and freedom of hunter-gatherers is the special ecological quality of the hunting and gathering lifestyle in which, if necessary, an individual or an individual family can survive for some time on their own. There are also networks of bonds that one can use. It is possible to see the hunter-gatherers' free engagement in fission-fusion as one effective means of resolving the paradox of acceptance and constraint that is inherent in bonds.

Whilst they are free, hunter-gatherer peoples want to align themselves with others, to do the same things as others; they are people who find value in being the same. They are people who live lives with double meanings: whilst hating restrictions, they hold dear bonds with other people. There is no other way for people to live a proper life—an independent adult life—apart from through being connected to and cooperating with other people. People seek out others and form groups; they exist as complete people by living cooperatively with others. People who wander about on their own are not at all complete people. Conversely, when people can no longer bear the restrictions that come with cooperative action, they seek out other bonds and move to another place. People's bonds are not limited to "here and now", they extend as far as the far off "distant world". "Distant bonds" guide people and lead them to new "here and now bonds". It is precisely these ambivalent and dynamic relationships between individuals and between individuals and groups, and the spread of the network of bonds that supports them, that shape bands as the way of life of hunter-gatherers.

Hitoshi Imamura uses the word "*social*" for the forces that connect people to one another and for the "social bonds" that act to resolve hostilities between people and bring about amity. He states the following regarding this issue:

> Compared to '*society*', which consists of relations of control and dependency, '*social*' is 'weak' mutual acts... Even if these only have a 'weak' function, at the very least they join people to one another and convert simple 'groups' (*mure*) into human groups (*shūdan*). This is how the foundations of groups are formed. Without '*social*' as the power of social formation, social order would not be able to maintain itself. (H. Imamura 2000: 30, translated by Terashima)

Borrowing Imamura's phraseology, let us sum up this chapter. Hunter-gatherer society is one that lives on the basis of social bonds that comprehensively reject a society composed of relationships of control and dependency and exists for the sole purpose of social formation. Viewed from the reality of the annihilation of practically all hunter-gatherer societies following the appearance of agriculture and stock farming, they may well have been "weak" societies. However, they are societies that give real form to the peculiarly human nature of groups in which independence and dependence are permitted to coexist without any constraint of others or being constrained by others, where links and equal relationships between people, as the realization of *social*, are central to people's values. These types of societies have supported humanity for an astonishingly long period of time. Although they may be minute in number now, the fact that they still maintain their vitality is surely worthy of special mention.

9 Perceivable "Unity": Between Visible "Group" and Invisible "Category"

Toru Soga

"Unity" and "cultural category"

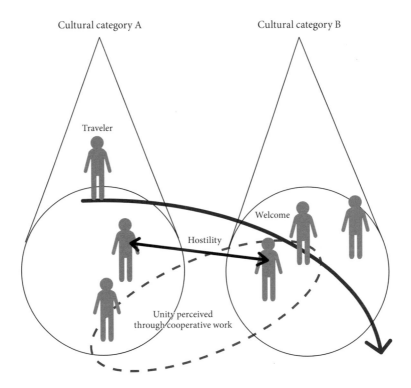

Humankind has the unique ability to segment the world according to symbols and create cultural categories. At times, humans find themselves constrained by cultural categories such as "ethnic groups". However, humans also possess a capacity they share with non-human primates, and by using this ability in a very human way, they can cast off the yoke of cultural categorization. This is the sense of "unity" people can feel with those they interact with, for example, while offering help to unknown travelers or working with people from other ethnic groups. I argue that the fundamental characteristics of humans, who succeeded in creating complex societies, lies in their ability to perceive "unity" with new people based on the personal experience of interacting with them, despite the impact of cultural categories.

Introduction

One evening in January 1994, I was in Northern Kenya conducting fieldwork in a pastoralist village of the Gabra, enjoying a conversation with Mr. Elema, as I always did. It was a pitch-black night without moonlight. Suddenly, a man appeared. He begged, in Kiswahili, to be allowed to stay the night because he was extremely tired. After listening to his story, Elema said to him that he could sleep by the livestock fence and told his wife to make some tea.

Kiswahili is the national language of Kenya, but it was never used in the Gabra village. People used Kiswahili in town only when they needed to communicate with those from different ethnic groups. Why, then, did the man speak Kiswahili in the Gabra village? I asked Elema and he replied, "He is a Turkana. He is on his way back home from town but is too tired to move on".

The Turkana were another pastoral people and were in a hostile relationship with the Gabra. There was a major armed conflict between the two in 1992, and a widow who lost her husband to the Turkana was still present in the village. I had never seen a Turkana near the village, and the two communities had no everyday communication whatsoever. Therefore, I could not imagine that a Turkana man would visit the village in the middle of the night, but I was even more surprised at Elema who permitted the man to stay. Why would he allow his enemy to sleep on his ground? Elema answered, "Because he asked very respectfully".

I always recollect this scene when I ponder the pastoralists' social relationships. We, as humans, sometimes find ourselves in a situation where we need to rely on those we consider our enemies. The exhausted Turkana man chose to sleep in his enemy's village rather than camping out in the wild. In the village, he would be safe from the hyenas and would not have to fear being attacked by a Gabra while sleeping in the open. Feeling empathy with the tired man, Elema responded with a minimum level of hospitality. People do not flatly refuse others who are dependent upon them, even if they are their enemies.

When humans come face to face with others, they have the ability to show deep empathy for each other. This is true even if they do not have kinship ties or are from opposing ethnic groups. People have the ability to put their group affiliations aside and immerse themselves in interacting with those in front of them. Having said that, people can also choose not to empathize and simply kill the other as an "enemy". In the latter case, people base their judgment on their opponent's group affiliation; in other words, their actions are determined in compliance with their respective groups.

Some of these behaviors in relation to groups are based on abilities humans newly acquired through evolution, while others may be founded on those shared with non-human primates. For example, uniquely human attributes include the ability to represent themselves as "ethnic groups" or clans, to have control over such representation and to decide how to behave accordingly. However, it is also true that there are many other activities that are based on abilities that we share with non-human primates. Being able to become immersed in the interactions with those before them, without being too constrained with what they represent, can be said to be one of these faculties. We tend to focus on newly acquired human abilities. However, it seems that at times it is just as important for us humans to express the abilities we share with primates in a very human way.

Based on field data from the East African pastoralist community, this chapter examines how humans use the abilities that they share with non-human primates by looking at behaviors connected with groups. To be specific, we will focus on the pastoral people of Gabra who live on the border of Northern Kenya and Southern Ethiopia. I conducted fieldwork in Kenya from 1990 to 1999 and in Ethiopia from 2000 to 2009. The population of Gabra in Northern Kenya (Gabra Malbe) is about 30,000, and they live in the arid Chalbi desert where the average annual rainfall is approximately 150mm. They are pure pastoralists, mainly breeding goats, sheep and camels, supplemented by cattle and donkeys. The neighborhood of Gabra Malbe is clearly segregated from those of other local ethnic groups. They are surrounded by the pastoral people of Turkana, Samburu and Rendille on the Kenyan side and the agro-pastoralists of Boran, Arbore, Hamar and Daasanetch on the Ethiopian side.

Conversely, the Gabra in Southern Ethiopia (Gabra Miigo) live in a relatively moist area with an average annual rainfall of approximately 400mm. The population figures are unclear, but are estimated to be around 20,000 to 30,000. They are pure pastoralists mainly raising camels but also a very small number of cattle, goats, sheep and donkeys. The neighborhood of Gabra Miigo overlaps with those of the other ethnic groups. Until 1964, the eastern side of Gabra Miigo's neighborhood overlapped with that of the pastoral people of Garre, and the western side with the pastoral people of Boran. Since then, the Gabra Miigo have been repeatedly forced into refugee or internally displaced status. Their internally displaced status continues today, and thus it is difficult to identify exactly where they live. However, during the 2000–2007 period when the research for this chapter was conducted, they lived in Yabelo area. The northern side of the neighborhood largely overlapped with that of the agro-pastoral people of Guji, but on the southern side they lived segregated from the pastoral people of Boran.

Visibility of a group

Now, let us first discuss what "group" signifies. When debating with primatologists, I sometimes cannot help but feel insecure about the terminology used in the field of anthropology on which I stand. "Group" is one of such terms. What is a group to a primatologist? "Group" to a primatologist refers to a visible "troop". A "troop" is an observable company of monkeys, and primatologists have been discussing *groups* based on their visibility.

Let's take chimpanzees for example. Though Kaori Kawai has touched on the experience of Junichiro Itani and Akira Suzuki in the Introduction to this volume, I would like to revisit it here as it is a key point in the argument to follow in this chapter. In the early days of chimpanzee studies, it was believed that chimpanzees do not form a stable troop because the frequent fission-fusion had deceived the eyes of the observers. Jane Goodall, who had been conducting research in the Gombe Stream in Tanzania, assumed that there was no stable troop among the chimpanzees other than the ties between mother and child. However, one day Junichiro Itani and Akira Suzuki happened to run into a large troop of chimpanzees. The incident happened in 1965 in Filabanga, Tanzania. In September of that year, nearing the peak of the dry season, the two researchers came across forty-three chimpanzees crossing a ridge in a linear fashion. They recorded the sex and age of the individuals and were convinced that this multi-male multi-female troop indeed comprised their Basic Social Unit (BSU) (Itani 1987b: 179–182). Later on, Toshisada Nishida, who succeeded in feeding the chimpanzees in the Mahara massif in Tanzania, identified the membership together with Kenji Kawanaka and was finally able to confirm that this large multi-male multi-female troop was their BSU. Once the BSU was confirmed, they were then able to observe features such as their diachronic structure, internal group structure, inter-group relationships and so on.

Now, compared to the primatologists who base their discussion on groups on visibility, those that the anthropologists deal with are not always visible. Roger M. Keesing (1975) pointed out that social groups and cultural categories were often confused and emphasized the importance of keeping the two separate. According to his definition, a social group is constituted by "actual warm-blooded human beings who recurrently interact in an interconnected set of roles" (1975: 10). This would be almost equivalent to the "troops" that primatologists deal with. On the other hand, cultural categories "exist in people's *conceptual* worlds", and "are sets we *draw mental lines around* in particular contexts" (Keesing 1975: 10). For example, though unilineal descent groups or ethnic groups such as clans and lineages do form "groups", they are in fact cultural categories that segment people by concepts. They differ from social groups comprised of actual living human beings. The unilineal descent groups or ethnic groups may be referred to as invisible "groups".

What makes things complicated is the fact that, even though a cultural category may be different from a social group, by belonging to a certain cultural category one will be charged with various rights and responsibilities. People interact with each other according to their cultural category. For example, if someone fails to carry out their responsibility as a member of a clan, the other clan members may gather to impose sanctions upon this person. As such, when living human beings behave in a certain way, the cultural category may appear before their eyes in the form of a real social group.

This confusion of cultural category with entity, despite Keesing's warning, seems to be the reason behind the instability I experienced regarding anthropological terms when I was talking with primatologists. The error, however, is inevitable. Motomitsu Uchibori in Chapter Two of this book astutely observed that efforts to identify the differences between groups and categories would most likely fail. In fact, Keesing himself seemed to be confused by the two concepts, stating at one point that the descent unit (especially at clan and moiety level) can no longer be called a group but is rather a social category (which serves to define the outer frame of exogamy or the minimum lineage responsibility) (1975: 32). He almost seems to be indicating here that the descent unit is connected at the real social group level to the cultural category level on the same dimension. If a kin group specialist like Keesing finds it difficult to differentiate, it is important for us mere mortals to set up a simple principle to prevent us from falling into the trap where we "go for wool and come home shorn".

The simple principle is as follows. If people are visibly gathering and interacting, we define this as a group. According to Keesing (1975: 10), "groups can be distinguished from forms of aggregation, such as crowds or gatherings, whose interaction is temporary and limited". However, in this chapter we also regard this temporary and limited aggregation as a group. Such aggregations may not be significant in the conventional anthropological or sociological context, but they would become an important index when discussing human ability that facilitates the group phenomenon (evolutionary historic foundation), as it is extremely difficult for primates who do not know each other to aggregate as a crowd. This is another indication that such ability is something unique to humanity.

Alternatively, though I admit that cultural category can become a motive for group formation, I would basically like to exclude this invisible concept from our group phenomenon. If we are to amend Keesing's confusion in his statement on descent unit, it would be that "families, lineages, clans, moieties and ethnic groups are all cultural categories and people sometimes gather and form groups based on the rights and responsibilities that these cultural categories charge them with". I am not sure if this process has prepared us well enough to avoid confusion of the two concepts, but let us return to the principle when in doubt and continue with our discussion.

Temporary groups formed by synchronizing behaviors

By setting our definition as follows, "if people are visibly gathering and interacting, we define the gathering as a group", we can now examine various "group" phenomena that anthropology and sociology have so far been unable to (or have chosen to ignore) and compare them with those of primates. Furthermore, it is important to note that we can now include gatherings, which have not been identified in the context of cultural categories, as group phenomena.

Primatologists have researched all visible gatherings. For example, there are many groups formed within a stable troop of chimpanzees (BSU). However, because these groups lack a stable structure, they are regarded as temporary unions and are labeled parties or subgroups. By defining visible gatherings as "groups", it is now also possible for anthropology to focus on such various types of groups.

This chapter also focuses on the behavioral synchronicity as a key function to form these various types of groups. I mentioned at the end of the last section that "people sometimes gather and form groups based on the rights and responsibilities that these cultural categories charge them with". However, the rights and responsibilities charged by cultural categories are not the only motive behind human group formation. Groups are more often created by behavioral synchronicity. Cultural categories are simply one of the mechanisms that create such coalescence.

Let us look at some examples of the types of groups formed by synchronizing behaviors.

Case 1

In 1991, I visited Itaru Ohta in his research field of the Turkana. Ohta happened to be out at the time, so I was dallying in the village waiting. Then, suddenly, I heard screaming coming from the bush. A number of men leapt from the scrub, grabbed a young village man and conducted him away. I was surprised and worried that something terrible may have just happened. When I reported what I saw to Ohta, he told me "Ah! That's a group sanction". The youths of several villages were captured and brought in front of the elders to be scolded. They were condemned for not properly supervising their young brothers grazing the cows. After the lecture, the youths had to offer their goat as punishment and they all enjoyed the treat of the meat together.

By reviewing this case from the perspective that "groups are formed by synchronizing behaviors", we can say here that a group consisting of elders and youths was formed to conduct a sanction. This was a temporary group that was established to capture the youths, lecture them and make them offer their goat. The group then dissolved after they shared the meat of the goat. It was a group that came into existence only through the act of the members conducting a "sanction" together.

Another example would be the gathering behavior of the women of the /Gui and the //Gana Bushmen reported by Kaoru Imamura (1993, 1996). Imamura emphasized the synchronicity of the women's behaviors. "Bushmen interact with others as proactive actors and synchronize their movements responsively or isomorphically" (K. Imamura 1993: 25). In this chapter, we understand that the women of the /Gui and the //Gana, while gathering, form groups through their responsive and isomorphic synchronization. Though these may be temporary groups, they are ones formed on a daily basis.

It is with reason that we purposely use the term "group" for such temporary and circumscribed phenomenon. Anthropologists and sociologists have been paying too much attention to "permanent groups".[1] Such a research position has produced the impression that the society is constituted from a number of permanent groups that cooperate, rival and integrate with super- or subordinate groups. In other words, it has given the impression that the world is made up of building blocks formed by permanent groups. Both the Nuer's lineage system (Evans-Pritchard 1940) and the Boran's *gada* system (Asmarom 1973) portray images of orderly societies built as methodically as some corporate organizations.

The issue with this view is that it may cause observers to overlook members' behaviors that have no relevance to cultural categories. In reality, behaviors that are irrelevant to cultural categories outnumber those that are relevant, and it is often these behaviors that drive people to form connections across cultural categories. The terms "group" and "cultural category" have so far been used in an overlapping manner, but in this chapter we intend to purposely isolate the former from the latter as we aim to focus on such a connection.

The connection across cultural categories can sometimes be referred to as a "network". However, the term "network" transcends time and space and would eliminate the "here-ness and now-ness" of the group as defined in this chapter. For example, though I am still connected through a network in my memory with an old friend who now lives in a distant locale, I do not form a group with this friend in the "here and now". Let us take these points into account and proceed with our discussion while being constantly aware of the "temporality" and "circumscription" of groups.

"Unity" perceived by overlap between groups

"Unity" that is neither group nor cultural category

I mentioned above that primatologists succeeded in identifying the BSU of the chimpanzees through observing a large troop, while struggling with their fission-fusion behaviors. However, it is not often that chimpanzees actually get together and form a troop as large as that observed by Itani and Suzuki. In their daily life, it is quite normal for chimpanzees to move in small unstable parties where membership constantly changes. The fact is that even the chimpanzees themselves do not often get to see the full outline of their own BSU.

Noriko Itoh (2003) set out to learn how chimpanzees living in their small and unstable groups (parties) were aware of the outline of their BSU. She did not assume the existence of a BSU as a given, and tried to observe how the chimpanzees themselves perceived their BSU through their daily interactions.

In their lives, chimpanzees spend their time foraging, grooming and traveling and while doing so, they form a party, or a "temporary unity", by synchronizing their behaviors with the individuals around them or by adjusting the discrepancies between them. Then again, sometimes they extend the inconsistencies between their behaviors. These individuals eventually break away from the "temporary unity" and form another one with other individuals. By repeating such behaviors, a chimpanzee spends time with most members of their BSU, or their "stable unity", on a given day. In other words, the chimpanzees form a number of "temporary unities" over the course of a day such that each individual eventually forms "temporary unities" with most members of their BSU. Here, the BSU vaguely reveals its outline as an accumulation of "temporary unities".

"BSU" is an analytical concept identified by the observer, but it is most likely that the chimpanzees themselves are aware of this "stable unity" as well. The "stable unity" is not a "group" as defined in this chapter, as the members do not gather on a daily basis, but it is not a cultural category either. However, such an extensive "unity", which is neither a group nor a cultural category but can be perceived as an overlapping of "temporary groups", also exists in human society. In the following section, let us focus on various groups formed daily in the Gabra society in Southern Ethiopia and discuss how a larger "unity" is being perceived through the overlapping groups.

Regional "unity"

There are numerous behaviors that form groups in Gabra society. Some of those that establish relatively large groups which can be seen regularly include making the elements for and constructing a new house, dancing, weddings, funerals, coffee ceremonies, hanging out at the bus stop, meetings and drawing water for livestock.

Among these temporary groups, some are comprised mainly of females (such as making the elements for a new house) and some mainly of males (such as meetings and drawing water). There are also groups formed at weddings and funerals where both genders meet, but sometimes the adult males and females then gravitate into separate subgroups. However, these groups are fluid and members come and go. Men are observed hanging around women making the parts for a new house, and women do their laundry next to men drawing water for the livestock. Here, it should be noted that even if people are doing different things, the behaviors or interactions of others remain within sight.

One of the "unities" that is faintly perceived through the overlapping of these temporary groups is that of the "region", though how this form of unity is perceived can vary.

Since migrating to their current land in 1992 as internally displaced people, the people of Gabra have been sharing the land with the agro-pastoral people of Guji. The Gabra and the Guji have hardly ever formed a group together, and the Gabra displayed most of the behaviors mentioned above within their own ethnic group. Occasionally, a very few Guji and Gabra friends visited each other's village, but most of the time the daily interactions between the two groups were limited to casual chatting at the bus stop. The Gabra and the Guji almost avoided each other in their daily lives. Given this environment, through the overlapping of the temporary groups the Gabra would have perceived a clear outline of themselves living in the region and a vague outline of the Guji who also resided in the same region.

Conversely, prior to 1964, the Gabra shared their land with the pastoral people of the Garre on the eastern side and the Boran on the western side. The Garre and the Gabra are culturally very similar and have many common clans. Inter-group marriage was common and men and women visited each other to work together, conducted coffee ceremonies and enjoyed dancing in joint groups. They were not conscious of their ethnic categories in their daily lives and it seemed that on the eastern side, they perceived the whole "region" as one big unity.

On the western side of their land, the Gabra formed various groups with the overlapping people of the Boran, even though their ties were not as strong as those with the Garre. They visited each other's villages to dance and perform coffee ceremonies together, and when there was a wedding, they visited in droves to celebrate. One of the most important groups of all was that formed to draw water for livestock. In this region, a herd of cattle had to drink water once every three days and camels once every nine days, and men regularly went to the well to draw water for these animals. As the groundwater level was quite low in this region, the well was forty meters deep. Drawing water from this deep well was hard work requiring six to twelve men to go inside the well to transport the bucketsful of water in relay. One could not draw water for their animals without the help of others, and in turn each had to assist others in drawing water for their animals. The Gabra and the Boran cooperated well in this water drawing activity (Soga 2007). It was the act of collectively drawing water that formed the temporary group and as such, new groups banded and disbanded at the well every day.

These groups also often became those for meat distribution. For example, when the Boran were to slaughter their livestock, they would invite the Gabra with whom they shared the task of drawing water to the village and the guests would slaughter the animal. The Gabra are Muslims and thus cannot eat animals slaughtered by non-Muslims. Therefore, the Boran would purposely invite the Gabra to slaughter their animal so that the Gabra could also eat the meat. The ribs of the animal were usually given to the Gabra. When the Gabra slaughtered their livestock, they also shared the ribs with the Boran. Sometimes, the villagers secretly ate the meat without sharing, but it seems that their shiny faces as a result of eating meat always gave away the truth. On finding out that they had been left out, the neglected people would grudge about missing the ribs, moaning "ah, my ribs hurt", or refused to draw water.

Photo 9.1 Boran youth drawing water from a deep well

Four youngsters are drawing water, in relay, from a well. The wells are generally much deeper, and
the Gabra Miigo and the Boran used to cooperate to get the work done.

Women also came to the well and were aided by men to draw drinking water. By
getting to know each other through these temporary groups, both the Gabra and
the Boran perceived the outline of the "region" that was formed around the well.

The unity known as the Hofte

"Region" is not the only unity perceived by the overlapping of temporary groups
in this setting. The next example looks at the people known as the Hofte who
suddenly appeared in the 1920s.

In reports written by the British colonial administrators in the 1920s, the people
of the Hofte appear along with the camel pastoralists of Gabra and the cattle
pastoralists of Boran. So, who are the Hofte? It seems that the administrative
officers could not decide whether they were Gabra or Boran. In the annual report

of 1922, there is a statement that says that the Hofte are a section of the Gabra, but they are later referred to as the "Boran-Gabra" in another statement (Hussein 2006). H. B. Sharpe, the administrative officer of the Northern Frontier District, wrote in his annual report of 1927 that the Hofte Chief, Sora Borsula, is a Gabra (Gabbra, Hofteh), but then from 1928 onwards he refers to him as a Boran (Boran Hofteh), evidence that he was unable to make up his mind as to whether he was a Gabra or a Boran.

Robinson, who reconstituted the chronicle of the Gabra living in Kenya (Gabra Malbe), indicated that the Gabra Malbe named the year 1921 as "Wednesday Year when the Hofte came fleeing", referring to the people known as the Hofte who came to Kenya across the border from Ethiopia (Robinson 1985). Robinson referred to the Hofte as the Boran and wrote that the reason for their escape was either because they feared the Abyssinians or that the drought that year was too severe, or both.

I have been very curious about these people, whom officials and researchers had difficulty identifying as the Boran or the Gabra, and for the last few years have been asking as to whether anyone knew who the Hofte were. The result was not positive. When I asked the Gabra Malbe in Kenya, one answered "the Hofte are the Boran" while another said, "I don't know". When I asked the Gabra Miigo, I was told "There is a song called 'the Hofte went extinct because they lied all the time (hoften, inkijibune, ba'de)'. The Hofte are the Gabra", or, "The Hofte are the pastoral people of Sakue", or, "In the Malbe region, there are people who are neither the Gabra nor the Boran but are a little bit like the Boran. I think they are the Hofte". However, nobody was able to reveal further details.

Leus, who compiled a dictionary on the Boran culture, described the Ooftee (Hofte) in the following way:

> Ooftee ... a term used for the Borana and Gabra living in Golbo and Malbe, down to Marsabit, so called because they move their animals together (loon oofani). In Kenya, where it is written Hofte, the name refers to the Borana who were the spearhead of the main migration south to Marsabit, around 1920. (Leus 2006: 497)

The Malbe referred to here is a region west of the Ethiopia Highway that runs longitudinally in the southern region of Ethiopia, while Golbo is a southern region of escarpment marking the Kenya-Ethiopia border. In 2006, I conducted an interview in Hobok at the northern tip of Malbe region and met two men, one of whom introduced himself as a Boran and the other a Gabra. However, the man who called himself Gabra actually lived in a Boran village in a Boran style house. In other words, he was a Gabra who was like a Boran.

Finally, these two men clearly told me what Hofte was. The answer reflected that written by Leus (2006). The word *Hofte* originated from the verb *ofa,* meaning "moving the livestock". Regardless of whether they are Gabra or Boran, people who moved frequently were called Hofte; however, there were no ethnic groups or clans by that name. The people in the region were called Hofte simply because they moved more often than others.

If the explanations by these men and Leus were correct, it can be said that the colonial government had created a kind of "ethnic category" from a unity that neither the Gabra nor the Boran recognized as the subordinate section. The colonial government had even appointed a chief in order to collect tax from the Hofte.

In the 1920s, a band of robbers by the name of Tigre[2] (most likely ivory hunters) were rampant in Southern Ethiopia, rustling livestock from the locals. People feared the Tigre and fled in flocks from Ethiopia to the colonial Kenya. Among those who escaped were the Gabra as well as the Boran. Such synchronized behaviors created a continuous stream of frequently migrating groups that was perceived as a "unity" known as the Hofte. The colonial government recognized them as a subordinate ethnic section by the name of Hofte.

What prompted the false recognition was this "*behavior*" of extremely fluid migration. People's behavior made it seem as if there was a "unity" called Hofte. When someone focused on the Hofte, it was only natural that they seemed both like the Gabra and the Boran.

The name Hofte last appeared in the 1929 annual report, when "Sora Borsula of the Hofte spent most of the year in Abyssinia" and was never to be seen again. We can assume here that the Hofte had returned to their native Ethiopia because the bandit group Tigre, who triggered their escape, was mopped up by the Ethiopian Government. The Hofte put an end to their intense migratory lifestyle and returned to their old moderate migration pattern. As the Hofte was a "unity" that was only perceived by their intense migrating behavior, its existence itself came to a halt when they stopped their characteristic relocations.

The outline of the "region" that the Gabra and the Boran perceived or that of "Hofte" that the colonial government recognized, were manifested through people's continuous interaction with each other. On this identifiable outline, Itoh (2003: 246) wrote "there is no guarantee that the parties concerned are paying attention to such outline or whatever it is that the concept of 'social' framework is represented as". This statement can indeed be applied to the unity of "region" and "Hofte".

As primatologists referred to this outline using the analytical concept "BSU", it is possible for anthropologists to use analytical concepts such as "regional group" or "ethnic section" to refer to such unities. However, for the parties concerned, these outlines were simply the bounds of something that was perceived beyond the overlapping of the temporary groups. In other words, the parties were never conscious of such outlines in their thoughts or actions. They never had the notion that they "belonged" to the "unity" of being in the same region or being a Hofte, and there were no rights or responsibilities that came along with either. However, even if the parties were not conscious of these "unities", they may still hold important meanings. In the next section, let us examine the conflict that may arise between cultural categories and "unities".

Conflict between cultural category and "unity"

Cultural category that precedes behavior

We have so far looked at temporary groups formed by synchronizing behaviors and "unities" perceived through the overlapping of such groups. Though there may be a difference in the degree, how they work is basically the same as the way in which chimpanzees form "temporary unities" and perceive "stable unity". Humans perceive the unities of "region" and "ethnic group" by using the ability they share with non-human primates. The ability to conceive cultural categories, on the other hand, is probably a uniquely human phenomenon.

Cultural categories impose rights and responsibilities on those involved and in that sense, they are socially meaningful concepts. Cultural categories can also drive people to take certain actions simply for the reason that they belong to the same category. On the contrary to groups and "unities", which are formed through established behaviors, cultural categories precede behavior. In other words, groups and "unities" come to existence through experience, whereas cultural categories are *a priori*.

If that is the case, how do cultural categories and "unities", perceived in everyday lives, affect each other when the pastoralists come in contact with other ethnic groups? We would now like to go on and look at some examples where "unities", perceived in everyday lives, transcended cultural categories as well as the reverse scenario, and discuss the pastoralists' social relationships.

Photo 9.2 Gabra Miigo youth and Guji youth

A Gabra Miigo youth is on the left with a Guji youth on the right, standing shoulder to shoulder at a bus station. They called each other by nicknames and appeared to be good friends, though they never visited their respective villages nor did they know each other's real names.

"Unity" that precedes cultural category

The first example we will look at is the pre-1964 relationship between the Boran and the Gabra. As stated earlier, the neighborhood of the Gabra and the Boran overlapped extensively. The wells in this region were located every twenty km, and the Gabra and the Boran cooperated at each well to draw water for their livestock. Both the Gabra and the Boran were aware of the regional sense of unity through such cooperation.

Now, the Boran occasionally killed men from other ethnic groups and severed their genitalia to show off their gallantry. When they succeeded in such activities, they were praised by people, offered cattle as gifts and loved by women, and thus the men of the Boran longed to wear a special trophy ring that constituted proof of a successful murder. The Gabra called this penis-seeking murder *gaada*, and made a clear distinction from *duula*, raiding for livestock and *olki*,[3] large-scale war between ethnic groups.

Gaada was often carried out by a group of Boran targeting a lone shepherd boy. It often happened at dusk, so Gabra youths and men often went to collect the shepherd boys around that time carrying spears. However, despite these efforts, Boran did occasionally succeed in murdering Gabra.

Case 2
In 1964, in the region around the well known as Irdar, a Gabra man by the name of Galgalo was murdered and had his penis severed. A young boy was also killed at the same time. The culprit was a Boran man from the Hidilola well region about thirty-eight km southwest from Irdar. In revenge for these murders, a Gabra man by the name of Durachi traveled to the township of Moiyale about fifty km south-southeast and killed two Borans near the town.

The important point to note here is the fact that the Boran who killed the Gabra came from outside the Irdar region, and the Gabra man whose family members fell victim to the murder also left the Irdar region to seek revenge. Let us look at another example.

Case 3
In 1946, in the region around the well known as El Leh, two Gabra men were murdered and had their penises cut off. People traced the footprints of the culprit but lost sight around the Irdar region about forty-four km west-northwest from El Leh. In revenge of this murder, four Gabra went to the Kenyan side about thirty-three km south from

El Leh and killed four Boran. On return to the village, the four cleansed their hands with milk scented with myrrh, quietly sacrificed a male goat outside the village and ate its meat. Years later, they built a ritual hut in a field, sacrificed a bull and ate its meat while singing a song about killing their enemies (*gob*).

Again, from this example we can see that the Boran who killed the Gabra came from outside the El Leh region, and the four Gabra who sought revenge also travelled all the way to Kenya to fulfill their mission.

When a Boran murdered someone in the neighborhood, the Boran living in the local region often feared revenge by the Gabra. At these times, people would continue saying "the land is at peace" (*Lafti nagaya*) and kept on drawing water together. The Gabra hated the Boran deep down, but kept chanting that the land was at peace and lived cooperatively with them. Despite their hatred, the Gabra did not attempt revenge within their local region.

The murderer who killed the Gabra was not a "specific Boran individual" with whom they shared daily tasks in the local region, but an "abstract Boran individual". The revenge carried out by the Gabra was not targeted towards the specific Boran individuals with whom they worked in the region, but towards the abstract individuals living elsewhere that fitted in the cultural category of Boran. These examples show how people hesitated in carrying out murder within the regional unity, but showed no empathy in the killing of members of another ethnic group that fitted the same cultural category.

Cultural category that precedes "Unity"

Next, let us examine an opposite case, where cultural category preceded "unity". As stated earlier, since 1992 the Gabra have shared the locality with the agro-pastoral people of Guji. However, the two hardly ever formed temporary groups. The Gabra and Guji avoided each other in their daily lives.

Though the Gabra and Guji lived together peacefully most of the time, since I started my fieldwork in 2000, there were occasional incidents of murder. Sometimes, it was Guji killing Gabra and in other cases it was the reverse. These homicidal incidents took place within the local region and the culprit was always captured immediately and sent to prison.

What was eerie about these murders was the fact that there were no apparent reasons behind the killings (such as conflict over resources). The investigations revealed how the murders took place, but never why. The only thinkable reason was that the victim was from a different ethnic group.

These murders must have taken place purely based on cultural category, i.e., the name of their ethnic group. The victim was not deemed as a specific individual person living in the local community. The Guji and Gabra had hardly any interaction in their daily lives and did not form a group together. Such lack of contact may have made them look at their murder targets as nothing more than members of a certain cultural category, a process of dehumanization.

It should also be noted that the framework of "ethnic group" began to have a more significant meaning in Ethiopia after the ethnic-based federal system was introduced in 1994. Ethnic groups, which are cultural categories, began to exert strong control over people's actions and this domestic politics may have played a role in why the Guji and Gabra were not able to develop a deep relationship. At any rate, these examples indicate to us that cultural categories may take precedence in groups whose members lack interactions with each other.

Conclusion

Humans interact with other humans and form groups of real living human beings. By overlapping these groups, they perceive a larger "unity". This faint "unity" may not always be something meaningful to the members of the group. However, humans may refer to these "unities" by the name of a region, such as Irdar or Hidilola, or by the name of the ethnic group such as the Hofte. When these faint "unities" begin to take clear shape, some of these may turn into cultural categories and begin to control human behavior. The emergence of a cultural category becomes evident when they begin to set out rights and responsibilities.

When we interact with people in front of us, we begin to feel empathy. By drawing water for livestock, enjoying dancing, sharing meals or enduring hunger together, one can overlap one's own experiences with those of others and deepen the empathy felt for them. In other words, these empathies are created through experiences.

Furthermore, by repeatedly forming groups with various people, we begin to perceive a larger "unity". This is a "unity" of people who have nurtured empathy for each other. Neither the Boran who sought to kill nor the Gabra who sought revenge did so within their local community. I believe this was because they had perceived the "unity" of empathy. Those who have shared experiences are excluded from the group of abstract "others" whom the cultural category specifies as the target of murder.

However, once this "unity" began to take clear shape and was recognized as a cultural category, empathy began to be vigorously controlled by it. While the category demanded "unconditional" empathy for fellow members, even if they had never met each other, it forced them to deny any empathy for those who belonged to other cultural categories, creating an environment in which even murder was possible.

The ability to represent "unities", create cultural categories and control such distinctions is something that is unique to mankind. However, I believe what is more significant for us is the fact that, while we create cultural categories and use them to control people's behaviors, we also manage to open new possibilities which allowed us to explore new relationships across the boundaries of cultural categories by using the ability we share with non-human primates.

Going back to the story I introduced at the beginning of this chapter, I labeled the Turkana man as an "enemy" as soon as I heard the name of his ethnic group. I tried to understand the human being in front of me simply by categorizing him into the ethnic category of Turkana. The Turkana man and Elema, however, put their ethnic categories aside and immersed themselves in interacting with the live person in front of them until they eventually managed to empathize with each other.

To be able to immerse themselves in the interaction with the person before their eyes, even under the control of a cultural category, and perceive a "unity" of a new kind—I believe this is a key feature of us humans who have managed to create this complex society.

10 The Small Village of "We, the Bemba": The Reference Phase that Connects the Daily Life Practice in a Residential Group to the Chiefdom

Yuko Sugiyama

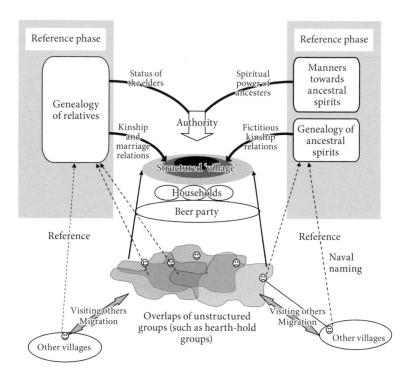

Having a "reference phase" is a common phenomenon in human groups, but what characterizes the chiefdom of the Bemba, a society of shifting cultivators, is that their "reference phase" is multi-layered. The Bemba residential group forms the outline of its overlapping non-structured groups. It is formed through the daily interaction of individual members, as its base, and provides structure to members by referring to the genealogy of kinship. The other phase referred to in Bemba society is that of ancestral genealogy and the custom of honoring ancestral spirits. Ancestral genealogy connects people directly to the Bemba chiefdom. While it legitimizes the village headman's authority, imbuing him with the supernatural power of the ancestral spirits, it also functions as a mechanism that allows people to travel over extensive regions.

Introduction

In the evolutionary history of mankind, the expansion of group size has always been noted as an indicator through which to evaluate the developmental stage of the group's ability on a scale that eventually leads to modern human beings. In order for group size to expand, it is necessary for the group to develop a political and economic system that secures a stable food supply for its population. However, it is also just as important for the group to undergo a process of changing its awareness towards others and creating a mechanism that nurtures "we" consciousness. In order to develop such a mechanism, people would need to have the ability to presuppose the existence of others, not just those standing in front of them but also those in a distant location as well as those whom they have not even met. By inventing numerous mechanisms based on this ability, human group phenomena have come to show a rich diversity, unparalleled by any other.

The Bemba who form the focus of this chapter are shifting cultivators living in the savanna woodland regions of Africa.[1] Their residential groups, villages, are very small in scale but together they form a chiefdom with a population of some hundred thousands to one million people.[2] Though these people are aware of their differing historical background, they all share the identity of being "we, the Bemba". By looking at the group phenomenon of the Bemba, we can explore the process of how various groups with differing historical backgrounds are reorganized by overlapping on multiple levels, which may give us a glimpse into the diversity of the human group phenomena. What we hope to achieve by looking at the cultivators of modern Africa, while briefly examining their history, is to see how the small scale residential groups of *villages* are managed through members' interactions and how the everyday experience of the residential groups are woven into the concept of "we in general", which even incorporates the existence of invisible "other" beings. In this chapter, we will not regard groups as a given, and would like to begin by reviewing them by focusing on how the interactions between members lead to the formation of the group.

When I first began living in a Bemba village, I did not know how to respond to the villagers who kept visiting me, regardless of the time of day, just because they had not seen me the previous day. They did not seem to have any business nor were they close friends of mine, but they came anyway and spent time chatting with me about nothing in particular. These visits made me feel embarrassed at the time, but I found out much later that they were in fact part of necessary social behavior for village residents, as well as acting to maintain friendly relations (Yuko Sugiyama 1987).

The village was formed with the matrilineal kin as its core and most villagers were somehow incorporated into this genealogy of kinship. However, being a member of the kinship group alone does not guarantee a friendly relationship, as blood ties alone are not regarded as forming tangible bonds. In the village life, people were always together at the same time and space. Their tendency toward always being with each other was deeply embedded in the villagers' day-to-day behaviors, and it was through these practices that people experienced intimacy and recognized each other as "fellow members". The village as residential group was supported by the continuation of direct interactions and was formed by residents' ongoing practice of "being with each other". Conversely, this meant that members had the option to leave the group by not choosing to "be with each other".

The reason why the Bemba do things this way is closely related to how the groups were formed while maintaining their mobility. The shifting cultivators of Bemba settle into a village lifestyle, but they have a cycle that incorporates fission-fusion in the long term and the population of the village hardly reaches more than seventy households. This method, which requires villagers to constantly confirm each other's intention to "be with each other", was created through this unique settlement process that ensured its mobility.

Below we examine various forms of "gatherings" that take place in village life and discuss what the residential groups of *villages* are and how they are incorporated into the chiefdom of the Bemba. In order to recognize group phenomenon flexibly, we will widely assume gatherings of people as "groups" and treat both temporary and long-term unity as such. However, to avoid confusion, we will use the term "gatherings" for temporary congregations and "groups" for longer/permanent ones, and try to separate the two wherever possible.

The structure of the Bemba chiefdom and its villages

The chiefdom's history and its centralized political system

The Bemba are said to be descendants of people of the Lunda and the Luba chiefdoms that once flourished in the southern part of the current Democratic Republic of the Congo, who migrated eastwards around the sixteenth century. By the seventeenth century, they had established their base near the Kasama district in the Northern Province of Zambia (Roberts 1974, 1976). The region was covered with savanna woodland dominated by the leguminous tree species collectively known as *Miombo* in the local term. The land was contaminated by sleeping sickness which made it unsuitable for raising livestock, but due to poor soil quality it was not particularly suitable for farming either. Given the nature of the region, the Bemba developed a unique method of shifting cultivation while they utilized their powerful military force to create a chiefdom headed by the Paramount Chief *Chitimukulu*. Through repeated combat with the other ethnic groups in the region, they had expanded their influence over the whole of the Northern Province by the beginning of the nineteenth century. By the end of the nineteenth century, they had advanced southwards to procure salt, ivory and slaves for trading purposes and monopolized foreign trade with the Arab nations and Portugal. The region was colonized by Great Britain in the beginning of the twentieth century, but the Bemba used the unstable political conditions caused by colonization to further increase their influence, expanding the chiefdom's territory to the current Mpika district. The Bemba then further extended their residential area to the Copperbelt province and Lusaka Province through migrant workers who bore the task of developing the mining cities during the colonial period.

Given the history, the political structure of the chiefdom of Bemba bears a highly centralized nature. Three senior chiefs including *Chitimukulu* himself reign at the top of the hierarchy. The senior chief positions are basically passed down in a matrilineal kinship system. Only the males from the crocodile clan, the so-called royal clan, have the right to succeed the senior chief's position and among them, only those from a limited special clan are allowed to succeed to the throne of *Chitimukulu*.

Under the senior chiefs are fifteen junior chiefs, created while the Bemba chiefdom was undergoing its process of expansion. During the nineteenth century, when the Bemba was rapidly expanding its territory, the principle of matrilineal succession was temporarily loosened to allow the sons of *Chitimukulu* to rule over the new territories as chiefs. Since then, it is the "sons" of the successive *Chitimukulus* who inherit the position of junior chief. At one stage prior to succeeding the position, they are granted "perpetual son-ship" of the reigning *Chitimukulu* and inherit the position of the "son" of the senior chief. As such, the political structure of the Bemba chiefdom is created by fully utilizing such constructive parent-child relationships and kinship patterns (Roberts 1974).

Each chief has his own territory (*ichalo*), and the dominant village headmen of the territory support the chief regarding politics and rituals by forming a consultative council. The chief is a highly autonomous king who not only has jurisdiction over a wide range of matters but governs the rituals and politics within the territory. All the villages in the territory belong to the chief and offer labor and gifts according to the chief's demands, but the villages themselves are highly autonomous. Each village has a village headman who is in charge of the rituals that takes place at the village's ancestral shrine. This political structure was further enhanced when it was incorporated into the colonial governance structure and remained in place even after Zambia became independent in 1964. People often use the phrase "*fwebabemba*" (we the Bemba) and like to talk about being a member of Bemba and the presence of the Bemba chiefdom.

Small-scale villages

In contrast to the centralized political structure of the Bemba, which embraces a population of hundreds of thousands to millions of people, the villages that form the basic units of the chiefdom have hardly any hierarchical structure aside from the village headman, who holds a certain degree of political power. There is a fundamental discipline within the community that encourages those who have better and more things than others to share, and the village's leveling mechanism which functions to distribute food among other things helps to suppress economic disparity. The villages are generally small, with only ten to seventy households and often do not last for a long time; in fact the village site itself tends to migrate in a ten to thirty-year cycle. This is deeply related to the fact that the villagers' livelihood is centered on shifting cultivation. When they run short of Miombo woodland suitable for *Chiteimene* cultivation, the villagers start migrating in search of adequately regenerated Miombo woodland.

As for the structure, the village is formed with the village headman and his matrilineal siblings as its core. Uxorilocal residence is practiced. Most males other than those of the headman's matrilineal kin basically have external origins. However, it is quite normal for siblings from different mothers, distant relatives and others who entered into constructive kinship to join the village. Thus, the village is a residential group that embraces different sorts of members, though it is based on matrilineal kinship. It should also be noted that both the divorce rate and the rate of households headed by a female (hereafter referred to as female-headed households) are relatively high.

Village outline formed by accrual of "gatherings" and interactions

Household and subsistence activities

The basic unit for subsistence is households that are mostly comprised of nuclear families. Their subsistence activities have a clear gender division of labor. The difference in the household's gender composition is a factor that has a direct influence over the economic disparity between households. This becomes particularly apparent when it comes to the agricultural activities in opening the *Chitemene* shifting cultivation fields.

The cultivation process begins with men climbing trees to cut only the branches and leave them to be dried. The women then carry the branches to the center of the felled land and pile them up in a circle. The branches are then burnt to produce a cultivation field (*Chitemene* field, from now on), and women carry out most of the rest of the agricultural work. The *Chitemene* field undergoes a few years of crop rotation followed by a fallow period during which the secondary woodland regenerates. This woodland is then used for hunting and gathering wild produce. As far as the agricultural work is concerned, men only help by cutting the branches, but as a new *Chitemene* field must be opened every year, the male workforce is essential. They also go hunting and gathering to procure animal protein and play the main role in money-earning activities.

Due to the high divorce rate, ten to forty percent of the village households are female-headed households. Economic disparity between households headed by couples and those headed by females is inevitable due to the availability of male labor for subsistence activities such as cultivating *Chitemene* fields and procuring cash and animal protein. However, the disparity at the production level is evened out through the consumption process at the women's gatherings in addition to a leveling mechanism that functions through the brewing and sale of fingermillet beer.

Becoming familiar: Updating relationships and grouping via daily interactions

As stated in the beginning, the Bemba village is comprised with the matrilineal kin system at its core, and most villagers are somehow linked into the kinship structure. However, this relationship needs to be constantly substantiated through daily interactions. In order to maintain a day-to-day relationship with the members of the village, one tries to meet and spend time with the other villagers, and this process needs to be repeated daily so long as one lives in the village. However, the frequency and range of the visits varies among the villagers. Women in general have more opportunities to gather due to the nature of the work they do, such as farm work, cooking and food processing. Men, on the other hand, whose primary roles are cutting tree branches for *Chitemene*, earning money and hunting, only get to meet a limited number of people through their work because they usually carry out their tasks alone. However, they are often seen visiting and chatting with in-laws of the same generation (wife's brothers), relatives and close friends. Outsiders who join the village through marriage have an in-law avoidance taboo and thus their movements are comparatively limited. Young men during the bride service period and some women with external origins fit into this category.

When it comes to visiting others in the village, the participation of the elder women stands out. As clarified elsewhere (Yuko Sugiyama 1987), these women frequently receive visits from other villagers, but they also often move around and thus have contact with most of the village members. They are consciously trying to meet and spend time with other villagers. By contrast, elder men such as the village headman do not often go out to visit other villagers but tend to receive visits from them. As such the day-to-day act of meeting with others may vary according to the social status of the villager, and members they actually "meet and spend time together" with may also differ from person to person, but the act of meeting with others on daily basis is spread out through the village.

Married women play a particularly large role in creating opportunities for gatherings where people can substantiate and update relationships with other villagers in their daily life. The gatherings facilitated by these women are related to cooking, the processing of produce and food sharing.

When focusing on production activities such as agriculture, the basic unit for subsistence is the "households" that are mainly composed of nuclear families. However, when closely observing the consumption process of the produce, a different activity unit emerges, which mainly consists of married women and their children. The women of each household have a good knowledge of the when, what and how much to harvest from their fields, and usually decide how to use their produce at their own discretion. In the process of consumption, the barriers between households are lowered allowing the produce to flow between them through the sharing of food. The inter-household disparities in the production quantity are thus leveled, thanks to the consumption units temporarily made up of women and children, other than the households. Let us call this unit the "hearth-hold", following the footsteps of Ferricia Ekejiuba (1995).[3] However, in the case of the Bemba, a "hearth-hold" should be considered as a fairly loose group with interchangeable members, as it is quite common for relatives and friends' children to join in at a meal.

Hearth-holds have become the primary unit for food sharing. By joining several hearth-holds together, a large flow of food occurs within the village. This is strongly influenced by the fact that Bemba villages are matrilineal, where uxorilocal residence is normally practiced. In Bemba society, daughters retain a strong

Photo 10.1 Childen making preparations for fishing

Children pounding *ububa* leaves (*Tephrosia*: fish poison) make preparations for their planned fishing expeditions. Children living in nearby areas participate voluntarily. It is a routine of village life for them to gather and work in cooperation, and even kill time chatting together.

bond with their mothers even after they get married. The newly wed daughter will not have her own hearth from the beginning, thus she will continue to cook at her mother's hearth and rely on the produce from her mother's *Chitemene* fields. As she undergoes a number of rituals, the daughter gradually becomes independent, leaving her mother's hearth to become the owner of her own hearth where she will cook mainly for herself, her husband and children. Even then, the mother-daughter and niece-matrilineal aunt relationships remain strong, and villagers in these relations will often get together to work in the *Chitemene* fields and process food. The processing and cooking of food mostly takes place at the houses of elder women, such as the mother or the matrilineal aunt, using their cooking tools. Once they finish cooking, the women sit and share the food with the children. When the husbands are in the village, the children will deliver food to them.[4] Through working and cooking together and sharing the results, the produce harvested by the women of different households are shared and consumed by all those who gather (Yuko Sugiyama 1987).

Let us call these loose ties among the women seen at the hearth-hold, "hearth-hold groups".[5] Women from the matrilineal kinship groups and women of the same generation who married into the family form the core of the hearth-hold groups, but the members at each gathering may change according to the day and time due to varied circumstances. It should also be noted that villagers who are not members of the hearth-hold groups are almost always present at the food sharing that takes place during the gatherings. Children in particular are allowed to join in the meal at any gathering and as often as they like. Through the hearth-hold group's gatherings, women can enjoy spending time together and cement their intimate relationships. At the same time, the gatherings offer them opportunities to share food and for hearth-holds to loosely cooperate. Even if there are discrepancies in male labor or the size of the fields among the households, such deficiencies can be accommodated at the consumption level through the gathering of members from different households and as a result, sufficient food can be procured for all.

Also of note is that food sharing centering on the hearth-hold group takes place simply through "being there", regardless of membership, descent or generational differences. There is no begging or thanking involved in the sharing. Everyone who happens to be there naturally eats together. There is also the freedom to not partake if one chooses. What occurs here is a gathering in which the sole and most important condition for attendance is to be there at that very moment.

By taking the time to meet and spend time with each other and by joining the above gatherings on a daily basis, the villagers can nurture their intimate relationship with each other. However, what is even more important when considering the Bemba village is the fact that these gatherings also have an instrumental aspect. As explained elsewhere (Yuko Sugiyama 1987), people who intend to leave the village do not attend such gatherings and try to avoid unessential contact with others. The villagers will eventually sense their intention to leave through their behavior and will refrain from approaching them. Bearing that in mind, it is evident that gathering with other villagers is a means to express the intention to stay in the village. It may be the most basic way to substantiate and update village relationships while confirming each member's intention. The village life, as a result of the accumulation of such direct negotiations, is supported by the intimate relationships formed and updated by being together and sharing food.

Confirming the social "position": The beer party

In comparison to the villagers' mutual visits and hearth-hold gatherings which form groups through the act of "spending time together", the beer party, following the sale of fingermillet beer, emphasizes the members' position and difference in the genealogy while facilitating friendship and order among the group.

The brewing and selling of fingermillet beer is carried out by married women. It is particularly important for widows without a male workforce in the household to sell beer to earn cash income. The brewing and sale of fingermillet beer functions as a key feature of the village's leveling mechanism, in that it contributes to level out the disparities in cash income among households. However, when considering the group phenomenon of the Bemba, the beer party following the sale of beer is particularly important for the village men because it provides a stage for them to get in touch and sort out the order of the group through their interactions. As stated earlier, men in general do not gather as often as women do and they rarely have the opportunity to meet with unfamiliar faces. The beer parties offer great opportunities for these men to meet with men from differing origins and generations whom they rarely meet in their everyday lives.

Fingermillet beer is sold at the house of the women who brew it. It is the men, who have the cash, who come to purchase the drink. The men who buy the beer do not take it home but rather pour it into a large bucket in the foreyard of the house where the beer is sold. People gather around the bucket in a circle, but it is never the person who bought the beer who drinks first. The sponsor exchanges courtesies with the other men in the circle and sits quietly among them. Eventually, one of the elder men slowly stands up and drinks the beer from a cup placed near the bucket. The other elder men then follow. At one point, one of the elders drinks up from his cup, fills it up again and offers it to the junior man sitting next to him saying, "Have the drink".

With this act as a trigger, the junior men proceed to the bucket to fetch their portion of beer. At the beer party it is a very important custom to follow the order of "elder to junior". The character of a man is judged by his manners and it is absolutely out of the question for a junior man to drink before an elder, or to offer drinks to members of the elder generation, such as a wife's father or uncles and aunts, just because he purchased more beer than them. Junior men are expected to behave conservatively, but as soon as the beer in the bucket dries up, they become the targets of begging for another cup of beer. It is not easy to say no to the begging of elders. If they do, the elders will make cutting remarks on the spot, saying things like "You have not learnt the courtesy of the Bemba". If the juniors generously make the purchase, they are applauded with words like "You are a real Bemba man", but the in the next instance, people forget who bought the beer and the beer party continues, sharing the drink until it is completely gone.

Compared to the food sharing occasions of the hearth-hold groups, where there is no begging, thanking or confirmation of the social order, the beer party is comprised of persistent confirmation of rank between elders and juniors and according to marital status. There is also repeated begging and applauding. The act of pouring all the drink into a single container and sharing, regardless of who bought how much, is another characteristic of this party. The purchaser has no control over how his beer is consumed and is not even permitted to drink before others.

The beer party, characterized by this lack of freedom, is a place where the participants are strongly made aware of their principle in village life, which is "those who have are to share with those who do not". It is also a place where participants confirm the social order and relationships between members to express how they acknowledge each other. Though a certain level of informality is accepted at each party, interactions among members are repeated in a way that emphasizes the differences between elders and juniors.[6] There is also another axis where the differences between men in marital relationships or kin of different generations ("relationships where respect and fear is required"), non-relatives ("*lupwa*" (relative) but outsider) and completely unrelated guests are repeatedly emphasized. By repeating such interactions, the group structure begins to surface and the order and bond between the members of the residential group emerges.

Various gatherings and outline of the residential group

Among the various forms of gatherings in the village daily life, the household is a group with a relatively clear-cut outline and structure in that its membership is apparent and it forms the unit for production. Conversely, gatherings such as "visiting other villagers" and "hearth-hold group" gatherings are formed through the act of "being here and now together". In other words, these are temporary groups where interactions take place with a footing in the "here and now". Its members may include those other than household members, relatives and hearth-hold group members and may change depending on the day, but regardless of who comprises the group, all the members present are treated as "fellow members", or one of "us". To be more precise, members are not questioned as to "who" they are and in that sense it is a non-structured group[7] where the order simply cannot be manifested.

However, the interactions at beer parties, comprised mainly of men, unfold with a footprint on a more ideological phase, one that is clearly different from the "here and now". At the beer party, interactions begin by acknowledging "who" each member is, followed by behaviors that clearly indicate each member's relationship with the others present. In acknowledging the other members, a "reference phase",[8] such as the genealogy of kinship, is provided through which the participants can refer to each other's relationship. Their interactions are conducted based on this phase. Here, the difference between the individuals is emphasized and the relationship of "us" is expressed based on such differences.

When a village is located as a residential group from a synchronic perspective, layers of different gatherings can overlap. There are groups of a different nature. Households have a clear boundary and structure, while series of non-structured gatherings based on co-presence such as hearth-hold groups have no clear boundary. Also, there are gatherings that provide structure to the group through interactions based on the "reference phase" such as beer parties.[9] It can be said that the outline of the residential group is created by overlapping these gatherings of different natures. But how are these small individual groups linked to the numerous other "we, the Bemba" configurations? In order to understand this connection, we have to look at another aspect of Bemba residential groups where "village" is held together by a kind of authority. The genealogy in the "referral phase" plays an important role in this context. The village's developmental cycle that creates fission-fusion in the long term is also significant.

Fission-Fusion: The developmental cycle of the village and authority

The genealogy of kinship and the village's developmental cycle

When Bemba need to make a reference, for example when deciding what to do at a beer party, the first phase they point to is kinship genealogy based on blood and marriage. The family tree, in the Bemba villagers' history, only extends back several generations. However, when the transition of power takes place from the village headman to the next generation, positioning in the family tree plays a very important role. As with the Ndembu portrayed by Victor Turner (1972[1957]), tension between matrilineal uncle and nephew (sister's son) is embedded in the matrilineal genealogy of the Bemba. In times of generational transition, this tension causes intergenerational conflict and often results in the disintegration of the village. Bemba villages, in the long term, have a developmental cycle where repeated break-ups and establishments are incorporated but it is at times of transitions that these conflicts, stemming from kinship genealogy, surface during attempts to challenge the foundation of the village's political authority.

In the case of the Bemba, the cycle of generational transition and that of the deterioration of the *Miombo* woodland almost synchronize, and the shortage of woodland suitable for *Chitemene* cultivation gives the young group of men an excuse to leave the main settlement. The village headman's nephews (sister's sons), the heirs to the headman, are included in such a group. These young men establish their own temporary huts some kilometers away from the village. Their households move to stay there for several months. Often, men from other villages also establish their temporary huts together in a nearby area, creating opportunities for these young men and their wives and children to gather. Sometimes, these households bond beyond the boundaries of their villages to form a group that may eventually become the core for a new village. Meanwhile, young men from within the same village also have intimate interactions and nurture powerful bonds with each other through their everyday life in the temporary huts. They live in together for these months apart from the main villages, eventually building the foundation of a group. Such ties could one day become a political force in the village (Yuko Sugiyama 2007a).

At times of generational transfer, conflicts between the group of the headman's matrilineal nephew's generation and the group of the headman's generation who founded the village tend to surface in the public domain. This is a process in which the villagers are reorganized into groups with differing interests, according to their position in the kinship structure of the "reference group". In many cases, these restructurings are triggered by disagreement in the treatment of a witchcraft case, where, in most cases, the village headman and the elder men of his generation are accused of sorcery (Yuko Sugiyama 2004, 2007a).

Fission-fusion: Witchcraft, authority and the supernatural powers of ancestors

Accusing the elders of the headman's generation of witchcraft is equivalent to questioning the legitimacy of their authority, because the source of the chief and village headman's authority and the power of witchcraft are one and the same. It is believed that the ancestors inhabit the *Miombo* woodland and have control over the fertility and wellbeing of people's lives. The ancestors' anger manifests as calamities and diseases in order to punish those who do not honor them with the proper rites. The chiefs and village headmen are said to have full knowledge of these proper ways to honor the ancestors. That is why they are granted supernatural powers. In fact, this is the very foundation of their authority.

They are said to use their supernatural powers to not only protect the villagers from witches but also to punish those who disturb the order of the village and disobey the rules. However, it is believed that sorcery to protect personal interests is also possible through manipulating the ancestors. To accuse the village headman and elders of sorcery is equivalent to condemning them for unjustifiably using their supernatural powers, the source of their authority, for personal gain. It is a serious denunciation that questions the legitimacy of the authority of the accused.

Furthermore, the supernatural power, which is the source of the village headman's authority, has to be updated annually by conducting ancestral rituals that require the villagers' cooperation.

The village headman is granted the ability to exercise influence over others, as a legitimate manifestation of his authority, through the supernatural powers of the ancestors, only while he has the support of the villagers. As soon as he loses their support, he will no longer be able to maintain the ancestral rituals that in turn deprive him of the headman's authority, as this would require the support of the ancestral spirits. The supernatural power that the headman exercises will be demeaned as powerful sorcery and he will be forced to leave the village. The village will disintegrate and villagers will be scattered. As such, the authority of the Bemba is a double-edged sword built on a fragile foundation[10] whose legitimacy falls apart once the direct support of the villagers is lost (Yuko Sugiyama 2004). The cleavage in the genealogy of the matrilineal kinship system and the way the authority is established incorporate the disintegration of the residential group in the village's developmental cycle, prompting migration in the long term.

Fusion: The ways of ancestral spirits that necessitate restoration of relationships

The village's developmental cycle is embedded with a pathway that prompts the disintegration of the residential group and separation of its people. It is also very important to note that there is a route that restores the relationships between the disintegrated people and necessitates that the regeneration of a new village is also incorporated. What allows this to happen is the passing down of the ancestor worship rites as well as a different layer in the "reference phase", which is ancestral or the Bemba chiefs' genealogy.

Villagers separated by the breakup of a residential group take shelter in other villages, but eventually one of the men steps up and proposes forming a new village with the separated matrilineal siblings and members of the household groups with whom he had spent time in the temporary huts. These members potentially include those from other villages. In order to become a village headman, one has to have a special knowledge of the customs to honor the ancestors, part of which is kept secret. To do so, a young man must restore his relationship with the former village headman and men of his generation from whom he had once parted. The special knowledge about honoring the ancestors can be acquired by "buying" it (Yuko Sugiyama 2007b). Theoretically it can be bought from anyone, but it is said that the one who would teach the most reliable and effective ways is the man's mother's brother (maternal uncle). By passing down the knowledge involved in approaching the ancestors, the maternal uncle becomes the supporter of the man who is to build the new village. He is supposed to possess the supernatural powers granted by the ancestors, which is necessary to lead the village members. For the young man who is to build the new village, the matrilineal uncle is usually the village headman or a man of his generation of the village he once lived in.

Therefore, the young man sends a messenger to the matrilineal uncle who lives in another village and requests the restoration of their relationship. He talks over and amends the story of the witchcraft incident and then receives the teaching on the customs to honor the ancestors. Sometimes, this prompts the matrilineal uncle to move into the village founded by his nephew as a village headman. This adds credibility to the new village headman among the villagers as it proves that he is a character who can retain a desirable relationship with a matrilineal uncle. The new village is then reorganized with the matrilineal kin of the headman as its core, as if the power had been smoothly passed down from the matrilineal uncle to his nephew without incident. As such, when the genealogy of kinship that once separated men from different generations into different groups is referred to once again, it provides the opportunity to reunite the separated people and secures the continuity of matrilineal kinship over generations.

As explained earlier, the ways to honor the ancestors, who would support the village headman's authority, are acquired by "buying" the "manner" in a proper way. The reason why the knowledge is thought to gain effective power simply by purchasing it is because the ancestral spirit is said to dwell in every single Bemba. In the following section, we will discuss how this links people together.

Multi-layering of the "reference phase"

"Ancestral genealogy" which overlaps that of genealogy

What is characteristic of the "reference phase" is that it is multi-layered, in which the genealogy of kinship overlaps the genealogy of ancestors. The word "ancestral spirit" (*mfumu*) in the Bemba language generally refers to the successive Bemba chiefs, starting with its founder. It is said that all Bemba members have the ancestral spirit of one of the line of Bemba chiefs dwelling within their body. Each Bemba can be linked to the genealogy of the chiefs through the name of the ancestral spirit dwelling within. What makes this possible is their faith in ancestral spirits and the practice known as "navel naming" (*ishina lyomutoto*).

When a Bemba woman becomes pregnant, the ancestral spirit of the chief is said to enter the fetus through the woman's navel. The pregnant woman or her matrilineal kin such as her mother or aunt will know which ancestral spirit had entered the child through a dream. When the child is born, it will be given the spirit's name as its "navel name". Thus, every Bemba bears the name of one of the Bemba chiefs and can be positioned along the genealogy originating back to the founding chief. The way the villagers are to behave with each other can be found not only in the "reference phase" of kinship based on blood and marriage ties but also in another overlapping "reference phase" based on the genealogy of the Bemba chiefs, which the villagers gain access to through their navel name. The reason why the headman and elders' manners honoring the ancestral spirits in a certain way have a strong effect on the villagers is because the Bemba people are all directly connected with the spirits.

By having an ancestral name as a "navel name", each individual is given a life as a spirit, on top of that of a specific individual who is present in the "here and now". This is a mechanism that allows the "ancestral spirits", who collectively exist in the "reference phase", to appear in the reality of daily life through the lives of individuals who are given their names. "Spirit medium" (*ngulu*) meetings,[11] "loud announcement" (*mbila*) which is repeated every time there is a source of dispute (Yuko Sugiyama 2006), fortune telling at net-hunting events and healing illnesses are other opportunities in which the ancestral spirits can show their presence in everyday lives. Villagers have the opportunity to get to know the characters of individual ancestral spirits through these events. They learn the personality of the ancestral spirits of their "navel names" and their relationship with other ancestral spirits through experience. In this way, the ancestral spirits become a very real existence known to the villagers through personal experience.

Interestingly, ancestral spirits dwell in villagers regardless of their gender, so the spirit of a female Bemba chief in her lifetime may dwell in a male villager, and visa versa. Therefore, one man may turn out to be the "mother" of another by tracing the ancestral genealogy from their navel name.[12] In addition, it is important to note that villagers have proactive access to the "reference phase" because they know these ancestral spirits as specific individual characters they can see and listen to through their regular appearance in the village's daily life. It is this proactive access that prevents the multi-layered "reference phase" from becoming absolute standards that restrict the villagers' relationship. The ancestral genealogy is a flexible tool that helps people to connect with each other and can be manipulated as necessary.

The overall picture of "we, the Bemba": Ancestral spirits as an image model

I mentioned above that when a Bemba man forms a new village, he gathers the dispersed matrilineal siblings and members of the household groups with whom he had spent time in temporary huts, including those from other villages. Here, the genealogy of ancestral spirits that can be referred to through the navel name plays an effective role. Frankly speaking, a new village is formed through the accomplished facts of people gathering in one location with the will to live together, and the reality is built up through the accumulation of day-to-day interactions, as previously stated. However, for a residential group to become a "village" as such, the people need a legitimate reason for "being there"; in other words, the group needs to be structurally positioned. Most of the villagers are related by blood or marriage, even though they may only be distantly related, and begin their interaction from the standpoint of kinship. However, those among the core household groups who do not have direct kinship ties with the village headman can refer to the ancestral genealogy through their navel names and link themselves with the headman along the lines of the ancestors (Yuko Sugiyama 2007a). Through this process, these household groups can easily be connected to the headman's genealogy of kinship as constructive relatives and become proper members of the village, or one of "us" in a localized sense.

Ancestral genealogy also plays an important role when the villagers individually migrate. The people of Bemba travel in a geographically extensive region regardless of their age or gender, but one of the mechanisms that allows them to move freely and extensively is ancestral genealogy and the constructive kinship that can be formed using this system. In most cases, the villagers who travel individually will find contacts at their destination through their network of friends and relatives.

However, when they cannot find anyone or when they are travelling in places where there have no connections, they form new ties with the people at the destination point by referring to ancestral genealogy through the navel name system.[13] For those who are not related by blood or friendship or are not even Bemba, a smart tactic of naming them with one of Bemba's ancestral names is often used in order to form constructive kinship ties.

Giving strangers local names in order to integrate them into the existing social system is a practice reported in many regions, starting with the case of the hunter-gatherer group of the San (Lee 1979). However, in the case of Bemba, the fact that the names they are given are ancestral adds extra flexibility. It is important to note that the relationships formed through direct interaction can be preserved because this flexible ancestral genealogy coexists with kinship genealogy. Once one acquires a Bemba ancestral name, one can automatically work out their relationship with others based on their names in any Bemba society, while maintaining the existing social relationship and kinship order and still being able to manipulatively refer to them as necessary.

By acquiring ancestral genealogy, people do not have to make the effort to figure out where they stand when they meet a new Bemba, and can start interacting straight away. As repeatedly explained, it is the accumulation of the actual interactions that builds the foundation of residential villages and substantiates "us" as fellow village members. In this sense, the system theoretically allows anyone to be integrated into "we, the Bemba" if they so wish.

"We, the Bemba" involves people from an extensive region, despite the fact that their residential groups are small in scale and thus there is a limit to the connections that can be traced through kinship. There are a number of reasons for this. The first is the fact that people have a flexible ancestral genealogy as a recognition tool that can be used aside from their experiences in real life. The second is that the authority that legitimizes the formation of a village is supported by the people's faith in their ancestors. The characteristics of this mechanism with multi-layered "reference phases" is that while it maintains the existing relationship within the residential group, it also makes it possible for villagers to incorporate strangers and people with no blood ties as one of "us" without discomfort. The ancestral genealogy is a tool that allows the flexible expansion of "we, the Bemba" because it opens possibilities for anyone to belong. Another important aspect to take into account is that the ancestral spirits whose names are used in the genealogy are limited to less than twenty chiefs who all belong to a few generations since the founder, and because of this people can share the image of the whole family tree.

The image of this ancestral family tree offers a model which evokes the picture of numerous "we, the Bemba" spreading over the extensive Bemba chiefdom.

When Bemba meet with each other, they all have multi-layered "reference phases", the genealogy of direct kinship and that of ancestors, and share an image of the ancestral family tree which evokes an overall image of "we the Bemba", no matter how small the population of their residential group may be. These features function to turn non-structured groups formed through daily interactions into a structured group of *village* and the villages that are then integrated into the Bemba chiefdom through these phases. In this sense, the mechanism that small-scale residential groups use in order to form *villages* is the same as that used by a far larger population, the chiefdom of the Bemba.

From a historical point of view, this incorporation mechanism has contributed to the expansion of the Bemba chiefdom by allowing people to take root and blend in when they migrated to other areas and come in contact with other ethnic groups. Such migration has allowed for further movements of people and goods to an even wider region, enabling the groups to congregate as "we, the Bemba" while creating opportunities to interact with people from a wide region. As a livelihood strategy, the system contributed to regulating the impact on the natural environment by maintaining small-scale residential groups in its involvement with the environment on a day-to-day subsistence basis, but it also provided a safety net which allowed people to endure challenges such as climate change and political conflicts through the extensive "social security" network of people living in a geographically wide region. Despite a few conflicts along the way, the process in which people from different backgrounds were reorganized and integrated into the framework of "we, the Bemba" while maintaining the group's autonomy are similar to how other chiefdoms in the region have expanded their territory (Ndaywel 1999; Phiri 1999).[14]

Conclusion

Kaoru Adachi, who discussed "groups" while focusing on Junichiro Itani's concepts of "structure" and "anti-structure" in Chapter One of this volume, focused on a phenomenon called "mixed species associations" as an example of a "non-structured" group. Adachi states "Monkeys that form mixed species associations hardly engage in special social negotiations … they rather coalesce and separate repeatedly in space, and exist in a weak, borderless union that is continuously changing", (p. 33) and indicates that the phenomenon of "groups without clear membership or outline", similar to the mixed species associations of the monkeys, can also be seen as a mixed group phenomenon in human society. This mixed group phenomenon which can be characterized simply as different individuals "being together" is something that has not been seriously studied until now, but Adachi's statement, that this perspective may also be useful in understanding human groups, is important. It is a view shared by Noriko Itoh in her discussion in Article One of this book.

Kaoru Imamura (1992, 1993), who discussed "sharing systems" through the analysis of the gathering activities of the women of hunter-gatherer San, stated that "intimateness" nurtured by the act of "being together" in the daily subsistence activity is the key factor that connects people together. She also stressed that it is this intimateness that supports the sharing system that characterizes hunter-gatherer society. The same can be observed in the shifting cultivators of the Bemba. It has been repeatedly stated that the loose groupings, such as those seen in the mutual visits in the village, hearth-hold groups and gatherings at temporary huts, form the foundation of the Bemba's residential groups and have supported the mobility of people and flexible changes in the group structure. However, while the people of the San stay together because they "share" the *activity itself*, the grouping of Bemba is far more intentional. As I stated at the beginning of this chapter, the Bemba seem to be consciously making efforts to see each other personally and spend time together, even if it is for very brief moments, because they are fully aware that it is the accumulation of such acts that nurture an intimate relationship. In that sense, the Bemba may be more conscious of the instrumentality of the act of "being together".

What is interesting about the case of the human is that the act of being together plays a unique role when combined with its orientation towards structuring. The importance of "being together" in forming groups is also seen in our daily life in Japan, but the way that the Bemba have developed is much more flexible and sophisticated. Bemba can integrate strangers who have no connection to the group by forming constructive parent-child relationships based on ancestral genealogy. Once the relationship is structured in accordance with the "reference phases", the interaction can begin from the standpoint of the given structure. However, by accumulating the interactions, the "relationship" is then backed up with "intimacy" and the structure will eventually be retrofitted with the substance. It is interesting that while mechanisms like this made a sort of "mix-up" possible, "being together" is still regarded as a basic factor when recognizing each other as fellow members. It is these factors that make the structure and scale of human groups look so diversified and versatile.

In the case of the Bemba discussed in this chapter, when someone needed to legitimize "friendly" relationships with strangers in front of them, they manipulatively used the "reference phases" of their ancestral genealogy. By doing so, the residential groups managed to join a larger community that shares the "we, the Bemba" consciousness without losing their autonomy or being unilaterally integrated into the chiefdom.

As Hideaki Terashima discusses in Chapter Eight of this book, the formulation of human groups presupposes "distant bonding" in addition to "here and now" bonding, regardless of the group size or degree of political integration. Various studies have identified that a mechanism that gives structure to a group through a phase separate from the experience of the individuals, such as that discussed in this chapter as a "reference phase", is essential in order to realize this, and human beings have made great efforts towards creating this mechanism of structuring. It can be assumed that through the emergence of a "reference phase", humans have acquired a fabricated reality and managed to use this as grounds to make certain actions. The "*a priori* cultural category" that Toru Soga discusses in Chapter Nine of this book is one aspect of this ability.

However, I would like to emphasize here that the way to recognize each other as fellow members by repeatedly spending time in the same time and space is just as important and forms the foundation for the other measures, where members relate to each other through the reference phase and interact accordingly.[15] The formation of a group by concomitantly using different recognition methods is a universal human phenomenon. Having acquired the room to form groups through people's movements or as a result of such movement, by combining these methods this allowed humans to create a variety of group phenomenon. How these two recognition methods appear in the mechanism that the society creates will certainly play a strong influence in the format of social integration in a given context.

Article 3

The "Group" called the Kenya Luo: A Social Anthropological Profile

Wakana Shiino

Introduction

How can we view the 100 years from the twentieth to the twenty-first century in the long process of human evolution? From a social anthropological perspective, especially in relation to the evolutionary basis of human groups, we can see the historical transition of the formation and maintenance of human groups and changes in its forms. With the unfortunate colonialist legacy behind the development of this science, contemporary social anthropology has adopted one major role after undergoing a period of self-reflection; that is, to reconstruct the histories of illiterate societies and to analyze social change to the present in the context of colonization and grounded in local people's experiences. In this article, I would like to briefly describe how the people called the Kenya Luo, who live in the south-eastern region around Lake Victoria in East Africa, have formed groups as "Luo". I focus on the historical transition of their identity as a physical aspect as well as their sense of belonging.

Group organization called the "Luo"

The Kenya Luo are part of the Nilotic Luo who belong to the same ethnolinguistic branch as the Nuer society studied by Evans-Pritchard from the latter 1930s in Southern Sudan, and from which he formulated his segmentary lineage system theory. As Evans-Pritchard wrote in an essay about the Luo (1965 [1949]), his apprentice, Aidan Southall showed that Luo society is a unilineal descent group with a segmentary lineage system by diagramming the breakdown of residential groups and the residential form of the society (i.e., members of compounds and the arrangement of houses) (Southall 1952).

When I began to study Luo society, which had been the subject of such orthodox studies of social anthropology, I was impressed by the orderly genealogical charts based on interviews and its residential form and unit which emphasized patrilineage. Remarks highlighting unilineal descent lines were widely observed in Luo social norms.

Present-day Luo form a residential unit called *dala* (comprised of a number of *delni*) based on the patrilineal extended family. A *Dala* basically consists of one married man as a hub and his wives, sons and grandchildren. Spatial arrangements are governed according to customary norms which show one's social role and status within the *dala* identified by gender, age and birth/marriage order. Wives and sons build their own house at an appropriate location and together form the compound and live within it. *Delni*, the smallest residential unit, are surrounded by trees or hedged with bushes and are scattered throughout the village.

Fastidiousness with such things as spatial utilization and the order of hostship form the centre of life within *delni*, as observed in the frequent reference to "Luo's traditional ways". *Delni*, the life and residential space of the Luo, are also a significant place signifying where members are born, raise a family and die. To return to Luoland and to establish one's own decent *dala* is a major purpose in life for Luo men working in remote lands. To be recognized as a "decent" man by establishing one's own *dala* is the dream and purpose of life for every Luo man. The possession of *dala* or being part of one is at the basis of Luo group consciousness and identity as Luo today (Shiino 2007a).

Physical changes in residential groups and British colonial policy

While studying the way of life of the present-day Luo villagers and interviewing them about the foundation of their village, I came to the question of when the currently observed *delni* had been adopted. After several months of living in the field, I came to notice the existence of stone-built enclosures where villagers used to live. These varied in size and were scattered around the village.

How had their ways of life and group formation changed? Prior to the arrival of white European settlers, villagers used to live together in a large group within a compound called *ohinga* (comprised of a number of *ohingni*) in Luo languages. *Ohingni* enclosed by stone walls are totally different from currently-observed *delni* surrounded by hedges. When I asked the names and homelands of the people who used to live in *ohinga*, I found out that they were not only patrilineal/agnatic descendants such as the founder of the village, his sons and grandchildren, but also included the founder's daughters, their husbands and children. There were also some Bantu Suba people, whose language belongs to a different linguistic branch, living together for a certain period. This indicates that *ohingni* housed a cohabiting collective based on loose relations, such as affinal ties or transethnic friendships.

While listening to old stories about the stone-built enclosures from village elders, I learned that *delni*, the residential unit as well as the most significant emotional anchorage for the Luo, are not really a long-standing tradition. In fact, those elders themselves were born and raised in *ohingni* in the 1910s.

The major change in the residential form, from stone-enclosed *ohingni* to hedge-enclosed *delni*, is thought to have resulted from the external factor of British colonial rule. There is a brief description of *ohingni* penned by a colonial administrator in 1906: "In shape the Ja-Luo village is circular, and is fenced round with Euphobia bush, stone or mud walls, or watling. In the centre is the cattle pen, around this are the store houses, and outside these again is the circle of huts" (Nyanza Province Annual Report 1905–1906). When residential areas were settled according to "tribe" and lands were assigned and registered in the names of married men with the introduction of the land readjustment system, lands came to be owned individually with almost no common lands remaining. With this, consciousness as a Luo or consciousness of family ties within the residential area was thought to have emerged.

In other words, the Luo's residential form, rules of life and notions of tradition seem to have changed or developed within the approximately 100 years from the commencement of British colonization. Patrilineage/agnation-centered thinking was strongly reflected in spatial utilization (e.g., burial sites or house construction), and exclusive remarks and behaviors against other ethnic groups observed in present-day Luo villages are thought to have been reinforced with the change in the residential form over this period.

Luo consciousness developed through colonization and independence

How has Luo consciousness been developed through their colonial experience and the change in the residential form in villages from residential groups based on loose solidarity to a more concrete residential unit called "*dala*"? As a result of the settlement of a residential area by ethnicity according to British colonial policy, the "Luo tribe" have been in a tense political relationship with the Kikuyu, the largest ethnic group in Kenya, since that country's independence in 1963. After independence, the Luo faced an increasing need to assert their specific ethnic characteristics in political relationships with other ethnic groups during Kenya's growth process toward a modern multiethnic nation based on tribes from the colonial period. The Luo are also thought to have put forward their "we-consciousness" in order to maintain their ground in urban areas where ethnically diverse people intermingle and coexist.

Let's turn our attention from Luoland to urban areas. There has been an upsurge in the number of Luo people who moved to Nairobi, the capital of Kenya. Some were brought for forced labor during the colonial period, while others came as seasonal workers after independence as the capital was growing into a commercial city with an advanced monetary economy. The city stretches in a circle around the government office district. There are slums on its outskirts where the streets are lined with row houses made of galvanized sheet iron. The city continues to grow from day to day. People living in Nairobi tend to gather together according to social class, homeland or ethnicity. Luo people who moved from villages live together in slums called "estates". The flow of people coming and going between towns and villages was facilitated as seasonal workers increased in number, resulting in the development of information networks between towns and villages and the formation of mutual aid organizations based on native clans.

For example, David Parkin, an apprentice of Southall, conducted studies in Kampala, the capital of Uganda, as well as in Nairobi from the colonial period until independence when so-called "tribalism" was coming to light. He illustrated how the Luo Union, a financial and life strategy based on friendships formed by the Luo people who moved to the cities between the 1960s and 1970s, was based on an extended kinship network deriving from Ramogi, a well-known ancestor of the Luo (Parkin 1978).

In the middle of the 1980s, there was a controversial case that highlighted ethnic identity in a modern multiethnic developing nation, attracting international media attention. This was the "S. M. Otieno Case," a court battle over the burial site of a Luo lawyer, Silvano Melea Otieno, fought between Silvano's relatives and his wife and her relatives belonging to the Kikuyu, the Luo's political opponent. A Kikuyu woman, Wambui, and a Luo man, Silvano, married and fostered a transethnic cosmopolitan life in Nairobi where they spoke English and Swahili. After Silvano's death, the couple's relatives disputed as to whether Silvano's burial site should be Luoland, his patrilineal homeland, or the compound in Nairobi where he lived a married life. The case triggered Luo's increasing tendency to emphasize "Luoness", coupled with the power relation between the Luo and Kikuyu in the political world.

Development of Luo group consciousness via presidential elections

The Luo tended to be shunned by the Kikuyu, who were at the centre of political life. Because of this situation everything, including regional development such as

Photo A3.1 An open-air market in Luoland

Raila Odinga, a presidential candidate from the opposition party in the December 2007 elections and the present Prime Minister, is the hope and a hero of the Luo. Posters of him are put up on the wall in almost every Luo residence. Calendars featuring the US President Obama and his family have also gained popularity recently.

road construction and electricity installation, tended to be neglected in western Kenya, the Luo's traditional settlement area, especially in southern Luoland which stretches around the shore of Lake Victoria. Compared to other areas with similar population density, NGOs began a late entry into this area only several years ago. Some studies claim that the present situation of extreme poverty where the area has been left behind by other regions of Kenya, even after fifty years of independence, primarily resulted from political conflicts around independence.

The Luo, who experienced many years of political and economical hardship, wrote the song *Malo Malo!* (rise higher and higher). This rhythmical song was easy for anyone to sing even without understanding the meaning of the lyrics, and became popular all over the country regardless of ethnicity from the latter half of 1990. The song embodies the aspirations for the Luo's prosperity on the theme of "we Luo" (Shiino 2007b). "Luoism" was increasingly intensified every time the

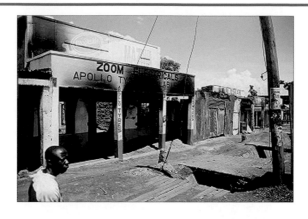

Photo A3.2 An aftermath of the Kenya Luo's political rebellion

In the December 2007 election, there were high expectations for the presidential candidate Raila Odinga among those, primarily Luo, who had been longing for Luo's reappearance on the political scene
for many years. His supporters had believed that Odinga would win by maintaining a lead throughout
the election campaign. When the present President Kibaki, who is from the Kikuyu tribe, declared
victory based on unreliable data, Odinga's supporters raged and raised riots across the country.

underlying structure of Luo-versus-Kikuyu political conflicts became prominent
in a presidential election, particularly that held in December 2007. The more vigorous the election campaign became, the more frequently the phrases "we Luo" or
"we Nilotes" were heard. A tendency toward excluding each other became manifest
even in ethnically diverse areas. The Kikuyu silently disappeared from Homa
Bay, a country town in Luoland, at the beginning of 2008. I was surprised when I
heard the phrase, "We don't want the Kikuyu", prevalent among local people (see
Photos A3.1, A3.2 and A3.3).

   This way, political factors played a crucial role in the development of and the
change in Luo group consciousness. In response to this trend, social science
studies about the Luo, including oral histories, have slightly increased in number
in recent years as exemplified in the title of J. R. Campbell's article, "Who are the
Luo" (Campbell 2006).

Beyond languages and borders: The future of present-day Luo

Some Luo people moved from Luoland to Nairobi for forced labor during the
colonial period or for seasonal work following independence, while others moved
overseas for study or seasonal work. As a more recent development, the Luo called

Photo A3.3 A scene of the Luo revolt in Homa Bey

In a small town, Homa Bay, people started to protest against the election results on January 1 and the police began to fire into the crowd on the following day. President Kibaki was said to have ordered the use of force in order to quell the rioting.

on acquaintances inside and outside the country via e-mail for cooperation in the campaign of a Luo presidential candidate, Raila Odinga, in the December 2007 election. I also received several such e-mails containing a large photo of Odinga. Luo people outside the country tend to connect with each other via the internet coupled with such campaigns. This is a phenomenon called "intentional grouping", which in this case aims for the formation and promotion of Luo identity.

The Luo are said to be a proud people by nature. T-shirts worn to declare oneself as Luo demonstrate this in a simple way (Photo A3.5). Luo politicians and frontline scholars usually use Luo names instead of Christian names in order to emphasize their ethnicity. Raila Odinga is one of them, and "Obama" is also a Luo name.

Since Barack Obama, whose father was Luo, was inaugurated as the President of the United States in January 2009, Luo people scattered all over the world have become proud of their ethnicity more than ever, and their movement to connect with each other on the internet has intensified. Activities of Luo's web communities based in England, the US and Nairobi have been vigorous. These communities hold forums for exchanging political views, organize so-called "offline gatherings" where members meet others face to face and help members find marriage partners (see for example such websites as www.Jaluo.com or www.luo-socialforum.org). How will Luo people overseas form real or imaginary groups via such websites in the future? On what occasions will such groups spring into action? Just as Barack Obama ran his presidential campaign by making full use of the internet, are the

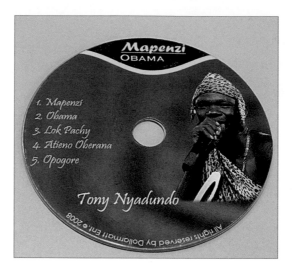

Photo A3.4 Tony Nyadundo, a musician of a form of Luo traditional music, Ohangla

Nyadundo released a song titled "Obama" shortly after Barack Obama became a presidential candidate in the US. The lyrics contain an imaginary conversation between Barack and Michelle Obama.

Photo A3.5 Luo pride

T-shirts with the phrase "Jaluo in the house!" emblazoned on the back proudly declare their wearers as Luo. The phrase "I biro ayie" on the front means that everyone in the house yields to the existence of Jaluo. "Jaluo" is the singular form of Luo.

Luo going to maintain Luoness and connections in the form of groups with those overseas for a variety of purposes by taking advantage of ever-evolving digital technologies or mobile phones?

Among Luo people born and raised in Nairobi, there are those who speak English and Swahili rather than Luo languages. They are known as "Nairobians" who are unfamiliar with Luo customs and speak Nairobi English, a mixture of English and Swahili. According to them, they are conscious of their ethnicity and tend to choose Luo people as friends or marriage partners, saying that people from different ethnic groups are ultimately undependable. Their consciousness as Luo is increasingly emphasized even though they do not necessarily speak Luo languages. I will continue to observe the establishment of connections among the Luo and the future of their group consciousness according to such variable situations inside and outside of Kenya.

Part IV
Towards a New Theory of Groups

11 Collective Excitement and Primitive War: What is the Equality Principle?

Suehisa Kuroda

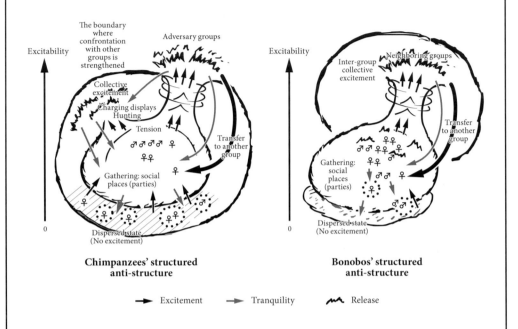

The boundary where confrontation with other groups is strengthened

Excitability

Adversary groups

Collective excitement

Charging displays
Hunting

Tension

Transfer to another group

Gathering: social places (parties)

Dispersed state
(No excitement)

0

**Chimpanzees' structured
anti-structure**

Excitability

Inter-group collective excitement

Neighboring groups

Transfer to another group

Gathering: social places (parties)

Dispersed state
(No excitement)

0

**Bonobos' structured
anti-structure**

→ Excitement   → Tranquility   ⋀⋀ Release

The unit-group (or community) of Pan, characterized by the individual nature of its members, is maintained through two mechanisms: acceptance of independent actions and collective excitement. In the case of chimpanzees, females tend to stay in a dispersed state where the level of excitement and tension remains the lowest. However, the males tend to aspire for dominance and fight for superiority by forming alliances based on the equality principle. Charging displays and hunting, in which males release themselves from the suppression required to maintain the equality principle, stir up collective excitement. A face-off with an adversary group functions to maintain group stability, as males go through such intense self-suppression that when released from it, they feel at one with a sense of pleasure. This is thought to be a developed form of excited community emerging through fear and aggression to counterattack when threatened by large carnivores. Conversely, the society of bonobos is a female led society where there is no male group formation or alliance. Groups are maintained through the solidarity of females, mother and adult son bonds as well as inter-group male confrontation. As there is little adversity among groups, collective excitement easily spreads from one unit group to another, leading to collective effervescence.

Introduction

Studies on the ecology and society of the two species of *Pan* have decades of history. Chimpanzees in the Gombe Stream and Mahale Mountains have been the subjects of research for almost half a century. The bonobos in Wamba, in the Democratic Republic of the Congo, have been studied for over thirty-five years, though intermittently due to civil war. However, there remain many behaviors with unknown functions that are yet to be explored. The phenomenon named by Toshisada Nishida (1977) as "collective excitement" is one of them.

> When a large number of chimpanzees converge in a limited space, they are often observed to be excited, uttering hooting and screaming, and drumming tree trunks and buttress roots. In such cases, the hooting is commonly emitted by all members of a subgroup (= party: terminology to be discussed below), and the chorus often attracts a number of other subgroups. When two or more subgroups come in close proximity, the excitement intensifies and induces chasing, dashing and various displays.
>
> When the majority of the members of a unit group are gathered in a certain area in seasons abundant with food, such collective excitement can occur several times a day, with each event lasting up to two hours. (T. Nishida 1977: 627)

These sprees, which occur when a large number of animals gather, have been reported at various chimpanzee research sites (i.e., Goodall 1986; Yukimaru Sugiyama 1981). Among chimpanzees, this is a phenomenon that occurs within a unit-group (or community). However, with the bonobos, cases of intense collective excitement involving two unit-groups have been observed (Idani 1990; Kuroda 1979, 1982). The social function of the collective excitement is difficult to understand and has not been studied to date. However, if this phenomenon instead involved humans, the matter would be easily resolved. Though there are no components of cultus, the excitement is almost identical to that observable at festivals. The shortcut to understanding this particular group phenomenon may be to refer to human festivals.

Émile Durkheim (1915) named the seasonal excitement displayed by the Aborigines in Australia during their religious cultus, "collective effervescence".

> When they are once come together, a sort of electricity is formed by their collecting, which quickly transports them to an extraordinary degree of exaltation. (Durkheim 1915: 215)

Such excitement would occur regardless of the religious cultus. In pre-modern society, when people gathered in tens or hundreds, they fell into unpredictable states of excitement (Elias and Dunning 1995). It seems reasonable to assume that the act of gathering is itself the stimulus that causes the effervescence. If so, even a gathering of a small number could induce members into a mild state of excitement, in preparation for an even greater sense of exuberance.

If we assume the same for chimpanzees and bonobos, their collective excitement begins to make sense. Then again, what is the grounds for this occurrence and why? How are these phenomena related in their society? In this chapter, we first build the hypothesis that *Pan* is an animal who is excited by the coexistence of other individuals, and begin our discussion from that point.

The forms of collective excitement observed in *Pan* are as diverse as those in the human world. In an incident Jane Goodall (1986, 1994) called the "primitive war" (detailed below), chimpanzees of an assault group gathered and, while in a state of fear, they raided their enemy territory as a group. What they shared in the attack was aggressive excitement mixed with fear. Could it be that they purposely entered the dangerous territory because they were intoxicated by the thrilling pleasure of an assault, just like humans would? (Please refer to Kawai's work in Chapter Seven of this book.) This discussion will form the other core theme of this chapter.

Such collective and aggressive excitement seems commonplace, when similar human group phenomena are used as analogies. In the case of humans, excitement accompanying group phenomena has been noted as an important aspect in various scenarios such as festivals and wars. Besides, our daily lives are filled with cultural apparatuses that prompt us to experience such excitement. However, once we set out to explore the issue, we soon find that we do not really know where this feeling comes from. Not even Émile Durkheim (1915), who discussed collective effervescence, Jean-Jacques Rousseau (H. Imamura 1982), who emphasized the role of festivals in the formation of human society or Roger Caillois (1994), who discussed war in the context of festivity has ever questioned the origin of the excitement closely related to the formation of human society. We may have perceived collective excitement as a given to humans and confined it to the field of sports on one hand, while on the other we worked on the art of its generation in the field of entertainment, but the truth of the fact is that we do not really know what it is.

However, if we look at *Pan*'s excitement in relation to their society as a whole, we begin to see another aspect to the issue. Could it be that the tendency to be excited, or excitability, was developed as a way for equal individuals to come together as a group? A group of individuals maintaining equality is what Junichirō Itani (1987b,

1988) referred to as a society based on the "equality principle". In other words, could it be that coexistence based on the equality principle is a state that can be achieved through states of excitement?

The two species of primates to be discussed below reveal a contrastive difference in their forms of collective excitement. The male chimpanzees' excitement, in particular, is associated with violence. In contrast, the bonobos' collective excitement either turns into sexual arousal and dissolves into chaos or remains as it is. The contrasting two cases are useful in relativizing the relationship between violence and collective excitement regarded as a given to humans.

Group phenomenon and excitement

Excitement arising from fission-fusion
One of the group phenomena unique to *Pan*—the formation which repeats fission-fusion—is deeply related to the notion of collective excitement. In general, fission-fusion is thought to be an ecological adaptation to suit their fruit-eating nature, as the size of the food patch and quantity available vary from season to season and year to year (see T. Nishida 2012). However, here we would like to discuss its social meaning in relation to excitement.

Chimpanzees in particular are known for their dynamic fission-fusion and it is very rare to see all members of a unit grouped together in one place. Behavior is gendered (see Article One of this volume) (Goodall 1986; T. Nishida 2012; Wilson & Wrangham 2003), in that females with young offspring usually tend to stay by themselves or with other close females, and only join the party when they hear males shouting upon finding food (food call) or commotion from hunting. Conversely, males tend to stay in a group (male group), forming the largest party with the alpha male as its core. Females in estrous often join this party, but sometimes lower ranking males draw them away from the group alone. It is reported that in Gombe, when such males are not found in the group they are often with females (Goodall 1986).

The bonobos' fission-fusion is different in that they spend their day in small parties and in the evening call for each other to form a large party at their sleeping place. Furthermore, parties often get together and all the members of a unit-group can often be observed at one place (Kano 1992; Kuroda 1979, 1982).

When fruit is scarce, chimpanzees disperse into parties of one to several animals. In these small parties, they seldom vocalize or display excitement. However, they do become animated when encountering other parties or when individuals join the group. They approach one another panting, showing their excitement, and hug each other. This is regarded as their form of "greeting" (Goodall 1986; T. Nishida 1977). When five or more animals arrive at a new feeding site, they spend five minutes or so screeching and causing commotion, engaging in behaviors such as charging displays,[1] chasing each other and branch dragging. When new members join in on hearing this noise, another round of excited greeting behavior begins.

Conversely, bonobos seldom exhibit the behaviors that accompany excitement as described above when in a group of less than ten animals. They do, however, respond when they hear calling from afar, even if there are only a few animals in the party (Kuroda 1979, 1980). When familiar animals individually join the party, there will usually be no greeting or displays of excitement. It is when several parties converge and form a group of ten or more that they fall into a state of excitement, frequently engaging in disorderly behaviors such as calling, screeching, charging displays, chasing each other and sexual behaviors.[2] When the group size exceeds ten, for about five minutes a commotion similar to that of chimpanzees arriving at a new feeding site occurs.

As such, it is the encounter, merging and group size that triggers excitement in *Pan,* which is caused by the group phenomenon of fission and fusion. It should be noted that it is quite normal for large groups or gatherings of mammals and birds to fall into a state of excitement, but such groups are formed when social units are disbanded to start seasonal migration or as temporary gatherings for mating. I would like to emphasize here that the way *Pan* get excited while maintaining their social groups is quite unique in mammals.

Freedom from the group and excitement upon joining the group

The fission-fusion phenomenon has not been seen in societies of most primate species. Members may disperse at times but they generally range and travel together as a group. In the case of the Japanese macaque, if some members of the troop (or unit-group) start making independent movements, it will be regarded as a sign for troop fission. If after being segregated for more than a month such individuals return, they would be attacked and fall to the lowest rank of the troop (Mito 1971). On the other hand, in chimpanzee and bonobo societies, it is common for individuals to go solo for one to several months. The group will become excited when they meet again but will accept these members back into the group. Male chimpanzees, during fights over ranking, may display fierce attacking behaviors if they have not seen each other for a while (Goodall 1986), but this does not signify rejection from the group.[3] *Pan* and Japanese macaques have a different social nature. Group membership in the case of the latter is not guaranteed by blood relations or memory, so their presence—being together with others—would be the condition for their coexistence (see Kitamura's work in Chapter Three of this volume).

In contrast, it can be said that the society of *Pan* is one that tolerates individual social independence.[4]

> A chimpanzee can almost always withdraw from an undesirable social situation and either remain alone for a while or seek a more congenial companion. In startling contrast to the vast majority of nonhuman primates, chimpanzees of both sexes can and do *regularly* go about their affairs with privacy. (Goodall 1986: 170)

Though they may not spend as much time alone as the chimpanzees, bonobo females have the option of refraining from interacting with other members of the party or keeping a short distance away from the others during the day in a semi-isolated state, and rejoin the group at their sleeping place. As with chimpanzees, all members of the bonobo group enter into a chorus when they form a large sleeping group. Even those who are on their own away from the group would join the chorus from a very young age of less than four years old (Kuroda 1982).

In essence, fission-fusion is a phenomenon formed by those who have the ability to keep themselves ecologically and socially detached from the group.[5] Such a state of self-reliance may be described as independence, or the "freedom" of individuals from the group. Further to these nuances, we may also call it "singularity" in the sense that *Pan* have the power to choose to be on their own (see Chapter Two, Uchihori's work, of this volume).

In relation to humans, while discussing equality in hunter-gatherer societies, Hideaki Terashima (2004: 30) stated "not being restricted and not having to depend on is something that cannot be discussed separately from equal relationships". "Not being restricted and not having to depend on" would be equivalent to the notion of "freedom" used in our discussion. The "freedom" of bonobos and chimpanzees forms a paired concept with the equality principle (to be mentioned later), as referred to by Itani, because those who are disadvantaged by inequality will simply "withdraw from an undesirable social situation" (Goodall 1986). It is not possible to bind self-reliant individuals to a group ruled by the order of dominance.

Such freedom from a group is not the freedom of the loser. It may be so in the case of chimpanzee males, who tend to stay in a group except when they temporarily go solo after losing a fight over their rank in the group, but certainly not with the females as it is quite normal even for the prominent ones to stay with a small number of other females whose offspring get along well (Goodall 1986). When such females go near the males, responding to their food calls and display of commotion, they undergo a psychological transition from a calm solitary state to one of coexistence with others, including males. The energy that enables them to achieve the transition is the "excitement" expressed in their greeting. This can also be corroborated by the fact that sick individuals often do not join the party, whereas females in estrous, being physically in a state of excitement, join the males without hesitation. Individuals with low levels of excitement may arrive quietly at the edge of the party but do not go as far as joining in the greeting.

Then, once the individuals converge in a large party, they must constantly maintain an idling level of light excitement, ready to enter collective exalted excitement as soon as they are triggered by some cause. In other words, in order to join a group, the individuals need not only the interest but also the energy to get excited. Both sexes of bonobos, and chimpanzee males, who have a stronger tendency for collectivity, are constantly in this idling mode in the company of many other individuals.

In summary, fission-fusion, which creates excitement, is a group phenomenon displayed by individuals who have the power to be self-reliant and independent from the group. When they join a gathering, they need the energy to become excited and once they do, they will remain in a preparatory state for excitement while they are in the gathering. Self-reliant individuals cannot be organized by the inequality principle (to be mentioned later) (or at least, not by itself), so if they wish to coexist, they inevitably need to form a society based on the equality principle.

The inter-unit group encounter and bonobos' excitement

Unit-groups of bonobos encounter each other through chases that can last for hours or days, in which it is hard to distinguish between the chaser and the chased, and while males will continue their skirmish in excitement, the groups eventually come together in what appears to be a single party (Idani 1990; Kano 1991; Kuroda 1982). Juveniles and young females in particular mingle well while males from different unit groups tend to remain in confrontation and do not interact to the same degree. However, unlike the chimpanzees, their relationship is not hostile, but rather "touchy".

When ten or more bonobos exist in the same space, they are easily excited by the voices or sounds of fellow bonobos in the distance and try to head in that direction. If the group consists of thirty to forty animals, responding loudly with displays such as branch dragging and the frequent drumming of buttress roots, huge commotions soon emerge. However, the inter-unit-group approach and encounter, accompanying the exchanging of voices, does not always take the form of a straightforward meeting where excitement calls for more excitement. Sometimes, groups seem as if they are going to meet but never do and simply continue on with their noisy transit, where they may continue to chase and be chased for days. It is as if they are scared to face each other or as if they are trying to extend their pre-engagement excitement.[6] As far as we have been able to confirm, such mobile groups are headed by males.

Equality principle, conflict and excitement

Two principles of coexistence

Itani (1987b, 1988), through the analysis of Japanese macaque and *Pan*, identified two patterns of coexistence among individuals within a unit-group and named them the "inequality principle" and "equality principle". The inequality principle is a form of coexistence based on the order of superiority, where the dominant-subordinate relation between individuals is fixed (ranking order). For example, in the unit-group (or troop) of Japanese macaques, there is a linear order of dominance among males and females respectively, and the higher-ranking individuals have priority access to food and members of the opposite sex. The inferior hardly ever overtake the superior in this rank order system.

Conversely, the equality principle is a form of coexistence based on interaction where there is no apparent order of superiority, for example, exchanging symmetrical and friendly behaviors (such as mutual grooming, symmetrical greeting and play), sharing food and nonexclusive access to members of the opposite sex. Generally, two forms of coexistence can be found in one species society, though two patterns of the equality principle—food sharing and granting non-exclusive access to members of opposite sex—are only found in a few species. Itani positions *Pan* society as one where the inequality principle had collapsed and the equality principle had emerged to replace it. Furthermore, Itani (1987b) claims that coexistence based on the inequality principle, or the order of superiority, can be made possible by the unilateral suppression of the subordinate individual, whereas a relationship based on the equality principle is being guaranteed by promoting a mutual approach. It implies that there is equality and mutual intention for coexistence among the participants.

Aspirations for dominance and the equality principle

There is a distinct order of dominance in chimpanzee male society. However, adult males usually interact based on the equality principle, especially in activities like grooming, greeting and sharing food, and tend to forage together. Despite the order of dominance, the occurrence of aggression often depends on the existence of allies and the assault may not be unidirectional. Only an alpha male in his stable reign can single-handedly maintain superiority in an attack. General mating is promiscuous but the alpha male tends to monopolize females in the ovulation phase. Even then, if a challenger emerges, the alpha male often starts permitting his allies to mate (de Waal 1982; T. Nishida 2012).

The chimpanzee alpha male struggles to maintain his position. While he often interrupts grooming among rivals to prevent them from forming alliances and conducts charging displays to show off his dominance, he also protects the females through fair fight arbitration and secures the support of the females and allies by sharing the game meat he hunts (de Waal 1982; Goodall 1986; T. Nishida 2012; Watts & Mitani 2002). According to Itani (1987b), who highly values food sharing as an objectification of the equality principle, an alpha male who often shares food with others is an embodiment of this principle. That is, his dominance is maintained by displaying behaviors that emphasize the equality principle.

These actions indicate that chimpanzees possess a degree of independence and the ability to view themselves with a certain level of objectivity, or self-awareness (de Waal 1989; Goodall 1986; Itani 1987b; Kuroda 1999). As discussed above, females have a level of independence and they can, when it comes down to it, join in the males' fights as supporters or adversaries or respond to unreasonable attacks by males. Female independence is not limited to the freedom to leave the group, but is also found in their social nature and their equality with males.[7]

However, things are a bit more severe in male chimpanzees. If the alpha male's dominance is shaken in some way, even if only slightly, males who are prepared to form an alliance with anyone in order to climb the ladder start emerging (de Waal 1982; T. Nishida 2012). This is followed by a continuous emergence of young males who keep challenging the alpha male, even if they are beaten (Goodall 1986). Chimpanzees have the ability not to be restricted by the current relationship or order of dominance, and thus males never cease to seek higher status.[8] De Waal (1982) calls this the aspiration for power. It is because of this that they need allies, care for others and act equally. The order of dominance among these individuals is, though heavily formalized, based on a temporary agreement, or a device hiding

Photo 11.1 A male bonobo at Wamba

Named Ten, the bonobo belongs to the E1 group. When I started observing the E group in October 1974, Ten was a child. He was estimated to be eight years old in 1979. After the breakup of E group into E1 and E2, he became an alpha male of E1 with his mother's support. I took this photo in 2006 when he was at the center of the party as a high-ranking male, together with Tawashi, three years younger and the alpha male at that time. In 2008, a young male occupied the top position, with Tawashi disappearing from the group. Ten, however, was there in good shape at the estimated age of thirty-eight.

Photo 11.2 Female bonobos at Wamba

These females belong to the E1 group. Sen is sitting at the center left and Haru at the center right. The bonobo females occupy good places for food collection, and only dominant males can access these areas. In February 1984, Sen supported her son, Ten (photographed in Photo 11.1), challenging an alpha male, Ibo, and his mother. This enabled Ten to be an alpha male, and Sen occupying the top position among females. Ten's brother, Senta (in the left of the photo), is stretching his arm to beg his mother for a sugar cane.

rivalry to maintain peace (de Waal 1989). Therefore, once the order collapses, the struggle for dominance can drag on for an extended period of time. The fight over power by three males at Arnhem Zoo lasted five years and only ended when two of the males collaborated to kill the remaining male (de Waal 1989).

Alternatively, in bonobo society, it is the females who take leadership roles. Bonobo males do not form alliances and sons depend on their mothers even when they are adults (Furuichi 1991; Kano 1992; Kuroda 1982). Furthermore, males do not fight over access to females. Young males may occasionally get into a fight but it is immediately taken over by their respective mothers. However, several dominant males can be assigned positions on a vertical scale, and there have been a few reported cases of the sons of dominant females challenging the alpha male for power with their mothers' support (Furuichi 1991). Superior males have certain benefits such as being allowed to enter the good feeding sites dominated by females and having a better chance of being chosen as a mating partner. Also, the more dominant the males are, the more often they tend to share food with the females (Kuroda 1984), but it seems that this is not necessarily done to maintain their order and is more because of their inescapable social position and their social nature—they just cannot ignore the requests of others (Kuroda 1999).

There is a vague hierarchic order in bonobo females, but interaction which would indicate a dominant/subordinate relationship is rarely observed in female adults and even if there is trouble, it is usually resolved through GG-rubbing (genito-genital-rubbing (Kuroda 1980)) (see Note 2). As such, behaviors that clearly indicate their equality or aspiration for dominance, often observed in chimpanzee society, are uncommon in bonobo society. The gentle nature of the males and the leadership and highly developed sexual behavior of the females together contribute to the creation of a tolerant and peaceful society based on the equality principle.

The equality principle and the unleashing of inhibition

The "rain dance",[9] the first reported case of a display by multiple chimpanzees with a strong degree of excitement (Goodall 1971), seems to be equivalent to Nishida's "explosive display" (by multiple males) (T. Nishida 1977) or "booming" (Reynolds and Reynolds 1965; Yukimaru Sugiyama 1981). The "collective excitement" Nishida (1977) refers to is a little different from these in that it is larger in scale, indicating a situation in which several parties simultaneously enter a state of excitement. However, when we follow Itani's hypothesis below, we can consider that all of these have a common function, categorize them into collective excitement and distinguish each if necessary.

Itani (1993: 270) explains the social significance of booming as follows:

[...] The maintenance (of the male bond) presupposes mutual inhibition in males. It is likely that the orderly structure of the chimpanzee groups that we are accustomed to seeing is one that the inhibition of males had created. Inhibition creates, or requires, various substitute behaviors. These include diverse and frequent greeting behaviors observed especially in males and the booming behavior that temporarily unleashes whatever the males have had to inhibit.

In other words, maintenance of male groups imposes self-inhibition on its members and booming or collective excitement is a way to release the resulting depression. (We name this interpretation as a "vent hypothesis".) The interpretation that acknowledges the excessiveness in males' greetings and views it as a substitute behavior of inhibition can also be applied to grooming. When the frequency of grooming among males and females was adjusted according to the number of individuals, it was calculated that male to male grooming occurred twenty times more frequently than its female counterpart (Kuroda 1980). It was also noted that when the order of dominance became unstable, the frequency of grooming increased nine fold (de Waal 1982), and even those who were used to choosing females as their grooming partners tended to select males during such time (Goodall 1986).

We can also employ France de Waal's words (1989) to interpret strong self-depression and excessiveness of interaction among chimpanzee males from another aspect.

[...] Second, the unreliable, Machiavellian nature of the male power games implies that every friend is a potential foe, and vice versa. Males have good reason to restore disturbed relationships; no male ever knows when he may need his strongest rival. Holding grudges may cause isolation, which within the coalition system amounts to political suicide. (de Waal 1989: 53)

In other words, the excessiveness is an obsession that forces them to maintain their alliances to facilitate their aspirations for power and to keep together, allowing them to monitor their allies because of the untrustworthiness inherent in their relations. At any rate, chimpanzee males fear isolation, inhibit hostility and repeat interactions to confirm their relationships. If the nature of the excessiveness of grooming is as such, the behavior may relieve tensions but would not truly alleviate their anxiety and tension. Male chimpanzees are destined to vent and explode.

When pent-up males gather on mass, it is considered that the desire to ventilate increases. The solution could be anything from booming to hunting or even promiscuous sexual behavior. The reason why males often go hunting when they gather could be because, in addition to the increased success rate (Hosaka 2002; Watts and Mitani 2002), it provides them with the opportunity to unleash their inhibitions.

Even so, game animals or females in estrous are not always available, and there will always be some who miss out on the game meat. The actions intended to release inhibition may at times cause further emotional conflict. However, the "enemy land" is always there. Sharing or monopolizing is not an issue in "hostile sentiments" and all can join in and be a part of the confrontation.[10] "Enemy" is a perfect outlet for the release of inhibitions that does not create emotional conflict.

Itani (1987b: 203–204) further claims that the "solidarity of males is achieved with the backing of mutual aggression of males in confronting groups", and "maintaining equilibrium through confrontation is the only way (for males) to secure their coexistence". However, that is not the only function of the excitement that accompanies aggression.

"Primitive war" and comradely unification

Let us now examine what happened in an incident that Goodall (1986) referred to as "primitive war".

In 1971, the group Goodall was following split into two (Kasakela community which was comprised of eight adult males and Kahama community made up of six adult males) and began having confrontations, culminating with the Kasakela males killing Kahama members in 1974. Chimpanzee males often cautiously walk around their territory border in a group. This behavior is called a "patrol" (Goodall 1986; Watts et al. 2002). The patrol troop stays silent, vigilant, smells the ground and/or vegetation, and members hug each other with fearful expressions upon hearing a sound while still moving forward. They relentlessly attack anyone alone or in a small group that they meet along the way, except for a group of tough individuals or a female in estrous. The patrol can last from tens of minutes to hours, and when they return to their territory they enter a state of collective excitement as if to release the accumulated tension in a burst (Goodall 1986).

Goodall (1994) describes that males often gathered, attracted by a female named Gigi in estrous, and went on a patrol of the periphery of their territory. This transformed into a series of lethal attacks on the adjoining group. There are several interpretations for invasion and lethal attack such as to acquire fertile females, to gain more food resources and/or to eliminate rivals from other unit-groups (Goodall 1986; T. Nishida et al. 1985; Wilson and Wrangham 2003). Itani's vent hypothesis, however, adds another important factor: to release the males from the depressive inhibition and maintain their group bonds in such a way that it may well be that the "specific target" of their ire actually does not matter.

It is clear that the members of patrol troops encourage each other to overcome their fear. It is quite common to see males hugging their nearby mates with a grin on their faces before pulling themselves together to return to the fray, and this also often occurs when they are fighting over their ranking order (De Waal 1982; Goodall 1986). Under these circumstances, the order of dominance within the group disappears and only the distinction between friend and foe remains. This state continues for an extended period of time when they are out on a patrol of some hours. The inhibition shown in their daily lives vanishes in the intense atmosphere of fear and offensive excitement and once the attack begins, all inhibitions are completely cast aside. This process of excitement influences the maintenance of peace among males in a dual sense in that it unleashes the inhibition accumulated through the practicing of the equality principle, and it creates equality through the comradely unification formed in the face of the enemy.[11] The fact that patrolling alone causes intense group excitement upon return to their territory indicates that, in a practical sense, all they need is a potential "enemy" with whom they could engage in conflict.

In the case of humans, behaviors that accompany excitement inhibited for long hours bring pleasure. The fact that the chimpanzees "chose" to repeat the act seems to indicate that they find some kind of pleasure in it, too. The young males' attraction to the hostile inter-unit-group encounters and the fact that they remain behind at the scene can also be seen as collateral evidence to support this supposition. One young male repeatedly headed the attack group and later returned to the scene of the meeting. Another young male had several triumphs, including one case in which he single-handedly put on a dramatic show of intimidation and kept throwing stones down on the enemy group who were screeching from the bottom of the valley. It is reported that he hit and stamped the ground and kept screeching provocatively even after he caught up with his group members who had fled before him (Goodall 1986).[12]

Communitas equality and *"structured anti-structure"*
The unification of aggressors amid fear and excitement, as we have seen in the previous section, does not fit with Itani's (1987b) category of the equality principle. Itani claims that the principle is to interact as equals by *denying dominance* and as emphasized in the self-inhibition of the male group, it emerges as selfish desire is inhibited. However, the unification seen above is nothing but equality, and this emerges in a situation that does not require self-inhibition.

Kuroda (1999) recognized two patterns in *Pan*'s food distribution—the pattern where food is shared through the food owner's inhibition of selfish desire, and that where generous distribution and division occurs after collaborative behaviors such as copulation or GG-rubbing, and cooperation—and called the latter "communitas distribution".[13] In "communitas distribution", the distinction between food owner and non-owner hardly exists and food is shared or divided as if it were common property. The relationship that emerges here does not require self-inhibition and both participants are almost unified.

Based on the above, we shall call the equality that emerges in the excitement and release of self-inhibition, the "communitas equality". In the case of humans, the typical example would be unification at festivals, but it can also be seen among soldiers in conflict situations. What we have confirmed in the case of chimpanzees' primitive war is that it is equivalent to the latter. In Itani's theory of anti-structure (non-structure), the communitas situation can be classified to the state of "anti-structure" (see Adachi's work in Chapter One of this book). Thus, Itani's vent hypothesis is connected with his theory of anti-structure. According to the hypothesis, the social structure on the base of the equality principle can be formed by each individual's self-inhibition, and cannot be maintained without a mechanism for unleashing inhibition. In other words, there is a contradictory structure in that the structure based on the equality principle can only be completed by the unstructured aspect (we will call this the "structured anti-structure"). In the case of the chimpanzees, the "anti-structure" emerges in the violent field produced by confronting "enemies", and functions to maintain their social structure. However, based on our discussion so far, the state of "dispersal" and booming would also become the structured anti-structure, though these are within-unit-group phenomena. Looking at this from another perspective, the structured anti-structure is a function connoted in the structure for its own maintenance and adjustment.[14]

Conversely, the bonobo society did not embrace the interactions, as seen in the chimpanzee society, in which aspirations for dominance and equality are emphasized. In other words, the bonobos' social interaction did not require a strong degree of self-inhibition and thus there was no need for release. Combat groups that paired up with "enemies" do not exist among males or females. The reason why they still manage to retain the outline of a unit-group is twofold: first, their highly gregarious nature with mother-son bonds and strong ties between females created through GG-rubbing at its core; and second, the existence of inter-unit-group male confrontation (but not "hostile confrontation"). As bonobo individuals do not often act individually, the role the state of "dispersal" plays as structured anti-structure is small. What plays a bigger role here is the widespread sexual behavior typified by GG-rubbing. Long lasting GG-rubbing provides intense excitement and communitas food distribution among females. Though not as frequent, males also screech in excitement as they indulge in furious rump-rump contact, where their testicles hit one another.

Then, what does chimpanzee collective excitement involving multiple parties or the inter-unit-group collective excitement of the bonobos mean? Thus far, we do not have enough data to interpret these phenomena other than to say that they are states in which the act of gathering itself becomes the stimulus and excitement calls for more excitement, and to position them as structured anti-structure. In the case of chimpanzees, the excitement is contained within the unit-group due to the presence of "enemy", but in the case of bonobos, the radius of excitement is expanded because they do not have the presence of a distinct "enemy". We could not identify the communitas equality in bonobos but if it exists, it would be among the excited males who head the inter-unit-group chase where they repeatedly chase and are chased.

Excitable animals that seeks unification

While many questions remain, we have managed to discuss the basic points thus far so from here, we shall conclude the argument by including a few points from an evolutionary perspective.

To begin with, the idea of independent individuals forming a group entails contradiction. In addition, we learnt that chimpanzee males aspire for dominance rather than equality. The aspiration for dominance necessitates the formation of a group. In fact, it is the absolute precondition in order for "dominance" to come into existence. That is why the chimpanzee males need to form groups and gather females as the object of advantage to facilitate dominance. This requires self-inhibition and the active embodiment of the equality principle. By releasing the inhibition accumulated in the process in the form of violence towards enemies, the males' groups become combat groups. The group that connotes the contradictive nature of aspiring for dominance while maintaining equality is thus stabilized and members will not be able to drop out due to its violent structure. If the formative process of chimpanzee society were to be modeled, it would look like this.

Having said that, even if the aspiration for dominance is present, equal and independent peers would not gather without particular reasons. We have to think of the factors that lead them to gather, perhaps some sort of outside pressure or an advantage that would be gained by group formation (see Uchibori's work in Chapter Two of this book). Itani (1987b) claims that a relationship based on the equality principle is one "that guarantees coexistence" by "promoting a mutual approach". Now, what factors would create a relationship which "promotes a mutual approach" among equal peers?

One factor that brings together peers who may not yet have aspirations for coexistence is the risk posed by large carnivores (Tsukahara 1993). When mammals are forced into a dangerous and inescapable situation, individuals of the same species often congregate and confront the danger as a "community". At this point, the other individuals cease to be competitors for food and members of the opposite sex. The confrontation of the "community" needs no intention. All that is required is for the individuals of a single species to simultaneously share both the fear of the carnivore and the excitement of counterattack. This would be a phenomenon where the simultaneity of behavior and emotion of multiple individuals separates "allies" and "enemies". This is the prototype of communitas equality. In such a scene, even when the danger has passed, if one animal becomes tense, the tension will be transmitted to their "allies" and amplified in an instant, while the act of being side by side has the effect of encouraging one another, providing an "unintentional mutual approach". The tension and excitement evident in the other individuals and the image of the foreign enemies further promotes the contiguity of a "mutual approach" and enhances the excitement. It should be re-emphasized that this occurs in the context of the fear and excitement of counterattack.

Now, we can see that the social structure of chimpanzees is surprisingly similar to the above "prototype". If the large carnivore is replaced with an enemy group, it becomes identical to the hostile inter-unit-group confrontation we have observed with the chimpanzees. The emergence of combat groups unified through communitas equality engendered by the state of fear and excitement of counterattack must also be an old pattern of behavior stemming from this prototype.

Fear is also evident in the offensive excitement of *Pan* males, indicating that it is an application of this old form of excitement. As the chimpanzee male becomes excited immediately before a charging display, his hair becomes raised and he sways his upper body. In the case of the bonobo male, this takes the form of a slight rising of his hair and a swaying of his body. It is common in all mammals for the hair to rise in situations of both fear and anger, but it becomes most intense when the fear and aggression is combined (Darwin 1921). The act of swaying the body also indicates anxiety. A male bonobo's intense display up a tree is accompanied by a grin, or the screeching of the inferior individual. The same can be said about the chimpanzee males' fights over dominance and attacks.[15] In other words, both their display of bravado and the actual attack are conducted in a state of excitement mixed with fear. They have to overcome their fear when venturing into such modes of display and attack (Goodall 1994).

This is a state of excitement embedded with fear and the need for support, accounting for why the displayer accepts the other individuals who run up to him with a pant-grunt, even if he is still in a state of excitement. It is likely that those who race up to him also react to such elements. Thus, we can argue that excitement seems to be basically a collective, centripetally psycho-physical state instantly shared by the group, and it seems that the individual charging display in itself entails a sort of inter-individual factor, in other words, an element to promote interaction.

The sharing and maintenance of this fear and aggressive excitement is an effect of primitive war. Everyone in the raiding group unites to become allies. This psychology of excitement might originate in the fear and aggressive group excitement that arises when facing carnivores, which was made continuous by the invention of an "enemy". This is a powerful mechanism that allows equal peers, struggling in aspirations for power, to rally while maintaining their equality. In other words, the existence of an enemy is what stabilizes the independent individuals as a group. This is what the evidence from the chimpanzees points to.

What about humans, then? The tremble of excitement we experience when we face our enemies can be said to be a form of aggressive excitement infused with fear. The concept of courage entails "to suppress fear and act" and we feel a thrilling pleasure when we do so. The excitement the African pastoralists felt when departing for dangerous raiding, as detailed in Kawai's work in Chapter Seven of this book, was exactly the same. The pleasure is not in the fight itself but in the continuation of the life-threatening aggressive state of excitement in a group. Various chronicles of war from around the world tell us that such situations often provide opportunities for the formation of eternally trusted alliances. Such pleasure calls for war, is systemized in sports and fascinates people. We live in the same terrain as chimpanzees.

Conversely, the collective excitement of the bonobos does not seem to involve "enemies". The shout of excitement attracts not only distant individuals and parties but also those from other unit-groups into an intense swirl of excitement. Males may confront males from other groups but never fight intensely. There may be exchanges of young females, but considering that this would occur even when groups are not in contact, it seems that the only reason for groups to stay together is to maintain the chain of excitement. This is a field that still needs exploring, but the discussion certainly indicates the possibility of an alternative method of embedding violence into the social system.

As we have seen, excitement is deeply related to the existence of equal peers. Durkheim (1915) and Clastres (2003) indicated in their writings that humans are creatures that use excitement and war in creating society, but they do not explain why or examine the nature of what the excitement actually is. If one wishes to venture to explore this field in any fashion, I believe it is necessary to begin by putting ourselves side by side with the primates who, as free and independent individuals, are living their lives with a sense of excitement.

12 Agency and Seduction: Against a Girardian Model of Society

Masakazu Tanaka

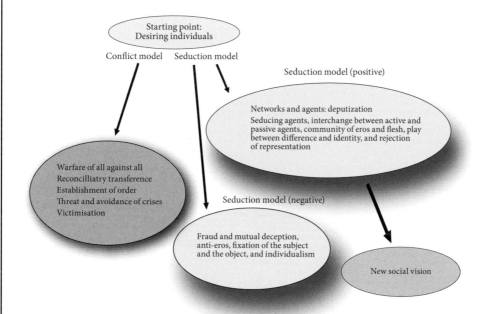

Conflict model versus seduction model

If the group theory of the "conflict model" is to connect individuals and collectivity by sublating the confrontation between "desiring individuals", the "seduction model" sees groups as a bundle of multilayered relationships formed through the infinite interchanges of activeness and passiveness by the above "desiring individuals". While accepting that the seduction model shares certain negative aspects with the conflict model, it is worth exploring as to whether we could create a new group theory that overcomes the problems inherent in the conflict model.

Introduction

We, as humans, need to cooperate with others and must come to terms with the world while we walk our paths in life. In doing so, how is it possible to do away with our personal desires and the interpersonal competition, conflict, violence and endless retaliation they give rise to? Thomas Hobbes was a thinker from the seventeenth century who tried to answer these questions. He wrote in *Leviathan* (first published in 1651) as follows:

> And therefore if any two men desire the same thing, which neverthelesse they cannot both enjoy, they become enemies; and in the way to their End, (which is principally their owne conservation, and some times their delectation only,) endeavor to destroy, or subdue one an other. (Hobbes 1914: 63)

> Hereby it is manifest, that during the time men live without a common Power to keep them all in awe, they are in that condition which is called Warre; and such a warre, as is of every man, against every man. For Warre, consisteth not in Battell onely, or the act of fighting; but in tract of time, wherein the Will to contend by Battell is sufficiently known: and therefore the notion of *Time*, is to be considered in the nature of Warre; as it is in the nature of Whether. (Hobbes 1914: 64)

As we can see from the above quotes, Hobbes had assumed a state of nature, a state of "warre, as is of every man, against every man", and discussed the need for "a common Power to keep them all in awe", specifically the need for an authority founded on a contract entered into by individuals according to their own free will. Here, we shall use the term "conflict model" to refer to the mindset that assumes a state of war or chaos in the primitive state—regardless of whether it really happened in that state or not—in order to discusses the mechanism of maintaining social order. René Girard, Hitoshi Imamura and Pierre Clastres have contributed to the development of this model in recent times. The purpose of this chapter is to critically review the conflict model and present a new alternate vision of society and group. As has already become evident, the conflict model has served a significant role in the realm of social thought. A number of the chapters in this book have been influenced by the works of Imamura, but here we will dare to critically review the conflict model. This chapter may present more of a social theory than a group theory. However, considering that social theory (or social philosophy) forms the foundation of group theory and that group and social theories are often in a parallel relationship, I believe that the discussion on society in this chapter will naturally contribute to group theory.

Conflict model: Girard, Imamura and Clastres

Girard explains violence in the state of nature as a series of "acquisitive mimesis" (1987: 20).[1] This is based on the principle that humans imitate desire and want what others are wanting. As a consequence, confrontation would arise between two individuals who both want the same thing. (See Hobbes' above quote, "if any two men desire the same thing".) The confrontation would then be followed by the imitation of violence and the repetition of retaliation. This is how reciprocal violence originated.

According to Girard, in order to avoid the critical infestation of violence, reciprocal violence among those in hostile relationships had to be unanimously directed towards a single individual. He claims that by doing so, inter-violence could be converted to collective violence and through this, social order could be restored. Girard (1987: 26) says that in the process of concentrating violence towards a single person, the "conflictual mimesis" comes into play in place of the acquisitive mimesis. At this stage, the target of desire is no longer present. What remains is simply the target of (conflictual) violence.

> Unification (of a community) occurs when fragmentation reaches an extreme, in other words, when the discord of imitation intensifies to an extent that retaliation is repeated endlessly and the community itself begins to head for disintegration. Individual to individual confrontation is suddenly followed by a one to all confrontation.
>
> A unity (of the community) that emerges from the moment when division is most intense, when the community enacts its dissolutions in the mimetic crisis and its abandonment to the endless cycle of vengeance. But suddenly the opposition of everyone against everyone else is replaced by the opposition of all against one. …there is now the simplicity of a single conflict: the entire community on one side, and on the other, the victim. …the community find itself unified once more at the expense of a victim… (Girard 1987: 24)

Religious taboo would prevent the imitation of interception that had created reciprocal violence. Then sacrifice is a repetition of the past collective violence, that is, "the opposition of all against one", which succeeded in restoring order, suppresses the endless repetition of violence and the chain of retaliation disrupting social order, and prevents the society from degenerating into the critical state (Girard 1977: 14; 1987: 28). It should further be noted that the victim would be believed to be "*sacred*" as the savior who had rescued the community from disruption. Girard called this "reconciliatory transference", in contrast to "aggressive transference" or victimization.

Sacrifice would eventually be replaced by the legal system, but in order to develop such system, a powerful authority (state) was required to support it. In short, that is to say that in the stateless society, order was maintained by sacrifice whereas in our current state society, it is maintained by law. This is the social evolutionary path that Girard has presented.

Imamura also tried to theorize the emergence of social order using the concept of "the third element issue" or "exclusion of the third element". From this point, we would like to focus on "the third element" issue using Imamura's *Bōryoku no ontorogī* (The ontology of violence) (1982) as our text.

Here, Imamura reinterpreted the story of Asdiwal analyzed by Claude Lévi-Strauss and identified two phenomena related to the emergence of social order: reciprocity and the third element issue. He regarded the first phenomenon, reciprocity, as that "developed with the axis of violence" or a "mutuality of violence". Here, reciprocity rules, in that violence cannot be exercised without suffering it. The social relationship is "constantly cursed with violence and conflict" (H. Imamura 1982: 30). Imamura then goes on to state, referring to Jean-Paul Sartre, as follows:

> Sartre described others as 'hell', but indeed this reciprocity is a hell-like reciprocity of violence and conflict. Whether physical or emotional, the mutual relationship between the subject and the object is obsessed by this kind of violence and cannot escape the hellish vicious cycle of violence and conflict. The *target activity by the subject* is sure to have an effect of *objectification*, similar to that of the head of Medusa. Here, mutual internal communication is fundamentally not possible, leaving no option but to keep responding with extreme reciprocation of *objectification* and *domination*. (H. Imamura 1982: 28, emphasis original)

According to Imamura, if two individuals wish to avoid a violent reciprocal relationship and maintain a harmonious one, the third element needs to be excluded (H. Imamura 1982: 29, 232). The third element assumed here is the sacrificial scapegoat.

The violence seen in the violent reciprocal relationship or the exclusion of the third element is not only the central issue in the history of European thought but is also a phenomenon universal to humankind. In fact, Imamura claims that the exclusion of the third element is none other than what Girard refers to as "reconciliatory transference" (H. Imamura 1982: 23–67).

The third element does not just take on a negative value. It may first be denied and excluded, but then undergoes a merger (or inversion) and is transformed into a positive and sacred being. This corresponds to Girard's "reconciliatory transference", introduced above. In other words, the third element can be a chief, king, god, fetish or even currency. The society that is (assumed to be) built by excluding the third element can be said to be one that is headed by something.[2]

Finally, we shall explore the discussion by French anthropologist, Pierre Clastres. In his book *Archaeology of Violence* (2003), first published in French in 1977, Clastres captures the uncivilized society as one where individuals are in a state of war against each other, as Hobbes has explained. However, he does not simply interpret them as being barbaric or aggressive. On the contrary, he claims that the endless war is preventing the uncivilized society from transforming into a centralized nation and is guaranteeing the diversity of said society. In other words, while Clastres acknowledges our view of the uncivilized society, he criticizes it for what it brings and why. According to Clastres, the uncivilized society is not one that seeks "a common Power to keep them all in awe" which Hobbes describes as the emergence of the first order in human history, but is in fact a society that opposes such authority. The society manages to sustain diversity and maintain its internal solidarity and order by continuing the war (Clastres 2003: 112). This is the originality of Clastres' discussion. However, Clastres made a key mistake in that, while Hobbes had assumed an individual-to-individual conflict, Clastres shifted this discussion by assuming an inter-group conflict. Consequently, his theory cannot explain how groups can unite at the stage when hostile groups do not yet exist. In this regard, his theory remains untenable, whether it is ethnographically correct or not.[3] Furthermore, it can be said that Clastres is presenting an imperfect version—a conflict model of groups rather than of society—that distorts Hobbes' conflict model and veers off the track from the discussion by Girard and Imamura.

The above conflict models by Girard and Imamura differ from that of Clastres in that they are attempting to answer the question from the individuals' perspective, the question of how reciprocal individual confrontation can be sublated (*aufheben*) to connect individuals to the society (collectivity).[4] The true value of the conflict model lies in the dialectic that "violence" would actually dissolve the confrontation and create a collective world. In this sense, the conflict model does not assume social order as *a priori*. It is a more dynamic model that entails the instability that could throw the society into crisis at any time.

Now, what formed the core of the conflict model was the desiring individual.[5] In this chapter, we will dare to further this view of the individual as it would not be productive to advance the discussion with a totally different take on this point. We believe we can conduct a more critical review by resting on the same premise, even if only hypothetically.

Network and relatedness: Conceptualization of interpersonal relationships

As with the conflict model, we can recognize criticism for the harmonious and static view of society in the backdrop of the network and relatedness concepts we are going to discuss here. These concepts also share another aspect of the conflict model in that they all assume a social model with individuals at its core. However, compared with the society and group that the conflict model assumed, this social model is milder and more versatile as it is not exposed to the threat of violence nor requires solidarity based on unanimous violence.[6] In fact, this is the very social feature that we should learn from network theory.

John A. Barnes, who had conducted field surveys of fishing villages in Norway, was the first to introduce the network concept to the fields of anthropology and sociology. He defined network as follows:

> Each person is, as it were, in touch with a number of other people, some of whom are directly in touch with each other and some of whom are not... I find it convenient to talk of a social field of this kind as a network. The image I have is of a set of points some of which are joined by lines. The points of the image are people, or sometimes groups, and the lines indicate which people interact with each other. (Barnes 1954: 43)

This article was published in 1954. At about the same time, another study on family and kinship in East London was published, whose analysis also focused on "network" (Bott 1955). By the 1960s, the network concept and its analysis methods were developed through studies in South African cities and industrial societies.[7]

Arnold L. Epstein, a distinguished member of the Manchester School of British Social Anthropology, who conducted fieldwork in the Copper Belt of Africa, summarized the network concept in three points:

1. "it is always egocentric"
2. "it is always personal"
3. "it is a series of links in a chain of personal interaction". (Epstein 1969: 109–112)

Some of the studies on network theory also entail theoretical aspects. Jeremy Boissevain advocated network theory as a criticism to the structural-functionalistic social view established in the 1940s and emphasized its individual perspective.

> Empirically, informants seemed to ask themselves the questions: 'What is the best for me and my family?' 'From what possibility will I derive the greatest benefit?' 'How much can I get away with?' as often as they did the typical structuralist questions: 'What is expected of me in this situation?' 'What is best for my group?' ...In short, it has become clear to me, as it has to others, that the static, structural-functional model of society does not work at the level at which real people interact. (Boissevain 1974: 4–5)

As is clear from this quote, network theory can be positioned as a criticism against group-centered social theory. Network is a concept that sheds light on personal and social relationships, in other words, human interactions that cannot be captured from the structural-functionalistic perspective that views humanity as a bundle of roles (rights and responsibilities). The relationships assumed in network theory are face-to-face dyadic relationships that are almost equal, borderless and indefinitely extending.[8]

Network theory was eventually succeeded by and developed in the fields of urban sociology, social network theory, graph theory, surveys and statistics. The individuals assumed here are the so-called *homo economicus*. As such, the characteristics of the individuals were never questioned. The relationships were quantified and evolved in a statistical direction. Therefore, I wish to emphasize that the current status of network theory does not necessarily overlap with the interests of this chapter. In fact, I believe that the significance of the early stage of network theory was actually succeeded by the concept of "relatedness", which I will now explain.

By the 1990s, a new concept to represent interpersonal relationships was advocated by Jannet Carsten (1995), which was "relatedness". According to Carsten, relatedness was a concept proposed to replace kinship in order to relativize the biological nature that this concept evoked in Europe. By doing so, a broader relationship that includes kinship comes into the scope of research. It also allows various "family relationships" created by today's reproductive technology to be researched from a cross-cultural perspective (Carsten 2000). However, what we find more significant for this chapter is that here the realities of the relationship (personal interactions), which are often overlooked in the network concept, are studied. This specifically refers to material exchange, sharing, activities such as labor and physical interactions between individuals.

> Here I focus strictly on notions about substance and the way it is acquired through feeding. My intent is to show how bodily substance is not something with which Malays are simply born and that remains forever unchanged, and to show how it gradually accrues and changes throughout life, as persons participate in relationships. (Carsten 1995: 225)

In other words, Carsten is trying to understand human relationships—including those that are considered to be biologically predetermined, such as mother-child relationships—not as a given but as something that is nurtured in daily practice, a kind of formation process. Even a group that appears to be static can be interpreted as a result of relations formulated through the daily activities of its members.

The concept of relatedness that Carsten proposed is further developed and amended in published papers and collected works,[9] but I do not think it is necessary to go into details in this chapter. Here, we would like to consider the relationship between the individual and the network and then move on to discuss individual interactions.

Agent and network

If network or relatedness signifies individual ties or relationships, one has to question as to who the individuals being connected are. Bearing such question in mind, I would like to introduce the concept of "agent". As already indicated, I adopt the view of individuals as the desiring subjects from the conflict model. Therefore, the agents introduced here would naturally be individuals who desire. However, before continuing the discussion, I would first like to explain the relationship between agent and network.

This chapter does not depend on a sociological paradigm that would discuss "agent" and "structure" in a confrontational scenario (M. Tanaka 2006). Agent is often used almost synonymously with "actor". What can be recognized here is the intention to create a new view of individuals, one that would replace that of those who are passively portrayed as members of the structure, society or group. The term "agent" can also be interpreted with a political intention that advocates that victims who are discriminated against and suffering violence because of their structural position should not simply be seen as helpless and passive beings, but rather as those who valiantly resist in an arduous daily struggle. However, no matter which view is taken, if the concept of agent is used simply to refer to an active being or one who initiates action, the modernistic view of individuals would be preserved in the concept of agent. I have no intention of denying the notion of the active subjects, but believe that the pitfall caused by only acknowledging the activeness in the concept should be avoided.

One of the oldest meanings of agent is deputy. Agent often refers to "a person who acts for somebody or deputizes" as in real estate agent, travel agent or CIA operative. These agents work for other persons or agents. Agency is the ability to deputize others and create a network. In a broader sense it is an ability for communication because the practice of deputization is inter-convertible in the relationships of a network. In other words, an agent is someone who deputizes and is deputized by another agent at the same time and is always closely connected to others. To deputize someone does not necessarily mean becoming submissive. An agent is a being who indicates a network of others who may become the object of deputization. In other words, "my existence" is embedded in relationships with others. However, it should be noted that this does not mean there is a one-sided master-servant relationship, as the relationship is inter-convertible. Even if the deputization is an individual one-on-one relationship, it creates an interactive and inter-negotiable space—a bundle of multilayered relationships. Agency indicates the existence of such a space, and is the ability to create such a space and to further transform it. Though it may sound paradoxical, the perspective that focuses on the agent does not necessarily value individual over structure. Instead, agent is a concept that makes others aware of a new network and the relatedness of which it is a part.

In social science, the question of structure or individual has been inextricably linked to the choice between looking at individuals either actively (i.e. individual-centered) or passively (i.e. society-centered). The concept of deputization was suggested here to indicate its convertibility. This was suggested mainly to overcome the dualism of structure versus individual. In the next section, I will discuss the nature of the mutual interaction between agents who are desiring subjects by referring to seduction. This is a rare concept that is not only closely related to desire but also has the ability to reverse activeness and passiveness.

Seduction: Mutual interaction between agents

What is the significance of focusing on the word "seduction" as a form of mutual interaction among agents?[10] There are two points to discuss in this regard: firstly, the "reversal of subject and object", and secondly, the importance of the body—the contingency and communicativeness it accompanies. Let us begin with the first point.

Seduction is embedded with the constant reversal of active subject and passive object (M. Tanaka 2009a). More than anything, seduction is the active approach of the seducer (subject). However, what the seducer is ultimately seeking is the activeness of the seduced. For example, French thinker Jean Baudrillard states that "...in a strategy of seduction one draws the other into one's area of weakness" (1990: 83). The role of seducer is to make the others become active.

However, seduction is not completed by an act of provocation on the part of the desiring side and a response from the desired side. In fact, by repeating such transaction, the relationship will proceed from the reversal of subject and object to the "inter-conversion of self and other" and eventually to a "fusion of self and other". A successful seduction does not leave a trace of seduction. Only when the attempt fails does the search for the responsible seducer begin.

It is evident that seduction is an extremely physical behavior. It is not exceptional for someone to seduce or be seduced against their will or without the intention of doing so. In fact, seduction is an act of contingency that occurs with the body as a medium.

Verbal maneuvering is one form of seduction, but even then we would like to focus not on the words but the voice—as an extension of the body. Seduction is a mutual interaction conducted by the body and not the head. Seduction betrays reason, and sometimes the body becomes uncontrollable. It threatens our autonomy, independence and the consistency that we strive to maintain in our daily lives. That is why we feel uncomfortable when we are being seduced. The discomfort is indeed a sign of the dissimilation of our daily life and natural attitude.

Seduction is an act that entices others into the world of Eros in that it is one which entails (the sign of) fusion between self and other that can be recognized in a good quality sexual act. The body in the field of Eros is a "communicative body" (Frank 1991) evident in performance such as dance and the act of care-giving.[11] This body awakens the sense of unity or bonding which connects the self not only with others but also with animals and nature in general. The world of Eros lurks in the very place where concepts criticized by modern rationalism, such as the reversal of subject and object, inter-conversion, contingency and the fusion of self and others, are concentrated. What leads to this location is seduction based on desire, the desiring body and the communicative body that responds to it.

That is not to say that everything is positive. Seduction is accompanied by risk. For example, the seducers' intention may be to simply satisfy their own sexual desire or to control others by providing sexual pleasure. Sexual intercourse can at times be simply an act of unilateral invasion. In this case, the body is simply an instrument to control (the body of) the other. The seduction model overlaps with the conflict model in this regard.

While we take into account such negative elements, the reciprocity of seduction in the model proposed here sits at the opposite end to the reciprocity of violence assumed in the conflict model. One of the reasons is that, though they both assume individuals and agents to be desiring beings, the seducer's desire is not directed at the target of the other's desire but is in fact, directed at each other.[12] There is no "acquisitive mimesis" that creates violence. In addition, the recognition of others in the seduction model does not consider "hell", as described by Sartre, as an inevitable situation. The gaze of the seducer does not cause the objectification of others, equivalent to death. (See the quote by Imamura above.)[13] On the contrary, the seducer is enticing the target to "to look at me". Seduction is a communication skill that does not exclude others, and such skill is referred to as "agency". What is required here is not how to act actively (in relation to others)—how to win the fight and dominate—but how to act passively in the mutual interaction of seduction. What the seduction model is calling for is *the power of being passive* or *the active passiveness.*

Having said that, one may counter-argue that while the relationship of desire may be dyadic, it may simply be ignoring the desire of a third party,[14] or the "acquisitive mimesis" assumed by Girard may still be applicable to the desire for things. I would like to note the following two points in response to this counterargument. Firstly, desire and the act of seduction are not closed in a dyadic relationship but are open to a third party and secondly, the interaction of seduction also emerges between humans and things.[15]

The community of flesh: With Merleau-Ponty

We have thus far discussed the seduction model as an alternative to the conflict model, but I would finally like to explore further possibilities of the seduction model by quickly examining the argument of Maurice Merleau-Ponty, whom Imamura has criticized. However, the reason I chose to discuss Merleau-Ponty is not just that cited above but because I can identify in his argument a more fundamental pattern of thinking. The discussion on the agent, representation and seduction so far was intended as a criticism against the conflict model. However, it is not necessarily suggested as an alternative to the primordial state of inter-violence assumed in the conflict model. We can only imagine what it was really like in the primordial days. If anything, the seduction model is a criticism of Imamura's view of society, to which the conflict model leads us, and an attempt to construct a new view of society. However, Merleau-Ponty's argument does not end there. He leads us into a fundamentally deeper realm. This is why Imamura could not ignore his work and had to critique him and why I chose to discuss him here, because I believe the conflict model can only be critically assessed in this phase.

Imamura's question is as follows:

> Merleau-Ponty believes that both self and other can simultaneously emerge as elements of intra-corporeality by a fundamental opening of perception, or a fundamental ecstasy, but could intra-corporeality really be formed so easily and smoothly? The perceptive opening is often called 'dehiscence' and the image of dehiscence is the aesthetic image of a bud blossoming into flower. Is not Merleau-Ponty's intra-corporeality achieved through 'perceptive opening' and 'ecstasy' portrayed with the beauty and smoothness of a flower? [...] While on one hand we have the horrendous phenomenon of social violence in our history and daily life, is it not a little bit odd that it has a completely irrelevant image of paradise, an image of dehiscence as its foundation? Should not there be at the fundamental level, a fundamental violence that connects with the various forms of violence in our daily social lives? (H. Imamura 1982: 216–217)

Here, I would like to mention the fact that Imamura seems to think that violence in our daily lives and "fundamental violence" are continuous, but I will not go into discussions of its validity here. Instead, I wish to point out that the seduction model presented in this chapter is similar in its standpoint to that of Merleau-Ponty, and thus against the position of Imamura.

Merleau-Ponty thoroughly criticized solipsism and focused on human perception and somesthesis in order to clarify the structure of the relationship between self and other, and that of self and the world from a phenomenological perspective. He wrote in his posthumously published *The Visible and the Invisible* on the sensation of the right and left hands touching each other (Merleau-Ponty

1968: 141).[16] When our hands are in contact with each other, we understand that one hand is touching the other by shifting our consciousness. However, it does not mean that both hands touch or are touched at the same time. The sensation is invertible (touch/touched) but not exactly the same. This applies not just with our hands that are parts of our body, but also when we touch things and people. As we touch, we are touched at the same time. This happens because both are "flesh", in other words, existence that can be sensed. Our flesh is a place where both subject and object are formed (Merleau-Ponty 1968: 147).

Merleau-Ponty calls this movement of formation "fission" or "segregation", and this is the process that Imamura questioned as overly idyllic.[17] However, rather than getting caught in discussing Imamura's critique, we instead follow Merleau-Ponty's thinking here.

In the case of the right hand and the left hand, the reversibility seems to be guaranteed as both parts are of the same body. However, there is no such great body, or "huge animal" in the case of (the right hand of) the self and (the right hand of) the other, when we are in contact with other things or people. However, Merleau-Ponty argues as follows:

(Organisms') landscapes interweave, their actions and their passions fit together exactly: his is possible as soon as… we rather understand it as the return of the visible upon itself, a carnal adherence of the sentient to the sensed and of the sensed to the sentient. (Merleau-Ponty 1968: 142)

What is open to us, therefore, with the reversibility of the visible and the tangible, is… at least an intercorporeal being, a presumptive domain of the visible and the tangible, which extends further than the things I touch and see at present. (Merleau-Ponty 1968: 142–143)

Through reversibility, we open our body up to others and the world as "flesh" that can be felt.

Like the natural man, we situate ourselves in ourselves *and* in the things, in ourselves *and* in the other, at the point where, by a sort of *chiasm*, we become the others and we become world. (Merleau-Ponty 1968: 160)

I don't believe I am the only one who is tempted to read this discussion of reversibility in an erotic context. In fact, I think it may actually be quite to the point.[18]

Furthermore, by understanding "carnal adherence", as in the quote above, as the "community of flesh" (what can be felt) (Matsuba 2008) which encapsulates the gap between self and others (the right hand and the left hand can never feel the same sensation at the same time (Kaganoi 2009: 281)) but denies representations or figures, it seems possible to connect Merleau-Ponty's philosophical thinking to socio-scientific discussions on society and groups.[19] In contrast to the conflict model that claims "first there is violence" and "first there is the war of all against all", Merleau-Ponty advocates "first there is the world of flesh which can be felt". The seduction model advocated in this chapter, though indirectly, supports Merleau-Ponty's discussion.

Conclusion

The conflict model's story of the emergence and expansion of individual inter-violence and the formation of a society (or group) through collective violence is an attractive one. However, is it not possible to assume a different sort of social origin, in other words, a situation that does not assume confrontation, violence or competition? Based on such awareness, what I suggested in this chapter, through our discussions on the concepts of network and relatedness, was the multilayered bundle of relationships (network) created by the infinite development of dyadic relationships. We cannot assume a clear and static border in the social world of network and as a result, it lacks or rejects the transcendental existence or an overarching representative symbol. One may argue that the connection of borderless people cannot be called a group or society, but the author disagrees. The conventional social view that assumes a border to separate self and other or the confrontation between society/group and individual is what needs to be questioned. We then moved on to discuss that agents and network are connected without being locked into a fixed master-servant relationship due to the deputizing nature of the agents.

In this chapter, the agents are the desiring subjects, as with the conflict model. However, what occurs between agents is not a violent confrontation but a physical and communal relationship based on seduction. As agents with desire seduce each other, a network of the seducer and the seduced is formed. What is important in this relationship bundle is how sensitive one can be to the delicate positioning of activeness and passiveness. In our modern society where self-reliance and independence are accentuated, the "activeness" of acting passively is what needs to be strongly emphasized. We focused on the concept of relatedness. What forms the core of relatedness is the mother-child relationship formed through acts such as breast-feeding. Here, we can see the delicate relationship between activeness and passiveness in the passiveness of the child (though the act of sucking cannot be said to be passive) or the passiveness of giving birth (though giving birth may be active behavior, the act is not actively intended as such). Returning to Merleau-Ponty, his philosophy of inter-corporeality teaches us the importance of this sort of relationship with others and the world. It may well be this passiveness that Imamura rejected.

In this chapter, we sketched out the seduction model as a replacement for the conflict model. It does not mean that there is no occasion for violence in this model. Such instances should obviously not be ignored,[20] and the seduction model is certainly not a rosy picture of human society. As the word "seduction" suggests, it entails various kinds of risks. Seduction too is one of the fundamentals in which we see the possibility of overcoming the limits of our modern society controlled by law and modern individuals with internalized discipline = training. From our beginning in the primordial state to the radical social thought of "flesh", we may have come too far while discussing the conflict model and the seduction model. However, what I believe is important is that the role of anthropology, which not only looks at the past but covers the future evolution of human society, should not stop at empirical analysis but go on to present a more fundamental image of humanity and human society for the future. This chapter is written as an essay with such intention.

13 Human Groups at the Zero-Level: An Exploration of the Meaning, Field and Structure of Relations at the Level of Group Extinction

Takeo Funabiki

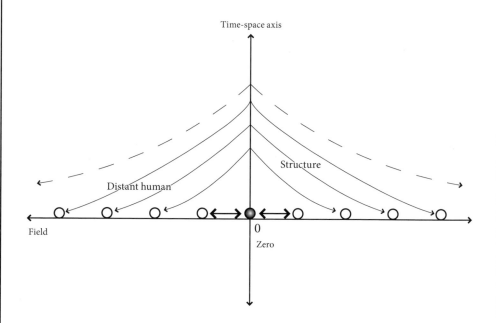

While human beings can form relations of mutual understanding with other human beings in spatial proximity, they must relate to those at a distance at the same time[1] in order to maintain the meaning and function that underpins life as members of a society. When we call the infinite expanse in which humans can gain mutual understanding a "field" and the aggregation of relations beyond the "height" of space and time punctuating it a "structure", we can posit that a human being cannot be a member of a group without such a structure with "height".

Introduction

The purpose of this paper is to answer what and how questions in relation to the fundamental level of human group formation. I first examine what happens when humans form a group and what makes it possible for them to do so. I consider the mechanism by which the group expands its scope to acquire expanse and height. Based on limit situations and hypothetical examples, I then discuss the situation when the level at which a group with such expanse and height (i.e., society) is formed is lowered, and question just how low can this level fall before the group ceases to exist. Using what transpires from the above consideration, I attempt to estimate the incipient point, or the zero-level, at which the group emerges.

However, any attempt to trace back to the embryonic point of group-formation in order to understand the nature of such a phenomenon is laden with methodological risks from the outset because, as Rousseau noted in his *Discourse on the Origin of Inequality*, investigations into the origin must be considered not "as historical truths, but only as mere conditional and hypothetical reasoning" (1978 [1755]: 200). In other words, the contents of almost all of the concepts we use in our discourse on groups such as "language" and "gesture" discussed below are guaranteed on the assumption of the ongoing existence of "society" as a human group or multiphase complex. We are unable to use these concepts to discuss groups once we have traced the levels of group formation to the point of extinction, and, we have not reached the incipient point as long as the group "exists". Therefore, the hypothesis we are trying to propose cannot escape logical tautology, at least during the dialectical process. Keeping this caveat in mind, we will develop our argument on the assumption that the analysis of the level at which the human group becomes infinitely close to zero and through which the human group "vaporizes" is in itself a worthy endeavor. Its value should be determinable at the end on the basis of the argument presented below.

The field and structure of human relations

Human beings are able to have relations of mutual understanding with others they meet and face. The possibility of understanding arises because humans can comprehend sounds made by others (language) and body movements (meaningful gestures) through hearing, sight and touch.

The question of how language and gesture were created and conveyed, i.e., the creation of human culture, is outside of the scope of this discussion. Here, we shall begin our exploration from the fact that humans are born into this world as beings that are capable of learning such language and gesture. The subject of this discussion is assumed to be human beings who are equipped with a shared sensibility regarding language and gesture. Similarly, the "relations" discussed here—while we recognize that culture and nature are inextricably linked in all human phenomena—refers to those in the cultural and social sphere rather than the natural biological "relations".

As mentioned earlier, human beings are able to form relationships of mutual understanding with others in spatial proximity to their language and gesture because of their ability to "comprehend sounds made by others (language) and body movements (meaningful gestures) through hearing, sight and touch". Here, they identify one another as human. Let us call the position in which such a relationship arises a "scene", and the infinite expanse of this position holding the possibilities of such relations a "field". Conversely, it is possible to say that the field is carved into scenes by the act of mutual understanding between humans.

Theoretically, however, humans must change their position in order to form relations with "distant" humans above and beyond the "close" humans on the scene with whom they have relations. We humans exist as horizontally-aligned individuals on the field, and when we relate with a certain close human, we are unable to simultaneously form relations with another human through our ability to "comprehend sounds made by others (language) and body movements (meaningful gestures) through hearing, sight and touch" if this other human is "overlapped and hidden", at a distance physically and metaphysically. Here, "overlapped and hidden" signifies physical reasons behind the difficulty in forming relations such as visual overlapping and concealment, and also social reasons such as existing relations preventing new ones.

One person (◐) may be able to share a scene with two or more others (○) in a radial fashion, but this does not expand infinitely. Empirically, one can only relate with a small number of people at once, say five or six or a dozen or so at most (Figure 13.1a). In order to confront and form relations with a person horizontally aligned and hidden behind another currently on the scene, one must move away from the person in the existing relation and shift one's position to the distant by "steering around" that person, physically and metaphysically (Figure 13.1b).

The new relation acquired by shifting one's position in this way, however, hides the previous relation in space and "degrades" it into a thing of the past in time. The previous relation becomes extinct spatially and temporally. If we assume that humans desire relations because they add value to the meaning and function

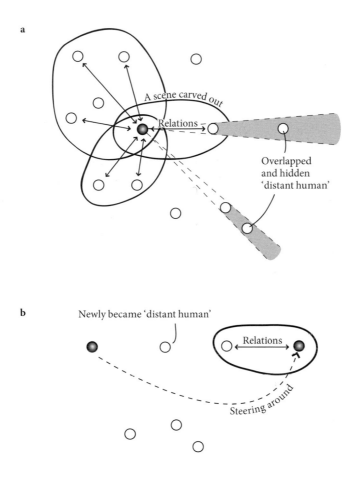

Figure 13.1 The field and structure of human relations

underpinning life, the degradation and extinction of previous relations due to the formation of new ones can be considered a "loss" incidental to the formation of the new relations. In this case, it is always possible that a human can seek to avoid the loss as they can gradually increase the value of living if they can preserve that of each relation he or she has formed in terms of meaning or function.

Now, when people wish not only to form relations but also to preserve the value provided by formed relations, how would they go about achieving this? If we think about it again, in order to prevent the hiding or extinguishing of existing relations by newly formed ones, we must be able to simultaneously relate to persons facing us and those behind them.[1] The main question is "how one can form a relation with a person positioned behind the scene on which one has an existing relation with another person", on a field where one can only form relations in a horizontal direction.

When we try to solve this question using figures 13.1a and 13.1b, we come to realize that it is possible once "height" to mediate relations is introduced into the field. In other words, by "jumping over" horizontally aligned people in space one can connect with others behind them, and by extending the time of the field in some way and "jumping over" the time on the field, one can connect relations into the past and future. We call the "aggregation of relations by accumulating positions in space-time" with "height" and "length" a "structure". The concept of "field and structure" contains a contrastive relationship in that "field" is infinitely expansive and non-temporal by nature whereas "structure" has a finite expanse and aims at an infinite length of time.

When we turn this concept into a model, the temporal and spatial axis has a direction perpendicular to the field and constructs a structure on the field. This structure can be explained from another perspective as a "cultural and social" structure that is stratified synchronically and diachronically. Thus, relations mediated by the time-space axis having a "length" and a "height" on the field can extend and carve out scenes to specific functional limits of the axis (Figure 13.2).

The question as to how the height facilitating such a structure comes into existence is also an evolutionary question of how human groups in which we presently live have gone beyond biological group composition, both qualitatively and quantitatively, and come to possess the scale and complexity of almost different dimensions. It can also be regarded as a question of human "institution". Although this discussion will not go that far, the discrete nature of the height and the structure will be referred to below.

Next, we head in the opposite direction from the analysis of height and structure, i.e., we descend to the situation in which one cannot even carve out a scene, to explore the zero-level of group formation.

The level of human group extinction

All existing sustained human groups have the height and structure mentioned in the previous section. While "height" varies in scale and complexity from one group to another, we will treat "sustained human groups" and "societies" synonymously in the following discussion.

A people called Mbotgote on Malakula Island, Vanuatu live in the central jungle region in dispersed dwelling groups or settlements, each consisting of several families[2] and having about 140 constituent members. They consciously maintain distance from each other in their daily living. For example, when one of them visits another settlement, the visitor uses a stick to carefully beat the bamboo fence of the pig pen on the outer perimeter of the settlement in the unique rhythm of the dance to which he is entitled in order to give the recipient time to prepare for the visit and let them know the identity of the visitor so that his visit is not treated as a "surprise attack".

My survey has revealed that when a whole group is faced with danger, they may go to the extent of disintegrating the dwelling group into individual families to live henceforth. Past examples of such dangers include disputes over property such as pigs, ritual rights and women (amongst men), epidemic diseases (presumably) spread by infection, a spate of earthquakes and so on. They disassemble and live at some distance in such cases. Members do not explain this behavior as a crisis avoidance measure for specific purposes, such as direct prevention of "infection" or avoidance of the risk of unexpected attacks from neighbors in the chaotic times after destructive earthquakes. It is possible to surmise that these times of crisis tend to reinforce the Mbotgote's belief that living away from each other attenuates relations and helps them avoid crises, even in times of peace as in the above example of visitation.

What we find here is a crisis avoidance measure in everyday life by intentionally lowering the "height" and the structure of the Mbotgote as a human group. However, this avoidance can be considered a measure to "maintain" the structure of the group by controlling the height of its axis rather than razing the group structure closer to the zero-level.

Now we shall examine two examples and try to simulate a descent to the zero-level. They are Shōhei Ōoka's novel *Nobi* (Fires on the plain) (1994 [1951]), and poet Yoshirō Ishihara's four short essays published in a collection of his poems and prose (2005).[3] They represent war-related experiences in the mountains of the Philippines and at a prison camp in Siberia, and depict the extreme situations in which humans are placed by war. We will attempt to find a clue to investigating the zero-level of group existence through these accounts.

The protagonist in *Nobi* is Tamura, who wanders through fields and mountains alone after being discharged from his platoon as a wounded soldier and forsaken by the field hospital. He was given a meager amount of food and a hand grenade, implying that he was expected to commit suicide. At this point, Tamura can be regarded as someone who is isolated from Japanese society after losing the mediating structure in the form of the Japanese army.

In the first half of the novel, the protagonist acts independently. American soldiers in the enemy forces are either stationed far away, flying or driving and never "confront" Tamura. The only people he meets are remnants of defeated Japanese troops wandering about just like him or Filipino farmers. However, Tamura tries to avoid meeting both soldiers and farmers because the former are his competitors in the search for food and the latter are potential enemies. Although he has consciously resigned himself to death, he continues to search for food. He sometimes shares his food with Japanese stragglers and sometimes steals from Filipinos.

Tamura feels the "freedom" of a singular man released from the rules of the army (group), but he has acquired this freedom because he regards himself as ontologically dead already, as suggested by his self-reflection while roaming through a forest: "A strange sense went by; the sense that *I would never walk on this road even though I was walking on this road for the first time in my life* (Ōoka 1994 [1951]: 9)". Monologues expressing his inner thoughts such as this signify not only that he is ontologically dead but also that this "presently-alive" man is "already dead" socially.

Yet, Tamura is also oscillating. As he says self-mockingly, he sometimes desires food, i.e., "life", and yearns for "encounters" as if he is living on the "field" above the zero-level. He has already sunken below the zero-level conceptually but, in a way, he can be considered alive ontologically while he still has a concept of his own death. As he is inside his monologues and without the language to carve out a scene with some other person, however, he is oscillating above and below the zero-level.

In the second half of the novel, the protagonist forms a group in a strange equilibrium with two Japanese stragglers he acquainted. Their only connection was a desire for survival. They cut up the corpses of their countrymen, who are "people at a distance" to them, and eat their flesh, then in the end go on to kill and eat other stragglers. This cannibalism does not necessarily mean that Tamura and the other two former soldiers are positioned below the zero-level. They still carve out their scene and remain as a group. However, their group disappears below the zero-level the moment the three men, who have supposedly met and understood each other, not only see "people at a distance" as mere food but also begin to kill and eat one another. This occurs even though Tamura shifts himself to a position of "angel" beyond field and structure in his thought at that moment when the other two men try to kill each other: "I felt anger. […] If I could vomit and get angry, I would no longer be a human. I would be an angel. I would have to express anger on behalf of God" (1994 [1951]: 123). Rather, it happens because of his intentional illusion of being an angel.

Yoshirō Ishihara depicts harsh living conditions at a prison camp in his essay "Aru 'kyōsei' no keiken kara" (From an experience of coexistence) (2005). It is the kind of experience that cannot possibly be "summarized" here. When two prisoners share their meager food from one bowl and sleep under one small blanket, it appears to be the human condition at the extreme or on the zero-level. Yet, as he writes that "[...] we learned over a very long period of time that this *distrust is the very thing that acts as a strong bond to enable people to coexist*" (2005: 94), there is still a group or a carved out scene. His sustained hatred toward his companion based on a sense of distrust founded the sense of togetherness. As he writes that "solitariness is never a state of singularity; solitariness is inescapably impregnated in togetherness" (2005: 95), language was not yet lost and he was not cut off from society as a "singular person" on the slippery slope from isolation, just like Tamura was in the first half of *Nobi*. Ishihara's worst situation comes next.

Ishihara was sentenced to twenty-five years in prison as a "war criminal" for committing "anti-Soviet acts" under the criminal law of the Russian Republic.[4] There was no reason for Ishihara to be tried under this law as he was not a Soviet or Russian citizen—he had nothing to do with its "height" or "structure"—but he was condemned as an enemy of "humankind". He was actually interned as one of the hostages by the USSR for negotiations with Japan. A Soviet interrogator suggested to one of Ishihara's colleagues who had met the same fate that he should become a collaborator. The prisoner responded to the interrogator with words of extreme pessimism: "If you were human, I wouldn't be human; if I was human, you wouldn't be human" (2005: 111). Here, a fundamental condition for creating a scene, recognizing one another as "human", is lacking between prisoner and interrogator. Here, both this prisoner and Ishihara had sunk below the zero-level as singular persons.

Ishihara loses a "longing for home" during his incarceration. It is replaced by a "bitterness towards home". "What I feared most was 'to be forgotten'. My home country, its new regime and its nation [...] one day forget about us completely. [...] Please remember once a day that I am here. If I die here, mark the place of my death clearly on a map. I continued to cry out from this maddeningly strong desire (70,000 Japanese prisoners died with their places of death unconfirmed)" (2005: 130). Then, it is followed by "obliviousness to home". "I was losing the view to see the Japanese as my 'compatriots'" (2005: 132). And finally, he becomes isolated from everything. "In the midst of a dense *taiga* forest, [...] I often felt envious of stands of trees voicelessly blocking the path ahead of me. They lived unconcerned with such matters as the self that is forgetting and being forgotten. I was envious of them for this reason and not in the least because they were "free"" (2005: 132–133). He not only becomes a singular person who has lost a companion to meet and confront but also becomes fully committed to being a singular person by firmly denying the possibility of having such a companion.

Let us look at some commonalities in these works of Ōoka and Ishihara.

When and how do Tamura's action in *Nobi*—or its author Shōhei Ōoka's thought expressed in it—and Yoshirō Ishihara's thought fall below the zero-level of group formation? It is when they lose the field of relations, or in other words, when they lose all companions to confront even in idea as well as in reality. To Tamura, it was when he accepted himself as a straggler who had been renounced by the army with only death in his hands. To Ishihara, it was when he rejected his existence as someone who was forgetting and being forgotten by his home country. It was when the solitary became the singular.

It is important for our investigation of the human group to confirm the difference between solitariness and singularity here. Solitariness occurs when one loses counterparts to meet or avoids meeting them while still on the field. Solitude may be an important mental state to the individual, but it has no significant meaning in a fundamental consideration of society or group. We must not think that "solitary" individuals are the "starting points" of the formation of a society or group. Individuals are "solitary" because of their sociality and collectivity, and therefore it is the "normal condition" of individuals in society and groups. Conversely, a singular person is someone who has already forfeited encounters with others. One exists ontologically but not socially. It means that only death awaits and that one's death will have no social implications because nobody else will acknowledge it.[5]

It is interesting that the notion of "freedom" appears at crucial points in both stories. Breaking away downward below the "field" or upward out of the structure on the field (army/Japan) leads to the freedom of the "singular person". Like "solitariness", however, this, a sort of absolute "freedom", falls outside of the scope of our present discussion.

The field of meaning and the height and the structure

It has become clear from our discussion thus far that sinking below the zero-level of group formation means losing the field and the possibility of carving out a scene. I wrote above that "human beings are able to have relations of mutual understanding with other human beings they meet and confront". I also wrote that such a relationship of mutual understanding "refers to the relations in the cultural and social sphere rather than the natural biological 'relations'". Now, what are the essential qualities of a relation and a scene thus formed? This question is equivalent to asking what "understanding" is. In order to answer this question, let us return to my earlier remark that "we assume that humans desire relations because they add value to the meaning and function for living".

I located "meaning and function" in parallel based on my own the hypothesis that "we recognize that culture and nature are inextricably linked in all human phenomena". If we construct a tentative model of this hypothesis, we can consider the meaning and function to be "the meaning as the purpose" of living culturally and socially and "the function as the condition" for living naturally and biologically in correspondence to the inextricably linked culture and nature. If we accept this hypothesis, we must stress the question of "meaning" in our investigation into group-formation by humans as cultural and social beings, in addition to biological group-formation. It is impossible to discuss this question thoroughly in this thesis, but it is clear at least from our analysis of *Nobi* and Ishihara's essay that a discourse on human groups is only possible when "meaning" could exist on a dimension different from the function. The straggler and the prisoner fell below the zero-level of group formation, the field of relations with others, because they lost the "meaning" to carve out a scene, not because they lost the function as the condition for living.

However, we have been receiving this argument, the importance of the "meaning" in human relations, like a cliché. Some would say that we have been investigating human groups in order to move the argument forward to the next stage. For this reason, if we are to reexamine this argument, it will be to redefine the old conclusion, not reveal a new discovery.

We do not treat "understanding" in a model by which two people's thoughts agree and reach one. Such a model would conclude that understanding is to exchange symbols reflecting a reality and to equalize the objects "meant" by these symbols as the same thing. All organisms have been performing this act since the beginning of their existence. This becomes clear, if we replace the equalization of the meaning with the reproduction of it, from the fact that reproduction is the fundamental condition for the survival of a group of organisms both at the genetic and biobehavioral levels. The "understanding" we should be thinking of in relation to human groups, however, is in terms of a model by which, from the two essentially different thoughts of two individuals, one meaning emerges between them. The meaning is newly created at each instance, not reproduced.

We should also notice that, if the newly created meaning is not a reproduction but rather something that is completely severed in the process of temporal degradation from the context of the aggregate meanings created in the past, it will only continue to suffer a "loss" as it appears, degrades and disappears on the field, as argued earlier. We know that, unlike other living organisms, human groups have not only found the meaning and function of relations on the field and carved out scenes, but also continued to "preserve the value of each relation (i.e. culture) he or she has formed in terms of meaning or function" within the "structure" against temporal and spatial degradation. From this point of view, it is perhaps more accurate to say that the "meaning" to human groups is not only newly created each time but also newly recreated.

Let us consider the "height" and the "structure" created for the purpose of preserving such value using the two aforementioned works as clues once again. The question of the height and structure in Ōoka's and Ishihara's compositions shows us a glimpse of their peculiar quality around the "fires on the plain" and the "cross" in Ōoka's work and the "Russian Republic and its criminal law" in that of Ishihara.

The protagonist in *Nobi* spots "fires on the plain" in the distance on several occasions while on the run in the mountains. Smoke plumes rising high from burning objects suggest the presence of humans below, probably Filipinos. They could be simply from fires burning the stalks and leaves of harvested corn or they could be signal flares for military purposes. In any case, they frighten him as a sign of the presence of the "enemy". At another time, he spots a cross over a seaside forest. He surmises it must be standing on the roof of a church. This also points to the presence of the fearsome enemy in the form of Filipinos and their settlements. In these occasions, the fires on the plain and the cross on the roof with a certain physical height symbolize the "height" supporting the structure in the form of Filipinos (villagers) or US troops.

While the fires on the plain signify the enemy such as Filipinos or US troops to Tamura, the cross also signifies, besides fearsome Filipinos, "Christianity" for which he had special feelings when he was young, although he now feels it is a "fallacy" (Ōoka 1994 [1951]: 44). He does not belong to the group of Filipinos but as a disenchanted Christian he does not consider himself as a total outsider to the structure. He enters the village he assumes to be deserted in which the church is situated where he is confronted by a woman and shoots her to death. We may interpret this event as follows based on the present argument, if we disregard its literary profundity. This episode suggests that the "field" has an infinite expanse and essentially no demarcation lines drawn to separate one person from another, whereas inside the structure formed by its height, there is a line that continuously excludes other persons which separates inside from outside. This line differs from the temporary lines that carve out scenes shown in Figure 13.1a. When the protagonist confronted the woman, the structure of "Christians" did not operate but the structures of "Filipinos" and "the Japanese army" came into play, mutually excluding these two people and depriving them of the chance to form a relation and create a scene.

For Ishihara, "Japan" was the structure that he specifically perceived as his homeland. And "Japan" was supposed to be that which would protect him against "the Soviet". However, he was abducted to the structure of the Soviet/Russian Republic and tried as its "state prisoner" for a crime against the state under its criminal law. Even stranger, the charge was replaced by one of "the enemy of peace and democracy" in the final indictment (Ishihara 2005: 124). When the hypothetical structure of "humanity", from which no one could escape, was imposed on him, Ishihara was turned into a "war criminal". At this point, his enemy was not only the abominable Soviet Russia that had condemned him to his suffering; the whole of humanity became his enemy. Theoretically, humanity regarded him as an enemy, a criminal. The only way to argue against this outrageous pretext was to repeat his colleague's words: "If you were human, I wouldn't be human; if I was human, you wouldn't be human" (Ishihara 2005: 111).

In terms of the model in our present discussion, it is perhaps obvious that "humanity" represents the "field" rather than a structure. If we consider the aforementioned episode in *Nobi*, height and structure have discreteness. In other words, a structure created through the medium of height has limits and a line drawn by the structure on the field, an outer border, separates inside and outside. Conversely, the field has no discreteness. The field is theoretically infinite and always open. If we consider the field to be a place to create the meaning of living for humanity as a whole, then the structure of "humanity" appears to rise up—but there is a trap. To see "humanity" as one structure is perhaps to limit it. It holds by assuming "the whole humanity" and excluding the "non-human", but in doing so, we tend to walk into the trap of erroneously seeing "those who are human as non-human".

I stated at the beginning that this thesis cannot completely escape circular reasoning. I am writing it despite this methodological flaw, because I anticipated that this exercise could identify which elements of the circular argument should be emphasized or disregarded. I believe this purpose has been achieved to a certain degree. If I can advance the argument at this pace from now on, I will address the question of "institution" and "law" in human groups, i.e., what is in operation inside these discrete heights and structures and how do they relate to one another, in my subsequent thesis. In it, "language" and "gesture" which function to carve out scenes on the field should become the multiphase equivalents of "ideology/ myth" and "ritual" respectively to function to create the heights of the structures (e.g. Nakajima 2007).

Conclusion
From "Groups" to "Institutions": Summary and Prospects

Introduction
Kaori Kawai

What have we been able to say about human society and sociality in the course of this book? Where do we go from here? By way of conclusion, let us summarize the key arguments and evaluate the outcomes.

This book is based on works from primatology and anthropology, but, as mentioned in the Introduction, we have not taken the approach of compiling arguments along the lines of the evolutionary lineages of both of these camps. It has been collated in accordance with the objective of attempting a rare example of collaboration between researchers on non-human primates and human beings, and is divided into four parts around separate themes. The reader will be left to judge as to whether or not we have succeeded in this attempt. There is, however, a need for a separate summary of the meaning and significance that this volume brings to the fields of primatology and anthropology. Accordingly, in the Conclusion, we have decided to synthesize the forms and particular characteristics of "groups" that have clearly emerged from human society in evolutionary perspectives under each of the three academic fields of primatology, ecological anthropology and socio-cultural anthropology. The respective contributors in charge of each of these fields are Suehisa Kuroda, Kōji Kitamura and Motomitsu Uchibori, and below these authors summarize the contributions from their academic fields.

Revealing the non-structured nature of social groups: From the field of primatology
Suehisa Kuroda

In this book, the works by Itō (Article One), Adachi (Chapter One), Nakagawa (Chapter Four) and Kuroda (Chapter Eleven) belong to the field of primatology. Although humankind are one species of primates, they construct such a complicated multitude of societies that they cannot even be compared to those of the chimpanzee (*Pan*) genus, to which humans are most closely related. There are unquestionably many people who doubt the ability of primatology to make any assertions regarding the phenomena of human groups. Moreover, the research from the field of primatology presented here focuses on non-human primates (abbreviated to primates hereafter) as their subject. Accordingly, what I would like to do is to proceed to an overall introduction of the field, keeping in mind the question of the contribution of discussions from the field of primatology to our understanding of human groups.

Writings from the field of primatology share not only an understanding of the structured concentration of primate individuals; they contain a viewpoint that perceives both decentralized and centralized conditions, including the non-structured aspect of groups, as group phenomena. This was stimulated by the work of Junichirō Itani, who paid attention to both intra- and inter-species gatherings that could not be seen as organized, and who advocated the importance of the theory of primate sociology comprehending the non-structured aspect. However, what brought greater clarity to the problem consciousness of the authors was a joint discussion with anthropologists who viewed both the non-structured gatherings and/or groups as usual. In this book, it is Adachi in particular who expands on and attempts to convey Itani's theory of anti-structure. Nakagawa carefully examines the dispersal and transfer of individuals from primate groups and revises Itani's theory of the evolution of social structure. Itō and Kuroda take the chimpanzee (*Pan*) genus as the object of their studies, designating their abilities to produce dispersal as autonomy and freedom from the group, and as the source of the non-structured nature of groups. In this volume, this perspective is shared by Uchibori's (Chapter Two) discussion on the human trend towards solitude and by the presentation on hunter-gatherers' group formation in Terashima's work (Chapter Eight). Let us turn our attention briefly to each of these writings.

Itō's work—with its objective of depicting female chimpanzees, who have come to be seen as "unsociable", as having an autonomous existence that is socially equal to that of males—is attempting to meet Itani's idea, to bring about a shift from "sociology of aggregation" towards a "sociology of dispersion". Itō attempts to represent groups from chimpanzees' eyes as a method of clarifying "group phenomena that come to nothing the moment one uses the word "group" to name them". In fact, she does her utmost to avoid conclusive representations of chimpanzee socialities, but shows observations that reveal the accumulation and inconclusiveness of the individual relationships of chimpanzees and describes "groups" from this basis. This may, on the surface of things, appear to be a simple matter, but there are problems of linguistic assumptions in terms of representing observations and—although Itō does not mention these—they raise questions about methods for responding to Donna Haraway's (1989) criticisms of primatology and are profoundly significant.

Adachi's work relies on Itani in viewing mixed species associations as non-structured groups, and gives expression to their formation and maintenance in a state of being unorganized by connecting Itani's phrase "a slight interest" with her own term "a slight indifference". Furthermore, Adachi adapts Itani's theory of non-structure to Imanishi's theory of *specia* (species society), and from this combination emerges prospects for the possibility of constructing "a sociology backed by an actual experience of life, for living things that encompasses all animal groups". Imanishi's society of all living things and the natural philosophy seem to lack *dynamis*m, but Adachi is formulating a new natural philosophy in which she supplements this with the theory of non-structure. Itō and Adachi have both already largely moved beyond the confines of Itani's framework.

The group phenomena dealt with by both authors have something in common. This is the loose and free associations that are expressed using social philosopher Hitoshi Imamura's term "*social*" in contrast to "*society*" signifying structured association. Terashima (Chapter Eight), working in the field of ecological anthropology, also borrowed the term to characterize the flexible grouping of hunter-gatherer societies. The groups that come into being via what Itani and Adachi term "slight interest" are maintained by basing the rules of coexistence on free relations between different species of mixed groups and the order of arrival. This is just like the classification into *social* that existed before the arrival of structured *society*. Itō and Kuroda's descriptions show that the connections of female chimpanzees to others also possess the characteristics of *social*. Just as Adachi identifies the hearth-hold groups in Sugiyama's work in this book (Chapter Ten) as this type of *social*, these primatologists expand Imamura's *social* into a concept that describes the condition of the coexistence in a group of living things that also includes humans.

Nakagawa's work is an elaboration of Itani's hypothesis on the evolutionary process of primate societies, focusing on the phenomenon of the dispersal of individuals away from their natal groups and areas. This work entails indispensable information to any future considerations of early hominid or its social forms. Nakagawa points out that though Carel van Schaik's socio-ecological model, one of the mainstays of this argument, explains an evolutionary process of the gregariousness and cohesiveness of female animals, it is not enough to account for those of male animals. In response to this, Nakagawa—examining the genealogy of patrilineal tendency by focusing on the unstructured aspect of male dispersal from the natal group—attempts to align and unify this model and Itani's hypothesis along the three opposing axes of: aggregation/dispersal, matrilineal/patrilineal and ecological adaptation/genealogical inertia. Itani's idea that lineage and social structure correlate, and that social structure exerts a large influence on the individual's behavior, is original in the field of primatology. As a paradigm, however, we can roughly say that it fits within classical historicism. In response to this, ecology and evolutionary ecology have regarded individual animals as having been selected so as to maximize the reproductive success within the constraints of genetic and environmental conditions, and this has led to a view of individual animals as leading existences in which they behave rationally. In this sense, it is a utilitarian view. Nakagawa's work is an attempt to take this historicism and utilitarianism and integrate the former into a pivot, leading to a reevaluation of the importance of the legacy of evolution.

Kuroda's work identifies societies of the chimpanzee genus as those that produce equals, and argues that its groups are maintained by two types of unstructured phase: a zero and a maximum degree of excitement. He argues that the unstructured phase of zero excitement that is born of the solitary nature of individuals operates to guarantee liberty, and that, in the case of chimpanzees, the phase of maximum excitement, while simultaneously leading to a clarification of the boundaries of the group in response to enemies, sees the appearance of anti-structure (non-structure) in the form of a communitas type unification within the group. Because groups of the chimpanzee genus are groups of equals, which are prone to collapse, one of the aims of the argument that group excitement coevolved as a mechanism to support them is to explain the relationship between the formation of human groups and collective excitement.

The difficulty in contributing to understandings of human group phenomena from primatology arises from the fact that we only have "presently living" humans and primates from whom we can obtain concrete behavioral data. There are two methods for inducing both to interact. One of these uses the genealogical relationships of living primates and fossil data to reconstruct evolutionary processes, and by means of a process of then making assumptions about the social structure and behavior of early humans, considers these to have some sort of influence on the group phenomenon of living humans. This orthodox method is apparent in Nakagawa's work. However, because this is an operation in which one abstracts a so-called "prototype" of human society, when faced with objections from anthropologists who say that on the basis of the variety and flexibility of human social structures there is no "prototype", then it is difficult to mount an objection in return, and the extension of these results beyond the time of early hominids is dropped.

The other is the non-temporal (metachronic) method of letting them interact directly. In a broad sense, Itō, Adachi and Kuroda's works fall within this method. The non-structured primate groups on which Itō and Adachi focus their attention can be regarded as being of the same type as the loose and free human groups. In Kuroda's case, he deals with groups of the chimpanzee genus and human groups through collective excitement and equality, and suggests that human collective excitement, which has not been a subject of inquiry in anthropology, may be a trait displayed by groups of equals. This non-temporal method is one that in an extremely simple sense speaks to the same type of patterns between systems, and that, in principle, has no connection with the additional consideration of a time axis. In fact, Kinji Imanishi's adoption of the concepts of anthropology and sociology, in the form of "analogies", in primate sociology—along with Itani's theory of the principle of equality and his theory of anti-structure—are all strongly characterized by this non-temporal method.

However, to get general recognition of the efficacy of this method, we need to refine describing the overall image of the respective primate societies—including the choice of model—as social systems. The evolutionary ecological model used by Nakagawa, which underpins his arguments, is also a system model composed of ecological and social factors, but Nakagawa's argument is one that splits these two elements. When describing social phenomena, it should be easy to achieve consistency by painting an overall image depicting the main social factors. For example, if we can form an overall image of chimpanzee society that has the equality principle at its core and connections between principle factors, then, if we can find similar connections in human society, we may be able to make some valid proposals about human society from the perspective of chimpanzees. This method needs supposition that a newly emergent system in evolution does not annul the old system; it is compatible with the thinking in neuroscience that makes effective use of the old system while transforming it into a partial one. However, what must not be forgotten is the possibility that, although we refer to an overall image of chimpanzees, differing images can be created by region, individual groups or by researchers. There are also a number of problems, including the issue of how we are to evaluate it, that we ought to resolve in this model.

Combining these two models also increases their persuasive power. In the case of Itani's arguments about the equality principle, if we attempt to revise this to a question of how equals can construct groups, we understand that the anthropoid lineage is a coexistence built by equals: a series of group forms. This is what Nakagawa calls the succession of patrilineal social structure, and the issues concerning non-structure raised by Itō and Adachi also permeate this succession. For example, in the case of gorillas, unit groups are semi-patrilineal with females moving between groups. Fellow females construct a ranking based on the order of acceptance into the group, but there is practically no direct negotiation, it is just their offspring who play together. This is the same as the mixed species group form of coexistence. Moreover, female gorillas, just as is the case with female chimpanzees, are very independent. (And almost no researchers, including Dianne Fossey, had shown any interest in the "sociality" of female gorillas as was the case with female chimpanzees!) Human social groups are the extension of this series, and their traits of freedom, equality and a non-structured nature can be detected along the whole of the series. It is the non-structured nature of the *social* that supports society, which is structured by institutions amongst humans; and Kuroda asserts that the relationship between the *social* and *society* can be seen in the chimpanzee genus.

This viewpoint draws out the thinking that human society is not structured in the form of layer upon layer piled onto a historical evolutionary base by representational ability and culture, but that products of the past are also reinforced and put to use on the surface. The efficacy of this argument will be tested in the coming "Project on Institution".

"Drawing together activity" using "representation" as a guide: From the field of ecological anthropology
Kōji Kitamura

The works that can be categorized as belonging to the field of ecological anthropology are those of Kitamura (Chapter Three), Kawai (Chapter Seven), Terashima (Chapter Eight), Soga (Chapter Nine), Sugiyama (Chapter Ten) and Umezaki (Article Two). This section—following the model of the thinking in Kitamura's chapter which, with its focus on both non-human primates and humans, sought to understand both on the basis of their commonalities—takes on the task of surveying the whole of the state of the human group phenomenon on the human stage. This is carried out on the basis of a demarcation that corresponds to differences in livelihood—hunter-gatherer societies (Terashima), pastoral societies (Kawai and Soga) and agricultural societies (Sugiyama and Umezaki)—

and by contrasting these societies after having classified them according to their respective characteristics.

Kitamura's main arguments can be summarized in the following two points. Firstly, groups and the group phenomenon do not originate from the likes of a basic human nature which aspires to cooperation, or an animal instinct that attempts to create social unity; they are rather activities that are chosen after having confronted the necessity for compatibility, from the individual's point of view, between appropriate relationship building with the environment from the individual's point of view and appropriate relationship building with others. It is in this way that social unity in the activity of relationship building with the environment is organized. In this regard, it is argued that there are no essential differences between non-human primates and humans.

Secondly, in seeking the elements that give rise to the disparities that do exist between these two societies, Kitamura focuses on the issue of what provides a cue for choosing members of the social union. At the level of non-human primates, the differentiation of "peers/not peers" is brought about in response to the effect of the articulated concentration of the members of the species society that is created along with these forms of relationship building with the environment, and the reproduction of social union can be seen as proceeding on the basis of this classification of others. As a result, between the classification of group members and the composition of the group another relation come into existence—each with their own origins and consequences. In contrast to this, at the level of humans, these activities see the appearance of a way of organizing groups within a framework designated by a "representation (= decoupled representation)" that exists in isolation, and this becomes usual practice. Because, however, the "representation" in this case does not compel but simply justifies the choices of members of the groups, the phenomenon of groups in human society makes possible a more flexible response to the issues of the moment, and this leads to far more diverse developments based on relationships with the other fields of activity in each society.

The following is a response to this view and an attempt to locate each of the human group phenomena depicted by the respective works using a comparative approach. Let us begin by dealing with the first point in question, and attempt a summary of the characteristics of each of the divisions regarding the different livelihood styles. Terashima establishes as his premise, when thinking about the group phenomenon in hunter-gatherer societies, that "bands" are the groups that lead cooperative lives in the manner of hunter-gatherers. There have been numerous arguments surrounding the premise, based on a certain understanding of hunter-gatherer societies, that groups called "bands" exist. If we follow the line of reasoning that says that this group phenomenon is an activity that produces social unity and attempts to build an appropriate relationship with the environment now and then, however, then this standpoint is connected to the understanding of the state of hunter-gatherer society premised on the cooperation found in bands, as "groups of small numbers of people who live by moving around in keeping with the availability of food resources each season".

While Soga, on the one hand, recognizes that the group phenomenon in pastoral societies is one established via a reliance on cultural categories, he concerns himself with "temporary groups" that are to be differentiated from these. As an example of this, Soga focuses on the activity found in the drawing of water for livestock, which is indispensible to the pastoral livelihood, and also that of "frequently moving livestock" that produced the establishment of groups transcending the borders of ethnicity and created the "unity" identified as the ethnic category of "Hofte" by the colonial government. Kawai describes the case of the perpetrating group in livestock raiding, which is formed by transcending ethnic boundaries, also corresponds to this. We can understand all of these as belonging within the context of "activities attempting to forge appropriate relations with the external environment alongside creating social unity".

However, it is fundamentally feasible to also adopt this in cases where groups—as distinct from the examples given above—are formed with direct reference to the so-called cultural category designated by "representation". In either case, this can be seen as attempting to make some sort of response to the issues that directly confront them as they produce social unity. The coexistence of both the application of this kind of cultural category and the attempt to respond to the issue without it is most likely an indication of the feasibility of a more flexible response in answer to the questions that arise at this time. With regard to the contrast between Kawai's description of the formation of the livestock raiding groups that transcend the boundaries of ethnicity, and, in the case of the party being raided, the actualization of a choice based on a cultural category—the level of the ethnic group—as an unhesitating, instant response at the scene of the raid, we see that the raiding party represents group formation based on the ecological need to acquire livestock, whilst the party being raided is based on the sociological need to respond to an attack by an enemy. Further, we can see this as the immediate surfacing of the cultural category that makes possible the summary decision based on the "enemy/friend" distinction.

Compared to the above examples, in the case of agricultural societies, we must assume slightly more complex conditions. It is not, however, at all difficult to comprehend the situation in these societies. In the agricultural, as compared to the hunter-gatherer or pastoral livelihood, far more distinct connections are formed between people and land. In the case of the slash-and-burn cultivators who are the focus of Sugiyama's work, even people with long-term settlement cycles of fission-fusion accompanying in-built migration, moving as they say they do every ten to thirty years, show an incomparably clearer manifestation of settlement than do hunter-gatherers or pastoralists. Due to this, we can see as inevitable the formation of "residential groups" connected by the type of friendly relationships that regularly strive to maintain these friendly relations that allow for the coordination of the interests of adjacent parties in land use. Consequently, in these cases, even when there is no concrete opportunity for urging collective activities of relationship building with the external environment, the activities that aim to create "groups" linked by these friendly relations are inevitable.

It seems to be in the context of these sorts of settlements that we first encounter examples of "attempts to produce the group phenomenon in situations where we would not have thought it necessary to do so". Sugiyama expresses this as "villages of residential groups are created via people's persistent practice of mutually choosing to 'be together'". The same sort of thing occurs when "residential groups" are also temporarily organized in the case of hunter-gatherer and pastoralist societies, even though the maintenance of their unity is for considerably shorter periods of time, and we could even say that the differences are a question of degree. It is, however, certain that circumstances that encourage embracing illusions such as "it is precisely living in groups that have been constructed with people being tightly bound to one another that is the true nature of humanity" first came into being with the establishment of this sort of settled lifestyle. While, on the other hand, temporary groups are constructed in response to the needs of the moment also in agricultural societies, this means that agricultural societies do not decisively differ from all others. Sugiyama gives the term "hearth-hold group" to one example of temporary groups that constitute the important aspects of group living and analyzes its significance. The groups, described by Umezaki as being formed in an effort to assemble supporters at the time of "disputes", also come about as the result of the very aim of constructing a "group" in order to fight another group. This is the type of activity in which these endeavors become the impetus for people then settling within that group. This, however, bears the indisputable character of having originally been social unity manufactured from the need to carry out a dispute with another group. In other words, along with the development of circumstances in which an agricultural society becomes settled, activity that is aimed at constructing groups itself becomes normal. Even in the group phenomenon found in these sorts of agricultural societies, however, we should see in their most fundamental aspects activities attempting to build social unity and to make relations with the environment at that time appropriate.

Let us now consider the second point. In the case of hunter-gatherer societies, there is a unique incompleteness regarding the use of "representation" that renders the differences with primate societies other than humans conspicuous, and this makes it tricky for there to be any simple understanding. According to Terashima, kinship bonds that extend bilaterally, with oneself and one's family at their core, provide the basis for the choices made by members of the living group known as a band. Expressed in anthropological terms, the classification of others that occurs in choices of group members at that time are thus "patrilineal kin", "matrilineal kin" and "in-laws". Consequently, if group members are selected on the basis of this type of criteria, these are all selections using "representation", which means that identification on the basis of third party criteria is possible.

Thought of in this way, shared characteristics with societies living other types of livelihoods can also be observed in hunter-gatherer societies, but in order to come to this conclusion, one reservation needs, in a sense, to be made. This is that even if they select group members on the basis of identifying them as third party criteria, the decision as to which criterion to use is entrusted to the parties concerned, and this is, ultimately, a selection using relative classification—along the lines of "related people"—that is unlimited and that originates in the perspectives of the parties concerned. This can even be associated with the recursive decision-making process found at the non-human primate level: it is precisely those people who have previously formed groups who become partners in forming a new group. If we focus on the differences with the example of non-human primates in order to locate this way of doing things, we get the following result. Namely, while forming groups in an attempt to deal with the issues of the moment, they are engaging in a way of doing things that concretely produces new, hitherto non-existent, "bonds" using the network expansion of "bonds" that originate in familial cooperation as a guide. This can be thought of as a way of selecting and renewing group members by referring to the third party criterion, which one should, in a sense, share with one's peers: the "distant group" that can be seen as the maximum extent of the spread of each individual's network.

Regarding pastoral societies, let us take the example of East African pastoral peoples as discussed in this volume. In these societies, it is extremely common to have a situation in which cultural categories established at various levels of patrilineal kin groups or cultural categories of ethnicity are referenced in the formation of the social unity that accompanies relationship building with the external environment. However, the way of deciding one's action about relationship building with some object by referring to these types of cultural categories is typical in cases involving relationship building with the external environment, but it is not so in the case of relationship building with others. Even where, as a consequence of certain cultural categories, the composition of conflict between groups is established in terms of "inside" and "outside", relations with those of "outside" do not always automatically reflect this sort of rivalry. It should not be forgotten that when an individual person is faced with issues, even though relations operate within these sorts of cultural categories, the scope for individual choices in relations is strictly maintained.

As mentioned previously, in the case of agricultural societies, activities emerge which are aimed at the very production of a group, but at that time, in order to make the formation of the group more of a certainty, particular social categories are referenced. In addition to that group's framework being decided on the basis of these, the group becomes a certainty as a structure. Conversely, however, linked to attempts to make the framework of the group a certainty, there will inevitably be attempts to maintain affiliative relationships of the sort that can coordinate the mutual interests between each member and repeated mutual affirmations of the "will" to perform these kinds of relations. Because of these, the composition of the group clearly has a character of having been chosen on the basis of the will of the respective individuals within it, and the endpoint of this situation can be thought of as leading to an orientation towards the present in which the composition of the group phenomenon itself becomes something that is entrusted to the free choice of individuals.

"Representation" that indicates the criterion for the choices of the members of a group in the "collectivity of activity" in which the parties concerned produce social unity alongside adjusting relationship building with the external environment can be considered to be performing the function of something that can be used to justify the choices of these members. This is not, however, the type of function that enforces choices originally based on it. Also in relation to this point, as something that can be differentiated from the domain of the collectivity of activity, we can assume a domain of control and integration of people's action guided by the workings of "representations" that justify the choices via the requirement that everyone should make certain specific choices.

Accordingly, note that when choices that have been justified in this way are not made this leads to a certain kind of "non-order", and the problem becomes one of stressing the realization of the value of avoiding this "non-order" and organizing order, as well as highlighting the achievement of the value of restoring order by making choices justified in terms of what ought to be done in order to resolve the "non-order" that has been brought about for some reason or another. In this sense, we can see an "institution" that operates to control and integrate people's activity as the next theme for investigation in efforts to probe the development of human society that displays this course of choosing actions guided by "representation".

Towards social assemblage based on the ability to symbolize and institutions: From the domain of socio-cultural anthropology
Motomitsu Uchibori

What kind of new viewpoints, ways of thinking and approaches have the arguments that took place during the collaborative research seminars (Groups Research Project), upon which this book is based, brought to research on groups within the framework of socio-cultural anthropology? The results of the seminars are not limited to the collected papers that it produced, but should also be recognized, both implicitly and explicitly, in the works published in this volume. Accordingly, as part of a review of the contents of these works, I would now like to undertake once again an appraisal of the degree of the novelty of this seminar, also encompassing elements that have not necessarily left a clear mark on these contents. This is because I feel that it is precisely by undertaking this task that we can provide a clear direction as to what the next stage of research will (and ought to) be.

If we look for the distant origins of present day socio-cultural anthropology in the various evolutionary trends of nineteenth century thought, then there are echoes that I would be tempted to call almost atavistic in the expression of "human society in evolutionary perspectives". The theoretical topics that have come up repeatedly in socio-cultural anthropology from the late twentieth century to the present can, I think, be said to have practically all been included in Marcus and Fischer's book *Anthropology as Cultural Critique* (1986), which may be called a "retrospective and prospective" sort of masterpiece with full inventory of worthy topics. When considering that the entirety of Marcus and Fischer's book has been written with a total disregard for the biological viewpoint regarding the current state of humankind, we must conclude that the idea of evolution, as at least a "new" possibility, has been thoroughly excluded, at least in Japan, from mainstream socio-cultural anthropology from this period onwards. The socio-cultural anthropologists who participated in the Groups Research Project were surrounded by an environment permeated by this inadequate level of thinking; to say nothing of the horizon of theoretical arguments on evolution, which—as one might expect from their distance from the possibility of empiricity that was guaranteed by social/cultural anthropology's participation in fieldwork—rendered the thinking environment even poorer.

If using the word "evolution" within the framework of "social evolution" such as that of neo-evolutionism, which played a big part for a period in American cultural anthropology, there might have been a certain degree of reduction of this potential discomfort, although a feeling of being a bit out of date remains. However, we should (re)confirm that although the arguments related to "social evolution" use the word "evolution", the time scale involved therein has generally not been that long. The goal of "social evolution" theory was clearly different from the way of framing questions such as: what kind of traits does man—as a member of the primates—and his sociality, compared to other types, possess, and how are these manifested as forms of group characteristics? Its goal was to take the various social organizational forms that occurred, either historically or ethnographically, and arrange them along a (pseudo-) time scale by following a certain type of logic or, in more theoretical cases, to construct this sort of logic. When seen in terms of the biological sense of a scale of evolution, these various forms of social organization discussed therein become nothing more than contemporaneous phenomena because they are evidently mere variants manifested among animate beings belonging to the same species called humankind or, more precisely, Anatomically Modern Humans. This may be a shift away from the normal usage of the word "history" by the criterion of the presence or absence of written documents, but what becomes problematic in "social evolution" is phenomena or events in the historical time that includes prehistory rather than in the evolutionary time scale. As the ambition of the Groups Research Project was to argue the whole matter in the length of evolutionary time range, socio-cultural anthropologists faced the question as to the extent to which their arguments would develop away from a historical time scale. That was a challenging and certainly difficult subject to resolve.

My own reflection is that it was, in fact, quite difficult to respond to this challenge with a neat discussion on the problems of time scale. Let us simply list the central themes (theses) related to the logical moments and opportunities for forming groups—excluding the ethnographic details—from all the works in this book from the field of socio-cultural anthropology. We can summarize these as: "steps towards an assembly from solitariness and individuality as the reverse aspect of groups" (Chapter Two, Uchibori); "sharing so as to acquire bonds that overcome mutual distrust" (Chapter Five, Ohmura); "violence that consolidates groups and violence that tears groups apart" (Chapter Six, Tokoro); "from the strife model to the allurement model as opportunities for group formation" (Chapter Twelve, Tanaka); "human relationships and group structure as seen from the standpoint of the dissolution of groups" (Chapter Thirteen, Funabiki); and the short argument on "groups and the process of ethnic identity formation" (Article Three, Shiino). What we detect here as a common point is the logic of group formation in various layers of human societies rather than the phenomenon of evolution that contains successive stages along a definite time scale. The questions asked here are: through what kinds of moments and opportunities do the entities that we regard as groups, with their variety of scales and contents, form and manifest themselves in human societies? On this point, we can say that these questions and arguments have an orientation that may fundamentally transcend time.

This orientation towards a metachronic way of thinking has not been at all popular in the recent mode of socio-cultural anthropology, which expressly emphasizes the historical formation of each particular society and culture. Amongst the above-mentioned works, Shiino deals with history in terms of how it weighs on the present, but the others, whilst not ignoring it, do not venture to push it to the fore. The significance of this in terms of what the development of the work shows is that socio-cultural anthropology will tend towards—or inevitably take on—a metachronic directionality when it builds in the ultra *longue dureé* of evolution, and tries to surpass the short-to-medium term timeframe of history. The metachronic stance that has thus been taken differs fundamentally in nuance from the anti-historical—or history-skeptical—perspective that was almost the main trend amongst British social anthropologists for a while. This is not skepticism dependent on the degree of provability; it is a theoretical stance that is seen as necessary in order to talk about the "present" in a time scale of evolution that actively transcends historical time. I imagine that Hitoshi Imamura, as a social philosopher, most likely felt no sense of unease regarding this point in particular during the seminar, and in fact took a pivotal part in the discussion; and also because of this, both primatologists and anthropologists were able to accept the various points in his arguments without any discrepancies. As Nishii says respectfully and affectionately in this volume, we have learnt a great deal from our communicative coexistence with him: it goes without saying that one such thing is the very concept of sociality, but this learning extends to human relationships such as violence and trade and further to the issue of ideation—representation and imagination.

An additional point that I would make in order to avoid any misunderstanding is that discussing groups in this metachronic fashion does not imply a lack of discussion of the creative *dynamis*m of the formation and dissolution of individual groups. On the contrary, it is from the very midst of this creative *dynamis*m that the special characteristics of groups explicitly emerge, and the fact that primates, including humans, have acquired the many characteristic moments and opportunities of this *dynamis*m in a species-specific manner through each one's own process of evolution is most likely recognized and accepted by the majority of advocates who, for all that, discuss it not in historical time, but on the, as it were, metachronic ground level of logic. Tokoro talks about attraction to and repulsion from violence in the formation of human groups, Tanaka and Ohmura explore the spaces where there are incentives that draw people together and unite them; and even the revival of the ancient *aporia* of group and individual that results from Uchibori's examination of the transition from the solitary phase to the group phase, and from Funabiki's analysis of the opposite phase of transformation, all are born of the desire to locate the creative *dynamis*m of groups.

I feel that it was precisely because of this orientation that researchers in socio-cultural anthropology were also somehow able to put forward the same arguments as primatologists. It is also certain that the existence of researchers in ecological anthropology, whom Kitamura summarizes in this chapter, acted as a mediating factor. Since all of the ecological anthropologists in this volume contribute studies that could be also called socio-cultural anthropology, this could be regarded as a fair degree of mediation. During the seminar series of the Groups Research Project, however, most developed arguments that faithfully reflected their scholarly lineage as Junichirō Itani's pupils.

However, if we are seriously to grasp the evolutionary time scale in a substantial sense, then the firm crossing of swords by primatologists and anthropologists— whether socio-cultural or ecological or, furthermore, physical/morphological—it cannot be denied that the optimal moment exists only in the earliest stages of hominization. All and the best that we were able to hold in the discussion during the seminar was to bring the circumstances of these real sequential moments into relief; we were able to make them stand out boldly only by shading the area surrounding them. As a method of discussing the evolution of forms of groups, which do not leave any physical remains, from two present day existences (that of humans and non-human primates), this is the only means for the present. And yet, in order not to leave these matters to a simple table discussion, it is necessary to always keep in mind the existence of real substantial time, which remains out of our reach despite all our efforts. It is likely that this very consciousness will then, as a matter of course, form the basis for discussions regarding "perspectives" when we talk about "evolutionary perspectives".

Conversely, we are compelled to say that the longest range of an evolutionary time scale that can most reliably be obtained by socio-cultural anthropologists is shorter by far when compared with the initial period of hominization. The second focal point that follows on from the initial period is the birth of the Homo genus from the Australopithecus genus, but this is quite beyond our capacities. After a long era further, we look to the period in which modern humans appeared as the main players, taking over the earlier Homo represented by Neanderthal man. The question as to what extent theories that stress this take-over as critical—those that strongly assert the so-called full-scale big bang of Homo sapiens—are valid, is what divides the arguments. However, socio-cultural anthropology should be able to speak more easily from this point onward. This is because leaving the stamp of the development of the symbolic potential and symbolic actions that have burst into blossom amongst the Anatomically Modern Humans, and locating the significance of these in light of all the various aspects of contemporary human lives, are adequately contained within the sphere of the defense of evidence in socio-cultural anthropology.

It is precisely here that we find the various social institutions that, in the midst of an imagination rich in symbols, are culturally diversified, and also here that contemporary anthropography and ethnography can argue using the same degree of empirical evidence as those of human natural history with its evolutionary time scale. The majority of the various studies in the anthropological contribution to this volume, including even those from ecological anthropology, point to the future of this thinking towards the human group's representational nature and the state of groups with their characteristic invisibility. It can be said that this orientation reveals the fact that human society is, in reality, an entity with its parts assembled by institutions. This is because while the workings of representation that makes those things that are invisible visible is certainly at play within the institutions, it is also at work constructing them. The task of positioning this argument correctly within the time scale of evolution is ahead of us, and within this task, extension, imbued with imagination, towards the future of evolutionary time may also become possible.

Epilogue
The Legacy of Hitoshi Imamura: The Macro lies in the Micro

Ryōko Nishii

A message from Professor Imamura

During my year-long stay in Thailand in 2005 I received a long email message from Professor Imamura. It outlined his thoughts about my draft introduction for a collection of research papers by a certain joint research project that had concluded the previous year and of which I was the head researcher. Together with his comments on the shortcomings of my clumsy argument, he gave me a homework assignment for the future. In short, we must take on the responsibility of demonstrating the extent of productive potential to which microanalysis focusing on the practical field of anthropology can bring to human knowledge. He continued as follows.

> Because 'Wohin geht es (Where to go)' cannot be articulated without a theory. Past theoreti-
> cal models, be they structuralism or Luhmann's systems theory, were very well formulated
> and able to expand human knowledge more than anthropologists had imagined. Due to
> their macroscopic nature, macro theories inevitably lack the micro domain and cannot
> always address human activity on an everyday level. I guess that is where the potential of
> microanalysis lies.

However, Professor Imamura warns that there are many hazards along the road. "The hazards will appear in the form of mental laxity, lack of conceptual thinking, deserting your theory and so on". These home truths were hard to hear then and still are now.

I believe Professor Imamura called the field of his specialty social philosophy. It is well known that he began introducing the ideas of thinkers such as Althusser and Baudrillard to Japan from the 1970s and led the country's contemporary thought from the 1980s. The program for the Hitoshi Imamura memorial symposium held in October 2007, several months after his passing, stated that he "squarely faced contemporary ideas such as Marxism and structuralism and enthusiastically attempted to construct a framework for new thinking beyond them in the field of the history of social philosophy and social thought".

When he sent me the aforementioned message, he was in the middle of writing *Marukusu nyūmon* (Introduction to Marx). He mentioned a methodology named micrology in his message, which he proposed in the book.

> Developing a technique to extract social structure from the micro is a better way of saying it. An individual, who is like a spec of dust, 'exists in reality' here and now. A microanalysis of this human is a detailed analysis based on the premise that social structure manifests itself in every action he performs. I name it 'micrology' in *Marukusu nyūmon* (published in May 2005). It means that the macro 'lies' in the detail.

He explains in *Marukusu nyūmon* that micrology is a methodology which is made possible by a mind that is sensitive enough to find and analyze minute differences, supporting Marx's thinking on the most profound level, and states: "When this sensitive mind combines with the analytical mind, the phenomenon to be observed and described will clearly reveal its action and true qualities by itself. [...] The crucial point is to sense various differences that are invisible and hidden and express them in language" (H. Imamura 2005b: 208).

I can say that this methodology also underpins Professor Imamura's thinking. Using this methodology, he contemplated his central themes of "labor" and "violence". Both involve a search for the roots or origins of the basic human existence. I shall venture to summarize his theoretical development using his own words below.

"Labor" in modern times brings to mind efficient instrumental acts based on economic rationalism. In this sense, labor is an abstract and mechanical activity measured according to time scales. By contrast, Imamura starts from the consideration of fundamental elements of labor and aims for its reorganization into a new form on the grounds that "labor was once aesthetic and play-like" and that "labor is the source of thought". This means that labor will become an activity that is aesthetic, play-like, rational and free from exploitation (H. Imamura 1988: 645–646).

In his consideration of violence, Professor Imamura developed his unique concept of "third term exclusion". He considers that violence is something that goes beyond the common image of "physical violence", something that arises from human existence itself and that is a peculiar force inherent in the relationship between people and between people and nature. That "there is" a person or a thing involves taking a specific place out of a pre-existing place. A violent cutting line is drawn in the incorporeal place. He calls the violence that enables the generation of the individual "original violence". The original relationship between individuals is of mutual exclusion or mutual hostility. There is no order. The mechanism to create order entails, in theory, the exclusion of a single individual. He calls this "third term exclusion". "The excluded third term" becomes the order itself and is consecrated. The force of social order formation is found in this process of order generation called "third term exclusion effect (= scapegoat effect)" (H. Imamura 1988: 565–566).

Prompted by the sensitivity to the minute differences in human existence he experienced on a daily basis and his discomfort with the preceding views, Professor Imamura developed his unique ideas into a macro-level theory that was applicable to all human beings. Perhaps this methodological thinking turned him towards anthropology, which appeared to be obsessed with micro-level phenomena more directly. From the latter half of the 1980s, he began to participate in joint research groups of mostly anthropologists at the invitation of Professor Shigeharu Tanabe of the National Museum of Ethnology (running from the 1980s to the 1990s). He stepped out of his study and traveled to Southeast Asia and China to gain firsthand experience of field research. Although he wasn't particularly confident about his own physical strength nor adaptability to a wide range of environmental conditions, he probably aspired to experience non-European societies personally through these joint studies with anthropologists, as he was already well acquainted with Europe and European thought through his studies in France and elsewhere. Professor Imamura once accompanied me to my study field of southern Thailand. I think it was around the mid-1990s. He was a little particular about what to eat and where to sleep, but I think he enjoyed many things more than I had expected. I once visited a Japanese alternative therapist in Pattani, southern Thailand, with him and a female Muslim friend of mine. When my friend removed her hijab to receive massage therapy for her head which she had injured in a car accident, I remember that Professor Imamura smiled joyfully, saying "Oh, lovely, lovely". As my friend was treated very kindly by him, she continued to reminisce and asked about him for years after that. Imamura's field trips culminated in philosophical travelogues, *Tai de kangaeru* (Philosophizing in Thailand) (1993) and *Chūgoku de kangaeru* (Philosophizing in China) (1994a), which are unusual titles among his works.

Professor Imamura's understanding of human existence

Now, how did Professor Imamura view human beings? I shall attempt to glean his view from *Shakaisei no tetsugaku* (The philosophy of sociality) (2007), published in the final year of his life. His words that left an impression in my memory from our research meetings with him were "particular human existence". He explained this notion as being distinguished from man as a natural being. "A man becomes a human the moment he comprehends and accepts the existence of a pre-ego self as given" (H. Imamura 2007: 40). One's self-consciousness begins with a sense of one's own existence as given, and a particular human being emerges when one senses the act of quiet giving in one's reception of the given existence. What is important here is that human existence is "being—thrown into", i.e., the thrown being, and that one is not necessarily aware of it but one always senses this original fact physically (H. Imamura 2007: 39). Professor Imamura got this idea from the concept of "thrownness" proposed by Heidegger, and identified the moment a natural man turns into an original human being in "accepting one's own existence through existential sentiment (so called emotions or feelings)". Existence and sentiment are indivisible in a human (H. Imamura 2007: 40). This sentiment is to sense one's own existence only as an "indebted" existence.

He thinks that this sentiment creates the sense of "coexistence", or a connection between a human and the environment through the human body (2007: 39). "To feel that one's existence is given is the fundamental way of being in this world for a human being" (2007: 40). A person as a coexisting being is one who receives and responds. Yet, the actual condition of human existence cannot be fully explained by the notion of a coexisting being here and now alone. For one to become a particular being, a unique twist must be added to the setup. I believe that Professor Imamura was trying to explain this using a ring as a metaphor.

Human existence as a ring

The ring metaphor was frequently mentioned at joint research meetings and Imamura sometimes used the metaphor of the donut as well, if I remember correctly. Why a ring? Because a ring has an empty space in the center and this void is created by the physical material around it. Let me quote again from *Shakaisei no tetsugaku*. "The metal part of the ring is in existence in this world (temporal and historical existence), and the ring cannot exist without the vacuum and the vacuum cannot be 'in existence' without the material ring" (H. Imamura 2007: 106). The vacuum is the original self-destructive element with a nullifying action which is constantly in competition with self-preserving existence and is indivisible from it. These are the two sides of human existence and the indebtedness is born out of the act of apprehension of this vacuum, or nothingness performed by the original existential sentiment.

The simile of the ring for a human being was used in slightly different ways on other occasions. The human being is perceived as a combination of the natural body and desire = the mind. The human being is also perceived as based on its physical body, i.e., no natural body, no mind.

> As we can see in the case of the void created by the ring, the round empty space would not exist without the material object (metal); but the void is not a natural material object. The void of the ring is something that is not found among natural material objects and it manifests as the *non-natural* within nature. Similarly, desire manifests as the non-natural (anti-natural) within nature on the premise of and in some relationship with nature. To nature, or to the body as part of nature, for example, desire is just *empty nothingness* unless and until it is fulfilled. If we call this kind of empty nothingness desire, then desire is something that nullifies nature within nature. And if we can call the action to nullify something intention or the mind as a non-natural action, then human desire must be the very *original* "mind". The original mind manifests itself as will or volition and volition aims for a specific object that can fill the void. (emphasis original)

Actually, the latter quote comes from a summary of his presentation at our Group research meeting on November 5, 2005 and the former from *Shakaisei no tetsugaku*. Chronologically, it is highly likely that the passage from the summary was written earlier than that from the book. While the metaphor of the ring in the summary from 2005 attempts to paint a picture of the human being that is non/anti-natural in the context of natural existence, the metaphor in the book published in 2007 focuses on the existence which has within itself a vacuum as a self-destructive nullifying action toward death, as opposed to the self-preserving existence to "survive". Assuming that the passage from 2007 was written later, this shift in relation to the metaphor of the ring may suggest a shift in his viewpoint from the division of nature in 2005 to a moral interest in the human being with an in-built orientation toward death in 2007.

What is the connection between the view of the human being as an existence carrying death inside and the sociality posited by Professor Imamura who was specializing in social philosophy? The inquiry into "sociality" is to examine the formation of society that is in fact occurring on a daily basis, the dismantling of the framework of the individual as a closed microcosm that is happening every day and the state of cooperation between us who live in reality. It was perhaps a way to extend beyond the proposition of "third term exclusion" in the discourse on violence, that is, the "transcendental" theory of the formation of society or community by victimization or exclusion, as discussed by Tanaka and Tokoro in this volume, and approach the philosophical anthropology of the original fact (H. Imamura 2007: 7–8). One possible course of transition from a human being carrying death inside to "being together" may be found in the fact that one cannot experience one's own death. Professor Imamura says as follows.

> My death is mine alone. And I cannot actually experience my own death [...]. My physical death is outside of my individual power and will remain with others. The corpse is merely a material object physically, but it is not a material object socially. The corpse has a social existence and continues to play a social role among other people; one still behaves like a living individual even after one's physical death. One is a social being not only while one is alive but also after one's death. (H. Imamura 2007: 108)

> And surviving while carrying the void of death in the center of one's existence gives rise to the indebtedness of living, a sense of 'guilt'. 'Being human' in itself feels guilty. It is not a mythological 'religious' feeling at all; it is because human 'existence' is originally constructed that way. (H. Imamura 2007: 105)

Here, I think I can see a path to "wakeful morality", which Professor Imamura planned to include in Part Three of *Shakaisei no tetsugaku*, which he did not write in the end. It is said that he defines "wakefulnesss" as the extremely rare and difficult, but not impossible, experience of understanding the existence of all things, including oneself, as they are, without any cover or filter, which becomes one with a feeling of joy (Sakurai 2011: 242). I wonder if this was the state of mind he attained at the point of his own death.

Homework from Professor Imamura

I can perhaps say that the continued presence of Professor of Imamura amongst us and the assignment upon us of the responsibility to respond to his presence are applicable not only to the members of various research groups who were encouraged, scolded and taught by him but also to all the surviving people who received his message from his writings. At research meetings, Professor Imamura pressed very hard, as if he was committing his whole existence, when he tried to get an important message across. So those who were pressed hard had to receive the message as their life-long homework. There are many people who received homework from Professor Imamura. He looked forward to our group research meetings more than anyone and he was always the first person to reply to the research meeting's scheduling enquiry by saying that he was "available to attend on any Saturdays, Sundays and holidays". He mentioned a certain research group of which he was a member at the end of the aforementioned email message to me in 2005, and wrote about the upcoming group research meeting. "We should have some primatologists and, together with anthropologists, find out 'how what we call society comes into existence' in the first place. I am participating in Dr Kawai's research group in preparation for it. To find suitable talent". In retrospect, Professor Imamura was also a great educator who committed his whole being to that task.

I shall tackle my homework as I keep his words engraved in my heart—"To sense various differences that are invisible and hidden and express them in language".

Notes

Introduction

1 Common among researchers involved in primate sociology or who major in human evolutionary theory.

2 A forum in which practice is created and sustained is referred to as a community, or a community of practice, to distinguish it from the traditional concept of a communal society (Tanabe 2002: 15).

3 The journal, *Asian and African Area Studies* (Kyoto University Graduate School of Asian and African Area Studies (ASAFAS)), was newly published in 2001 with the close involvement of established ecological anthropologists.

4 It goes without saying, but the socio-cultural anthropologists participating in the Groups Research Project are scholars who also at least have an interest in human evolution.

5 In a legal society, the individual is a substitutable entity.

6 More correctly, the Japanese macaque group is a gathering of matrilineally linked individuals, and while females never leave the troop into which they are born, males will depart their natal group around the time of sexual maturity and become solitary, ultimately joining another troop (transferring allegiance). It has been known for males to thereafter pass through several troops over their lifetime (Sprague 2004).

7 As in Kitamura's paper in Chapter Three, this book addresses both non-human primates and humans, and includes attempts to explore ways to understand both from common foundations.

8 The "solitude" spoken of in Chapter Two (by Uchibori) and Chapter Thirteen (by Funabiki) could probably be described as ontological modes unique to human beings. In this book, the theory of groups is developed to include how it is that the relationship with a group identified from an individual manifests in the individual. For example, in Article One Itoh asserts that for a single individual in a chimpanzee troop manifesting fission-fusion, it is still unclear as to how the overall troop (social unit) is understood by an observer.

9 At this level it is the ecological concept of a regional group of individuals. It points to a gathering of individuals of the same species. There is no need for that species to have formed a group, and the word "gathering" here refers to a set, as used in mathematics.

10 For example, a recent publication dealing with this major trend is Allen and Timothy (eds) (2000).

Chapter 1

1 A unisex gathering refers, for example, to a group only of males, as seen in gorillas and gray langur.

2 Competition within a troop can also include indirect competition (scrambles). Refer to Chapter Four by Nakagawa.

3 The discussion by Itani of the principle of equality relates to structure in troops rather than anti-structure, however, and differs from Turner's communitas based on the principle of equality.

4 In affordance theory, things in the environment are not simply there in a physical sense, but are there to cause an actor to seek affordance in the form of information linked to the acts of the actor (Sasaki 1991).

Chapter 2

1 The concept "*specia*" as used by Imanishi originally in 1941 (cf., *Seibutsu no sekai* (The world of organisms) 2002) asserts that there is a transcendent existence in contrast to individuals in a given species. When its multilayeredness is only seen with disregard for its transcendency, it would be placed at the top layer as an aggregate of all individuals of the same species. On the contrary, to think about a single species society does not eliminate the possible formation of certain groups or something group-like between different species. It is not logically incorrect to see an assembly of different primates or one of livestock and humans, as seen in nomadic groups, as a group of beings.

2 In my previous discussion (Uchibori 1989), I referred to societies in layers below a *specia* as "intermediate categories of the whole society" on the supposition that the whole society is smaller than a *specia,* but I also intended to indicate the logic that nation-states also fall into a real intermediate category with the growth of the whole human society on a global scale.

3 Constructed representations are often legitimated or substantiated through successive phases from past to present. This is especially true for the self-awareness that meets such representations. This is why ethnic groups are said to require (a) history (Anderson 1991; Shimizu 1992). However, in principle, the construction of a representation does not need to be shaped on a temporal phase. There may be a representation that is discarded as a *metachronic/*

absolute truth and may perhaps, in this way, have a stronger effect on belief formation.

4 I first learned about this view from Devereux (1967). This book also taught me that human culture and society should be understood according to the following terms: not simply based on groups; on the assumption that what is visible are individuals; and with researchers' introspection and reflection on their own deep psyche.

5 Graeber (2001) states his impression that, even in Merina Kingdom, a hierarchical country in Madagascar, its rural village societies are based on egalitarianism. When it comes to the issue of class or status-hierarchy, it is necessary to always be precise as to which social layers the framework in question is referring to.

6 Differences between evolutionary and historical times have been briefly argued (Uchibori 2006) and await further discussion.

7 For polygamy refer to Lévi-Strauss (1971). It may be more approximate to reality to restate the aforementioned assertion about human polygamy, as humans do not generally form polyandrous societies; in other words, humans do not often form multiple pairs with one female as a hub. Although polyandry in humans is seen to a very limited extent, it should not be disregarded. In the case of a small-scale polyandrous society in Tibet (Durham 1992), it is doubtful as to whether it can be referred to as pair formation or not. An ethnographic example of "multiple fathers" (a concept that multiple males can be reproductively involved in one pregnancy) as seen in some groups of indigenous people in South America may rather be one form of the instability of human pairs (Beckerman and Valentine 2002).

8 At any rate, it seems inappropriate to raise the issue of lineality here. Any lineality in human society is no more than ideology and requires the representation of temporal sequence of past, future and present (present as something to sever and reconnect past and future). Accumulated paternity can only be a problem among primates other than humans.

9 In relation to this, Cartmill (1993) points out the ideological nature of the hunting illusion/fantasy in European social history. Of objective significance is the issue of shifting the balance between the possibility of eating and the risk of being eaten in a new environment. The point is how the risk of being eaten changed following human evolution into hominidae, and the arguments of Hart and Sussman (2005) are not entirely convincing in this regard. The

authors of *Man, the Hunted* even say that in outer edge areas of forests are more informative than apes in forests in terms of investigating the behavioral characteristics of early humans as inhabitants of outer edge areas. Based on such analogical reasoning, regardless of its uncertainty, humans may be defined or caricatured as "macaque-like chimpanzees".

10 Though slightly varying in expressions and contents, Kuroda (1999), Sugawara (2002) and Kimura (2003) take approximately the same direction. Conversely, M. Nishida (2003) considers dispersive living in the form of a pair as a more decisive moment for human evolution than gregarious living in the form of a troop-like group. The current discussion is approximate to Nishida's view, except for his focus on the hominization processes in a factual sense.

Chapter 3

1 A term from Imanishi (1971) meaning a way of life in which a biological individual, as a living subject, is strongly tied to its place of living. Refer to Chapter One of this volume.

2 For example, the Japanese macaque forms extremely compact gatherings, or "troops", comprising multiple adult males and multiple adult females and their children. The membership of the group is considered to be stable in the very long term, but factors that can destabilise it are: (1) the phenomenon of entry and exit between groups by individuals, and (2) the sub-grouping phenomenon, in which some members of a group temporarily separate from others and range independently. However, these may be considered a kind of deviation from the normal state very generally recognised in the primate group phenomenon, and if accepted as a premise, then what is called a "troop" here, which is typical of the Japanese macaque troop, may be thought of as a group with stable membership.

3 Many similar stipulations have been reported in hunter-gatherer societies (for example, on the !Kung San, see Lee (1979), and on the Mbuti Pygmy, see Ichikawa (1982)).

4 This is not a deviation like that of the sub-grouping phenomenon in the Japanese macaque troop, but rather a normal state of ranging dispersed into distinct sub-groups (parties or sub groups) in which membership is fluid (Kitamura 1983).

Chapter 4

1 Photographs on the Diagram at page 80 include mountain gorilla (taken

by Juichi Yamagiwa), pygmy mouse lemur (taken by Shinichirō Ichino), Philippine tarsier (image taken from *http://blog.livedoor.jp/himan/ar-chives/50557195.hml* with permission) and other primates (taken by the author).

2 For the full version of van Schaik's socioecological model, see Isbell and Young (2002), Koenig (2002), Nakagawa and Okamoto (2003) and Nakagawa (2007).

3 Individual fruits and leaves that are the staples of primates are attached to individual plants and are generally not found elsewhere. In other words, looking at each individual species of plant, the fruit and leaves exist as a particular clump. These clumps of food are called "food patches", or just, "a patch". Normally, each individual tree may be considered a patch. For further detail, see Nakagawa (1999).

4 Refer to the section entitled "Selection pressures that form matrilineal, multiple female groups". For details, see also Nakagawa (1999).

5 In terms of van Schaik's social categories, this corresponds to 2) to 4) female resident categories.

6 In terms of van Schaik's social categories, this corresponds to 1) female dispersal category.

7 There is also another hypothesis that deems infanticide to be a selection pressure in the formation of multiple female groups, which evolved to enable females to cooperate in protection against infanticidal males (Treves and Chapman 1996). I would add that there are also hypotheses that deem sexual harassment (Fox 2002) and risk of predation (Hill and Lee 1998) as selection pressures in terms of the formation of two-sex groups, which posit that females formed groups with males to receive protection from these dangers (see Nakagawa and Okamoto 2003).

8 The need for infant care by males is not a selection pressure in the formation of one male-one female groups, but in some quarters it is thought that infant care by males becomes acceptable as a result of the formation of one male-one female groups. It should also be noted that at the least, there is no well-established theory that can be used to apply to all the relevant species, and debate is ongoing (for example, see Fuentes 2002; van Schaik and Kappeler 2003).

9 Normally we should now introduce the selection pressures that form fission-fusion and multi-level societies, but we will use author's licence and omit them. Please see Aureli et al. (2008) and Lehmann et al. (2007) for the former, and Barton (2000) for the latter, respectively.

10 In the thinking behind van Schaik and Wrangham's socioecological models the concept was that the matrilineal group evolved because it is advantageous in competition over food, the single most important resource for females, and it was therefore perfectly natural and logical to theorise the evolution of patrilineal society from the perspective of competition over females, and oestrus females in particular, the single most important resource for males. Unfortunately, however, the situation is that there is no theory that uniformly and without contradiction explains the factors in the formation of patrilineal societies, including Atilinae and the hamadryas baboon.

11 Previously this was separated as a suborder of the prosimian suborder, in the sense of them together being a taxon with primitive traits.

12 Previously this was separated as a suborder of the prosimian suborder, but it is now called suborder Halorrhini and includes Tarsiidae.

13 Refer to Hamada (1999) for the taxonomy of primates. For the Japanese names of primates, see Yukimaru Sugiyama (ed.) (1996) and in particular, Lemuroidea is in accordance with Aimi and Koyama (2006).

14 The Maximum Parsimony Method is one used to explore the traits of common ancestors in extant species. Characteristics for which changes in traits that happened in the passage to extant species are reduced to an absolute minimum, are surmised to be the same as in common ancestors (see T. Nishida 2001). See also Figures 4.8a and b, which present specific examples.

15 A gathering of individuals of the same species living in a certain region. The species does not need to have formed a group, as the word "gathering" here has the mathematical meaning of "a set".

16 The white-handed gibbon (*Hylobates*) of Khao Yai, Thailand, basically manifests a classical one male-one female group, but at the same time 21.2 percent form multiple male-one female groups, and because these gibbons have manifested diverse social structures, there are those who consider their diverse social structures to be in common with other apes (Reichard and Barelli 2008).

17 It has been reported that white-handed gibbon males in captivity play with infants, and also carry infants as old as five or six months (Clemens et al. 2008).

18 Recent molecular genetic methods have nevertheless given rise to reports of examples where there are no kin relations between adult males in a multiple male-multiple female group, but each have sired infants (Bradley et al. 2005).

19 Contrary to forecasts, it was not possible to obtain results from chimpanzees

from Gombe in Tanzania and Kanywara and Ngogo in Uganda, indicative of sex-biased dispersal (see Di Fiore 2003).

20 Compared to fifty km² in Mondika in the Democratic Republic of the Congo, it was possible to confirm a tendency towards male-biased dispersal when a survey was carried out over a broader range of 6000 km², and it was also discovered that among the western gorilla at least a portion of males disperse over a very long distance (Douadi et al. 2007).

21 Hypotheses about the social structure of early humans by primate researchers include: Mori (1990), Yamagiwa (1994), T. Nishida (1999, 2007), Furuichi (1999), Hasegawa and Hasegawa (2000) and Yukimaru Sugiyama (2002, 2007c).

22 What would normally happen here is that my hypothesis should be verified by the deliverables of ecological anthropology, which is another approach for unravelling the way of life of early humans. Itani considered the vililocal marriage of the Mbuti pigmies who live in Teturi in the Democratic Republic of the Congo, who were studied by his apprentices, Mitsuo Ichikawa et al. to be the human social ancestral type (Itani 2008: Vol. 4, 292). Vililocal marriage refers to a couple living after marriage under the same roof as relatives of the husband, and for that reason, when used as a primatological term where there is deemed to be female (social) dispersal and male philopatry, it matches my hypothesis. According to Colin Turnbull, however, who studied the Mbuti of Epulu only fifty km away from Teturi, the ancestral type is ambilical marriage, and can be described as neither vililocal nor uxirilocal. Both then succumbed to claiming the other to be something that formed secondarily as a result of social transformation (Ichikawa 1986). Thereafter, Hideaki Terashima, who conducted ecological anthropological studies into the Teturi pygmy, expressed the view that the Mbuti originally had ambilical marriage, and vililocal marriage was due to the influence of neighbouring agricultural peoples (see Terashima's paper, Chapter 8). In this way, the original human appearance is debated even between ecological anthropologists, and for that reason I considered it impossible for an outsider like myself to appropriately extract from several papers and publications the appearance of present-day hunter-gatherers appropriately applicable to early humans, and abandoned the idea. While it is a result that is at odds with my own hypothesis, I will nevertheless herein introduce the results of locational dispersal studies only using molecular genetics, which has formed the main focus of this chapter. In dispersal over a wide range on a global scale, sex-biased dispersal has not been identified, but in that on the scale of a particular region, female-biased

dispersal most typically manifests, and this is linked to typical vililocal marriage. Vililocal marriage and female-biased dispersal are more common in pastoral and agricultural peoples, in which there is a tendency for a son to inherit the livestock and land of the father and to cleverly succeed in attracting a woman as wife. What becomes an issue here, however, is the situation of hunter-gatherers who are considered the ancestral type. In hunter-gatherer societies, amassing resources could not be considered to be particularly important. In fact, in real present-day hunter-gatherers, either sex-biased dispersal cannot be identified, or at times only male-biased dispersal is evident (Wilkins and Marlowe 2006).

Chapter 5

1 As Ingold (2000: 52) points out in reference to Schutz's definition of sociality, namely "Sociality is constituted by communicative acts in which the I turns to the others, apprehending them as persons who turn to him, and both know of this fact" (Schutz 1970: 163): in the midst of people who behave in this way, whether in dealings between "humans" or between beings that are "human" and those in "nature", or then again between beings in "nature" whatever the involvement, communication founding sociality could be established. Communication as a phenomenon is not peculiar to "humans", and nor is the sociality that is constituted by communicative acts strictly limited to "humans". Accordingly, when we consider the sociality constructed via communication, we should not regard as given the modern western "human/ nature" dualism that sees the existence of a boundary between "humans" and "nature" as self-evident. However, as has been argued in detail throughout this volume, there are, of course, various differences in forms of communication, and a variety of groups will come into existence in response to these disparities in form. Moreover, as Adachi succinctly makes clear in Chapter One of this volume, the form of and borders between groups are not necessarily prescribed by biological differences between species. Even between diverse biological species, mixed groups emerge through interactions. As is argued from a variety of angles in this volume, it is the differences in forms of communication that prescribe the forms and boundaries of groups. While on the one hand mixed groups that transcend species differences are possible precisely because the members of the group can communicate with one another beyond biological species differences, on the other hand, differences existing in the forms of communication give rise to considerable distinctions in the form of sociality

that is constructed from these disparate forms of communication, and additionally in the forms of the groups that emerge as a result.

2 "Subsistence" is used here as "a term that indicates the totality of activities for acquiring, handling and consuming natural resources as well as the social relationships that accompany this. Even activities accompanied by financial gain, such as the use of traps for catching animals for their fur and pelts, where this profit is used in order to preserve 'traditional' lifestyles, are counted as subsistence, and differentiated from commercial resource use—fishery and forestry" (Stewart 1996: 126).

3 Social groups, which form the basis of social relationships in Inuit society, are the kindred groups called *ilagiit*; and within these, extended family groups, called *ilagiimariktut*—real *ilagiit*—are the units of everyday social relationships. Extended family group "refers to, not just those who are in extended family relationships, but people in close cooperative relationships through dwelling in the same place and engaging in the same economic activities; namely, people who form concrete social groups. This latter group, the *ilagiimariktut*, are ultimately largely made up of one's own parents, siblings, wives and children, grandchildren, uncles, aunts, grandparents and cousins" (Kishigami and Stewart 1994: 421). Associations are formed that have this at their core and that transcend kinship groups: quasi-kinship relationships, such as adoption; and voluntary associations, such as name-saking relationships and avoidance relationships. Complex social relationships are created with extended family at their core.

4 This is a summary of a story told to me by an elder on the 8th of August 2003.

5 What is important here is the fact that it is not because the Inuit are actually the weak party in relation to wildlife that they entice them. Rather, the Inuit entice the wildlife so as to place themselves in the role of the weak party. Also, as will be argued, the assumption of the status of weaker party with regard to wildlife by the Inuit is of decisively important significance in the construction of equal social relationships based on trust amongst the Inuit themselves. The enticing of wildlife and the intentional assumption of the position of weaker party is not essential ecologically; it has a social philosophical necessity for the Inuit to create equal social relationships. In this sense, as pointed out by Ingold (2000: 61–76), there is a logically necessary relationship between the state of social relationships and the form of skills/techniques. However, whilst Ingold sees the foundation of the form of skills in hunting and gathering subsistence in relationships of trust between humans and wildlife, as I will

argue in this chapter, the Inuit do not simply trust wildlife and merely wait for them to give their bodies to them, they actively appeal to wildlife using the technique of enticement. It is more accurate, therefore, to regard enticement, and not trust, as being the foundation of Inuit subsistence.

6 However, what is important here is that this tactic of enticement is developed between the body of the wild animal and that of the Inuit. Because this tactic is developed through the medium of bodies, the Inuit experience a sense of unity as "eros", in M. Tanaka's (2009a) words, with wildlife, and moreover with the whole "land". Tanaka says, "Enticement is not the type of tactic that is carried out while contriving to force the other party to surrender. Through the mediation of the 'body', which includes the voice, one acts against his own feelings and entices the other party, or alternatively, opens up the body to enticement" (M. Tanaka 2009a: 44). Enticement, "by virtue of the fact that it occurs corporally, gives rise to situations that one cannot control using one's own will" and "just who is being enticed? Is it me? Is it you? Is it your body? It is precisely the incidental contingency that does not allow for easy answers to these sorts of questions that is what we call enticement". He points out that, "what we have here are the signs of the fusion of self and other that are acknowledged in good quality sex", and he argues that these symptoms of fusion lead to eros which produces "a sharing of the world and an expansion of it not just by oneself, but oneself together with the other, and a sense of that type of construction of the world" (M. Tanaka 2009a: 46). It is, as I argued in another work (Omura 2008), precisely this "eros" that is the sense of unity that the Inuit feel with the "land" through the actual practice of subsistence. However, since the person employing this corporal tactic of enticement experiences a loss of personal autonomy and independence, it simultaneously induces fears about losing oneself. This is a "basic anxiety that is acknowledged in dealings with others" (M. Tanaka 2009a), and amongst the Inuit, tales are frequently recounted of the transformation of wildlife into humans and of humans into wildlife involving fears about losing oneself (Omura 2009).

7 This argument may provide the basis for sketching out a methodical understanding of "enticement" as a relationship with others outside the group and "trust" as harmonious relationships within the group.

8 If, as argued here, it is accurate that the form of the reciprocal act of gift giving can be secured via the transformation of the "enticement/command" relationships between the Inuit and wildlife, then Imamura's proposition that "the system of gift giving is a system of the reciprocation of defiance; a

reciprocity of conflict" (H. Imamura 2007: 406) might possibly be restated as follows: "the system of gift giving is a system of the reciprocation of enticement; a reciprocity of love". This transformation of the proposition, as argued by Tanaka in Chapter Twelve of this volume, brings a new perspective to the philosophy of society, one that transcends the "conflict model" of the creation of society, and one that has the potential to open up a new image of people and of society.

9 Up until this time, it had been thought that reciprocity was the driving force behind the creation of human societies (for example, H. Imamura 2007; Oda 1989, 1994), but based on the arguments in this chapter we can understand that the dynamics at work at the basis of reciprocity are concerned with the "sharing" of resources, in particular the "sharing" of food. Future avenues of investigation include the following: tracing the processes through which the various forms of exchange, such as reciprocal gift giving and also redistribution and equivalent exchange, come into existence through the dynamic that surrounds this "sharing" of food as the primary impetus; and, elucidating the process of the historical evolutionary logical development of human social groups even into today's capitalist society.

Chapter 6

1 On the distinction between the terms "society" and "group", see the article in Chapter Two of this work.

2 Refer, for example, to the following passage: "Hereby it is manifest that during the time men live without a common power to keep them all in awe, they are in that condition which is called war; and such a war as is of every man against every man" (Hobbes 1651 (trans. Mizuta 1992: 210)). Hobbes saw mankind in his natural state as living an essentially violent existence—"a war as is of every man against every man", and consequently he developed the doctrine that what was needed was the control of the state from above. We can, therefore, see that Hobbes assumed a trend from a state of mutual violence in mankind's original natural state towards the development of social order that was managed by a system such as the state—a vector, as it were, from a lack of order towards order in the history of human evolution. In contrast to this, Rousseau conceived of the history of mankind from the viewpoint that in his natural state man was originally in a peaceful state, the so-called "noble savage", and that it was rather with the emergence of the state and with civilization that war and violence spread amongst mankind. Consider, for

example, Rousseau's comment that "nothing is more timid and fearful than man in the state of nature; that he is always in a tremble and ready to flee at the slightest noise or movement" (1755). We could say that this is a "from order towards lack of order" vector—the exact opposite of Hobbes' position. Furthermore, for detailed accounts of the conflict between the Hobbesian and Rousseauian views of mankind see Kurimoto (1999) and H. Imamura (1982).

3 On the disputes surrounding the hunter hypothesis and criticisms of it see Ardrey (1961, 1976), Kurimoto (1999) and Baba (1999).

4 See Diamond (1992, 1997). However, Diamond himself also calls our attention to the fact that, in comparison to the violence found in non-western societies, the scale of violence carried out by civilized western societies is on a remarkably larger scale—for example, the acts of genocide carried out by western societies.

5 At a symposium attended by Junzō Kawada, Toshisada Nishida and others, Kawada spoke at length about this situation. During this symposium, in contrast to the arguments of primatologist, Toshisada Nishida (1999), who stressed the continuity between man's wars and chimpanzees' "wars", saying "war is to be found in hunter gatherer societies and also in pre-humans", cultural anthropologist, Junzō Kawada (2006), emphasized instead the aspect that "large scale war" had flourished "since the appearance of the centralized modern state". In general terms, these types of disputes originated in the problems surrounding the interpretation of fossils and ethnographic data, and also depending on the extent to which emphasis is placed on actual cases of violence in the data, there is a tendency to be led to different interpretations of the same cases on the basis of differing views of mankind. Finding a speedy and unambiguous resolution will be a difficult matter. On this point, see Kurimoto (1999) and Baba (1999).

6 In Hitoshi Imamura's social philosophy, for example, at the same time as stressing the discontinuity between man (homo sapiens) and other animals, there is also an implication of the opinion that precisely for this reason, human society has at its foundation a dark side; that it is fundamentally permeated with violence. On this point, see H. Imamura (1982) and also H. Imamura (2005a).

7 See Frans de Waal, (2005). In contrast to Wrangham and Peterson—and other advocates who single out chimpanzees from amongst higher primates, with their similar levels of evolution as man, and stress the continuous and inherent nature of violence between man and chimpanzees—de Waal stresses the

strong tendency in bonobo (pygmy chimpanzees), which are also extremely close to man in terms of evolution, to resolve the conflicts and tensions that arise in their society not by violence but by sexual communication. Thus, de Waal suggests that in terms of the nearest model to man—a model, as it were, of "our inner ape"—not only chimpanzees, but also pygmy chimpanzees are potentially important. Also, as de Waal and others point out, it is well known that chimpanzees display peaceful interactions such as "conciliatory behavior" alongside violent acts and that these are also important parts of chimpanzee society (see de Waal 1989).

8   Kurimoto, quoted previously, whilst stating that "one of the basic premises in war studies is recognition of the fact that war is a destructive act and one type of social pathology", also, however, simultaneously pays heed to the fact that "there is a creative aspect to war; an aspect that constructs societies and communities" (Kurimoto 1999).

9   In cases of the slave looting during the pre-colonial period and the abduction of hostages by present day pirates, the circumstances differ to some extent, but in these cases also the acquisition, as it were, of others as "things" is the aim, and although the use of violence against others may be a means to this end, it is not in itself the aim.

10 Toshisada Nishida points out that it may be that the germ of feuds and retribution behavior exists in chimpanzees, who are close to humans (See T. Nishida 1999). Also, in our Group Research Seminars, Naofumi Nakagawa commented on this issue that such feuds and retribution type behaviour can also be observed in some species of monkeys like the Japanese macaque. I would like to express my gratitude to him for having pointed this out.

11 See Nimmo (1972), Kiefer (1985, 1986), Sather (1997) and Tokoro (1996, 1999, 2003) for a geographic outline of the Sulu Archipelago and the ethnography of local groups such as the Tausug and the Sama.

12 Examples of the Muslim secessionist organizations that appeared after the 1970s' civil war include the MNLF (Moro National Liberation Front), the MILF (Moro Islam Liberation Front) and the ASG (AbuSayyaf Group). The MNLF, the largest of these organizations, concluded a peace treaty with the Philippine government, but subsequently, following considerable vicissitudes, sporadic outbursts of armed conflict continue to this day between the MILF and ASG and the Philippine National Army. In this chapter, I will not deal with this kind of conflict between the Muslim secessionist groups, with their political agenda, and the government. For a discussion of this issue see Tokoro (2006).

13 In addition to these there is violence carried out by a variety of armed groups—
the private armies of politicians, vigilante groups such as the CAFUG and
ICHDF—and there are also really quite vague areas around the borders of
narrow political violence and other types of violence, but the focus in this
chapter is on piracy and feud disputes.

14 "Pirates" in the context of current international law are mostly indentified
under Article Five of the "Law of the Sea" as "the passengers and crews of
privately-owned ships and aeroplanes, for private purposes—on the high seas
or in areas not falling under the jurisdiction of any country—all illegal acts of
violence, detention and looting carried out against other ships and aircraft and
also the people and property within these vessels" (Masahiro Nishii *Pirates
in International Law* 1998).

15 See the entry for "pirate" in the fifth edition of the *Kōjien*.

16 Other names that they may call themselves include the "Balangingi" or the "Aa
Bagigi", where "Aa" means people.

17 Moro National Liberation Front—a Muslim secessionist organization in the
Philippines.

18 It would be appropriate to use descriptions by Kawamoto, such as that below,
as an abstract and restated summary of the process discussed in this chapter
of separating groups through violence. In other words, "it is not the case of
a pre-established self-altering as one changes one's borders, it is the very self
that must be formed on every single occasion as the operation continues. In
this sense, the formation of boundaries in a system is simultaneously 'self-cre-
ation'," and "when it operates in such a way as there is no internal or external
part to the operation of action, then, for the first time, the boundaries of the self
are formed through the continuation of one's own actions… the continuation
of ongoing action at these times delineates the boundaries of the self and this
is where cyclic prescriptions come in for the organization" (Kawamoto 2000).

19 See Clastres (1974, 2003) regarding points of view on the afferent/efferent nature
of violence. Clastres advocates that war in primitive societies has a centrifugal
effect when waged against a larger state, a centripetal effect against individual
primitive societies, and that it serves to maintain the autonomy of primitive
societies.

20 Conversely, we can say that in cases where the hostile feud groups originally
belonged to the same village or ethnic group, the same feud dispute violence
would operate centrifugally as far as the original large group category is
concerned.

Chapter 7

1 Raiding never occurs within an ethnic group of this area, and in this regard, it could be used as an indicator that clearly identifies the boundaries of such. If raiding occurs among two different groups, they belong to separate ethnic groups and if not, they belong to the same group.

2 Imamura says that Lévi-Strauss is advocating "exchange monism", in that all social behaviors are exchange behaviors. In his definition, all mutual behaviors constitute some form of "exchange", and "social relationship" is the sum of these exchanges (H. Imamura 2000: 51).

3 H. Imamura (2000: 54) defines the term *koeki* (trade) as follows: "In Japanese, '*koeki*'s' '*ko*' is to intersect and thus means relational transition and a transitional relationship and *eki* means spatial and temporal transition of humans, materials, affairs, concepts and mentalities [...]. This view may contradict the conventional view, but by looking at it in this way, the terms of *koeki* can be expanded to cover the concepts of social or mutual behaviors, and even the concept of society itself".

4 "Excitement" has many synonyms such as exaltation, emotions, feelings and anger, as pointed out by Kuroda in Chapter Eleven of this volume. However, I will use "state of excitement" here, akin to the "heightened emotion" as defined in Kuroda's thesis.

5 The pastoralists of North East Africa in the southwest regions of Ethiopia and southern regions of Sudan that Katsuyoshi Fukui et al. have studied are known to have "archrivals" and "allies" (Fukui et al. 2004).

6 We shall define war, in accordance with Kurimoto (1999), as "organized armed conflict between political groups". Kurimoto has introduced Caillois' definition, "war is a collective, intentional and organized conflict" (Caillois 1974). In anthropology, we have seen many armed conflicts on the borderline of whether they should be called war or not, feuds and raids often among them (Kurimoto 1999). However, feuds and raids can vary in scale and degree of organization, from those equivalent to small-scale robbery to those that could well be called large scale organized military operations (Kurimoto 1999). The raiding groups of the Dodoth are usually small in scale with only several to several dozen members, but there have been rare incidents of organized armed groups of hundreds and thousands of members looting tens of thousands of livestock. In this chapter, we have small-scale raiding by dozens in mind, rather than large-scale acts of looting.

7 Let us give the example here of people of the Bodi in Ethiopia, in respect to the

aggressiveness displayed in the region. When a cow that a man deeply cared for dies, the owner of the cow attacks the neighboring cultivators and kills a man in order to alleviate his pain (Fukui 1979). The Bodi are known to go raiding as well, but the above act is clearly one intended to kill and has nothing to do with raiding for the purpose of looting livestock. Incidentally, there has been no report of such practices among the Teso-Turkana pastoralists.

8   Here is a classic example that shows how the movable assets of livestock come and go between two or more groups. The chief of the settlement I stayed at had a donkey named Karopeto, which he acquired at a stream named Karopeto by *akoko* (stealing). Soon after, the chief lost almost all of his livestock through *ajore* (raiding) by the Turkana, and Karopeto was among the animals that were taken. However, when I returned to the field a year and a half later and conducted a routine interview survey on the status of the number of livestock through identifying the animals, I came across another animal by the name of Karopeto. I thought that the chief had named a newly acquired donkey by the same name, but it was the actual Karopeto that was taken by the raiding attack. Apparently, the donkey happened to be among the animals they stole from the Turkana by another *akoko* (act of stealing). As such, the Dodoth and the Turkana's movable assets of livestock keep coming and going between each other. Thus, it can be assumed that the total number of livestock in this region, though it is not easy to set the exact boundary, would remain almost constant unless incidents such as severe drought were to occur.

9   In the case of raiding in the scale of hundreds, the elders plan the raid and each organizes a small unit of about twenty to thirty by gathering his comrades. This process is followed by the formation of a larger group, with these small regional units (or any other units) as subordinate groups. This raiding group would have a stronger militant aspect; for example, each unit may be given a systematic role in the raiding operation.

10 The Dodoth are pastoralists who place high value on their cattle, but with an annual rainfall of approximately 450mm, it is possible to cultivate grains with rainwater during the wet season and they actually grow pearl millet, sorghum, corn and sunflower. Their dependence on farming is extremely low, however, as it often results in no harvest due to drought. Nevertheless, they place certain value in farming as a livelihood activity.

11 The Dodoth live in a semi-permanent settlement called *ere*, in which the chief, married women, including his wife and daughters, their young children and

elders live with a small number of lactating cows and cultivate farming patches around the settlement. Most of the cattle are kept in livestock camps called *awi* away from the settlement and are tended and grazed by single and young married men.

12 Please refer to Kuroda's discussion in Chapter Eleven of this volume.

13 Annihilation, including the case reported by Goodall, is an extremely rare phenomenon. While the chimpanzees are in antagonistic relationships with neighboring groups, it is not as if they are constantly exterminating each other.

Chapter 8

1 See, in addition, Masakazu Tanaka's thoughts on network theory and relatedness (Chapter Twelve of this volume).

Chapter 9

1 I have used the term "permanent group" here on purpose, but according to the definition used in this chapter, such a group does not exist. In this chapter, a group is defined as a visible gathering of living human beings who interact with each other, and thus every group has a temporary existence.

2 They differed from the Tigray, who were the ethnic group in Northern Ethiopia.

3 In Chapter Seven of this book, Kawai pointed out that raiding for livestock should not be regarded as the root of war, and I tend to agree with this view. The Gabra distinguish different forms of fighting, such as assassination, raiding and war. Large-scale war is said to be a relatively new phenomenon. The Gabra have chronicled their history based on a precise solar calendar (Soga 2006) and according to this record, did not exist prior to the war between the Boran and the Garre in 1932. Conversely, raiding for livestock is a common phenomenon to the pastoral societies of East Africa and a practice actively pursued by the Gabra.

Chapter 10

1 In ecological anthropology, which tried to observe people living in the "natural society" and applied the evolutionary history of mankind to these findings, shifting cultivators were positioned as a bridge between natural and modern society. Itani (1986b), who interpreted the emergence of "egalitarian society" as seen in hunter-gatherer communities as an evolutionary stage unique to

human society and believes that human society has since chosen to return to an inequitable form, described the shifting cultivators living in the Miombo woodlands as "people who cannot complete the crossing of the threshold from egalitarian society to the world of inequitability" (Itani 1986b). Makoto Kakeya, who studied the shifting cultivators of Tongwe in Tanzania, pointed out that the mechanisms of envy and witchcraft played a big part in establishing order in the Tongwe community and unveiled the differences between the Tongwe and the hunter-gatherer society. Kakeya also stated that the Bemba community shared a common mechanism with Tongwe society (1991).

2 According to the Zambian government statistics office, Bemba speaking people now number more than two million. Population statistics by ethnic group have not been available since Zambian independence.

3 By assuming the concept of "hearth-hold", Ekejiuba tried to reconsider that of household, which was perceived statically in the patrilineal society, and to depict subsistence activities in a dynamic way.

4 These manners have hardly changed since the colonial days. Audrey Richards, who wrote a detailed ethnography of the Bemba in the 1930s, also focused on gatherings of Bemba women and stated that the collaboration of the women overwhelmingly improved the efficiency of the work (Richards 1939). Here, it should also be noted that this procedure establishes the pace of consumption for the whole village.

5 Richards (1939) termed these loose ties between the adult women "cooking groups", but here the concept of "hearth-hold" is embraced in order to clarify that it is not just about the women's behavior and clearly establishes the activity unit which consists mainly of women around their own hearths as well as their children. These groups are mostly seen during daytime when the adult women are engaged in processing and cooking food.

6 The elders are the generation with grandchildren, and juniors are young to middle-aged members without grandchildren.

7 The word "anti-structure" is used here in the same sense as discussed by Adachi in Chapter One of this book.

8 The concept "reference phase" used here is similar to that of "cultural category" as discussed by Soga in Chapter Nine of this book in the sense that it becomes a point of reference when people take action, but here we are focusing on a phase that exists outside the individual and is referred to when needed rather than constituting a category.

9 It should be noted that in the background of the beer parties, there is a mechanism that assumes the existence of power, a sort of authority that controls curses and the supernatural power of the ancestors, in a context different to that of the accumulation of the practice of daily interactions. Kakeya revealed in his discussion regarding "institutionalized envy" that fear of envy underpins the discipline of distribution to others (Kakeya 1983), but at the beer party, where Bemba men gather, the disciplinary distribution is conducted in an extremely elaborate manner. This can be apprehended by combining discipline with the underlying understanding that "if one doesn't distribute, one may cause envy and be cursed for it".

10 H. Imamura and S. Imamura (2007) describes such authority as "authority whose teeth are removed as soon as it comes into power".

11 Spirit medium (*nguhi*) meetings are events where spirit mediums meet and let the ancestral spirits "play". There are a number of spirit mediums in each Bemba village. The spirits that take possession of a certain spirit medium are almost always the same, and each medium has more than one spirit. They allow the spirits to possess them as necessary for fortune telling. At the fingermillet harvest ceremony, net-hunting and spirit medium meetings always take place together.

12 The chief's mother and sisters are included as ancestral spirits.

13 Please refer to Yuko Sugiyama (2007a) for details of these cases.

14 As Kakeya (1991) discussed, while it did not flourish into a chiefdom as large as that of the Bemba, the patrilineal society of Tongwe in Tanzania also shared the mechanism of creating an "almost public authority" while maintaining a functional leveling mechanism. This could be a key to categorizing and comparing the group phenomenon of the cultivators of the Miombo woodland regions.

15 The discussion of intercorporeality and co-presence touched on by Aoki (2009) also indicates the importance of "just being together".

Chapter 11

1 The "charging display" involves a flamboyant rampage with hair raised. When it accompanies shouting and a lot of noise, it usually takes the form of a rage without a particular target, but if it is a quiet rampage, it is generally aimed at one particular individual, and often followed by true attack (Bygott 1979). When the male stops the behavior, lower ranking males and/or females will approach and touch the male to calm him down (T. Nishida 1977).

2 Bonobo's sexual interaction occurs not only between different sexes, but also between members of the same sex. It includes male-female mating, mounting, genito-genital rubbing (GG-rubbing, Kuroda 1980) in which females hug and rub each other's genitals, male-male mounting, rump-rump touching where males attach their buttocks together and shake their hips, and male-male genital contact, called penis fencing, where two males hanging from branches make their erect penises meet each other like swords (Kano 1992). These behaviors are observable when they are excited or during play. Mating and GG-rubbing are often seen in children's play, too.

3 There was a case in Mahale Mountains and the Gombe Stream where the ex-alpha male, who had lost a power struggle, returned to reclaim his leadership and was fatally wounded by the male group's all-out attack (Goodall 1986; T. Nishida 2012). Individuals may not be allowed to rejoin the group depending on the circumstance and the character of those involved.

4 This is the general character of the society and thus individual cases may differ. For example, if a female chimpanzee who had just transferred to a group gave birth to a child, she would try to be around the most dominant and generous male wherever possible because a male child may be killed by adult males, or other females may pray on her child, regardless of its sex (Goodall 1986; T. Nishida 2012). However, if she remains in the group long enough for her son to grow and establishes her status within it, she would be able to be alone depending on the circumstances.

5 Needless to say, the independence is supported by the individuals' independent sustainability skills (= ecological independence). Adult primates exhibit ecological independence but in the case of Japanese macaque, for example, members do not have sociological independence, namely, they cannot to leave the group.

6 In September 2008, two unit-groups in Wamba, namely El and P repeated such approach for a week before actually meeting, confronting and splitting up.

7 Recently, at Tennoji Zoo in Osaka, an inferior female who was constantly attacked by a superior female and was protected by a young male teamed up with the female who was attacking her and began confronting the young male (Kyosuke Fukunaga, personal correspondence). It seems that females also have the ability to resort to alliances as a diplomatic expedient.

8 Goodall (1986) reported that there were two individuals among the Gombe males who showed no interest in dominance. In other words, most males do have some level of aspiration for power. It is noteworthy in relation to the

discussion later that those exceptional low ranking males were participants in the "primitive war" and were classed into the most brutal group of aggressors.

9 "Rain dance" (Goodall 1971) refers to the commotion seen at the beginning of a heavy rain event where several males conduct charging displays and enter a state of commotion where females join in as well.

10 See Note 8. Participation in inter-unit-group attacks is not correlated to ranking order in terms of frequency or role (Goodall 1986).

11 It is assumed that collective excitement in the face of an enemy creates group unification and promotes friendly relationships even when the conflict is over, but there is no data as of yet to verify this view.

12 Despite such braveries, young males cannot join the adult male group. This is also indicative of the position of the young human warrior. The male who threw down stones was a Kahama male called Sniff who was later killed.

13 Kuroda (1999) distinguishes between food distribution used to maintain a dominance hierarchy and communitas distribution diminishing hierarchy, and argued that the *dynamism* of the latter promoted the development of the former, contributing to the development of the food distribution system. The logical structure of this chapter as a whole is the same in that the equality principle that creates order is maintained by the communitas equality that destroys it (Kuroda 1999).

14 Itani (2008) positions the intra-unit-group infanticide carried out by chimpanzee males as an act in which the members "themselves transform" the social structure to achieve the result of stabilizing the male group and promoting their transformation into a combat group (*"war culture complex"*: Kuroda 1999). He intended to go on to discuss that chimpanzees also perform cultural behaviors to achieve their identities, but this act may also be an example of structured anti-structure. This is because, according to Itani's logic, infanticide occurs in a state of chaotic excitement, and the infants and juveniles, through the grotesque excitement and the fear that they could have been the victim themselves, are attracted to the violent adult males to acquire the violence to grow up into an adult who may one day perform infanticide himself.

15 In the actual physical attack, there are cases when they do not raise their hair or screech (Bygott 1979). This only occurs when the order of dominance within the group is clear.

Chapter 12

1 The discussion on Girard is based, in particular, on *Violence and the Sacred*, published in 1972 and *Things Hidden Since the Foundation of the World*, published in 1978.

2 Please refer to Imamura's discussion of interpellation (H. Imamura 1989a).

3 Santos-Granero (1991), who conducted a field survey in lowland Peru and has also discussed Hobbes, supports Clastres' argument, but I cannot help questioning whether the "uncivilized societies" in areas other than South America, for example the societies of Bushmen, were at war with each other. Please also refer to Chapter Six of this book by Tokoro.

4 *Totem and Taboo* by Freud, published in Gemany in 1913, may also be considered to be based on the conflict model.

5 Here the discussion is based on Girard in particular. Imamura's theory is not that simple when some of his other publications (H. Imamura 2000, 2007) are taken into account. Hobbes' conflict theory lacks thoroughness in that both the emotional man who promotes mutual violence and the rational man who overcomes such a state of nature coexist. For Hobbes' view on humanity, please refer to Arima (2002).

6 I will refrain from discussing specific ethnographies in this chapter but I would just like to note the Malay society in the Southeast Asian islands portrayed by Tachimoto (1996, 2001) as a typical network society.

7 Notable studies in this field are gathered in Mitchell (1969).

8 The network theory introduced in this chapter is proposed from a relatively logical and universal perspective. However, there is another interpretation that takes the historical perspective that a "network" is a form of flexible and egalitarian human connections that replaces groups with established boundaries and identities as they deteriorate during the modern and post-modern historical stages. Please refer to Katsuhiro Yamazumi and Yrjö Engeström (2008) for details.

9 Please refer to Carsten (2000), Viegas (2003), Bamford (2004) and Nyambedha and Aagaard-Hansen (2007). Myers (1986) also proposes the relatedness concept but Carsten has not referred to this work. According to Myers, "creating and maintaining relatedness requires mutual interaction, reciprocity and exchange" (1986: 163). And as is evident from this quote, "relatedness" here is very similar to Carsten's concept. I learnt of Myers' work from Terashima's article in Chapter Eight of this book.

10 Please refer to Omura's thesis in Chapter Five of this book for specific cases.

11 Commonality with resonance in Kitamura's thesis of Chapter Three should also be noted.

12 Althusser distinguishes between subjectification by interpellation of Subject and interpellation of subjects (Althusser 1971; M. Tanaka 2006). The former is the origin of interpellation and according to Imamura, this is the third element (H. Imamura 1989). The latter will not emerge without the interpellation of the former. Also, the former is not a mutual interpellation but is instead one-sided. If one only looks at the reciprocity of the latter then the similarity with seduction may stand out, but the difference between seduction that does not assume the former and the interpellation that does is huge.

13 Freud, in his thesis on Medusa (1997[1940]), wrote about the similarity between turning into stone upon being stared at by Medusa and an erection. Turning into stone = erection does not necessarily lead to the world of Eros. It can even result in an excessive and unrelenting objectification, in other words, death. This is what Imamura discusses as the hell of materialization, citing Sartre.

14 This is detailed in Girard (1965: 1–2, 1977: 146). When one enters a competitive relationship with others over a single target, Girard refers to "others" as "*internal mediation*" (1965: 9). The "self" imitates the desire of the mediator. On the other hand, when the mediator becomes the target of worship, competition will not arise even if the self desires the target of the mediators' desire. Girard calls this the "*external mediation*". They are both media but one reinforces jealousy while the other reinforces worship. In our current society, confrontation based on this internal medium will become dominant. Girard seems to think here that the phenomenon of mutual violence may become more intense in today's environment.

15 The relationship with things has been discussed in M. Tanaka (2009b).

16 Imamura mainly discusses Merleau-Ponty (1959), but as the content overlaps Merleau-Ponty (1968), we will examine the arguments in the latter from this point.

17 If I dare say so, "violent expression" as seen in dehiscence is not violence against others, but in that it is violence against the self, it is still an expression that entails a violent nature.

18 Regarding this point, Takiura (1974: 124) wrote "for Merleau-Ponty, *existence* was an invisible being, the unconscious Eros". Here, the flesh is regarded as the tangible form of the "existence". Please also refer to Michel Henry (2000) for his discussion on eroticism, though it is critical of Merleau-Ponty. Merleau-Ponty cited the example of a woman in the street feeling someone looking

at her breast, and checking her clothing (Merleau-Ponty 1968: 189–190), and such examples can be interpreted in the context of seduction. Though I could not include it in this chapter, I would also like to mention Naka (2008) for his criticism on reversibility.

19 In regards to flesh, please refer to Merleau-Ponty (1968: 259–260). Matsuba (2008) objected to Merleau-Ponty's claim that a community needs some form of representation, saying that this would be difficult for a community of flesh. However, this should not be regarded as a defect.

20 As already pointed out, this is because the sexual relationship itself entails the potential for anti-erotic development and upholds the homosocial bond that leads to patriarchy (Sedgwick 1985).

Chapter 13

1 To consider the state of "relate […] at the same time", we would need to further examine the meaning and content of "at the same time" and "relate". Nevertheless, we shall advance our argument here on the assumption that a broader definition is used.

2 The ethnographical present is the time of completion of my survey in 1977 (Funabiki 1981, 1982).

3 The four essays were published in the following years respectively: "Aru 'kyōsei' no keiken kara" (From an experience of "coexistence") (1969), "Peshimisuto no yūki kara" (Out of courage of a pessimist) (1970), "Bōkyō to umi" (Nostalgia and the sea) (1971) and "Shitsugo to chinmoku no aida" (Between a loss of words and silence) (1972).

4 The Russian Republic as a constituent republic of the Soviet Union.

5 For example, Robinson Crusoe did not stop being a social being despite his solitariness, even before he met Friday. That is why he was "able to meet" Friday. Robinson in *Furaidē aruiwa Taiheiyō no meikai* (Friday or the limbo of the Pacific) by Michel Tournier (1996), a novel based on *Robinson Crusoe*, is portrayed as a singular man who does not carve out a scene with Friday. As Gilles Deleuze explains, Robinson continues to live in "the world without others" (Tournier 1996: 299) even when he meets Thursday after Friday departs.

Bibliography

Abbink, J. (1993) "Reading the entrails: Analysis of an African divination discourse". *Man* (N.S.), 28: 705–726.

Adachi, K. (2003) "Kongun to iu shakai" (The society of mixed species associations). In M. Nishida, K. Kitamura and J. Yamagiwa (eds) *Ningensei no kigen to shinka* (Origin and evolution of humanity). Kyoto: Shōwadō, 204–232.

Aimi, M. and N. Koyama (2006) "Kitsunezaru-rui wa dono yōni bunrui saretekitaka" (A historical review of Lemur classification)". *Reichōrui Kenkyū* (Primate Research), 22: 97–116.

Allen, W. J. and E. Timothy (eds) (2000) *The Evolution of Human Societies: From Foraging Group to Agrarian State* (second edition). Stanford: Stanford University Press.

Althusser, L. (1971) "Ideology and ideological state apparatuses". In *Lenin and Philosophy and other Essays*, B. Brewster (trans.). London: New Left Books.

Anderson, B. (1991) *Imagined Communities: Reflections on the Origin and Spread of Nationalism*. New York: Verso Books.

Anderson, B. (1997) *Sōzō no kyōdōtai*. Japanese translation of B. Anderson (1991). T. Shiraishi and S. Shiraishi (trans.). Tokyo: Libro Port.

Aoki, E. (2009) "Shinmitsusei to karada: Fetishizumu genshō to jinruigaku no chihei" (Intimacy and the body: The fetishism phenomenon and the horizon of anthropology). In M. Tanaka (ed.), *Fetishizumu ron no keifu to tenbō* (The genealogy and prospects of discussion on fetishism). Kyoto: Kyoto University Press, 319–356.

Ardrey, R. (1961) *African Genesis: A Personal Investigation into the Animal Origins and Nature of Man*. London: Fontana Collins.

Ardrey, R. (1973) *Afurika sōseiki—satsuriku to tōsō no jinrui shi*. Japanese translation of R. Ardrey (1961). K. Tokuda (trans.). Tokyo: Chikuma Shobō.

Ardrey, R. (1976) *The Hunting Hypothesis: A Personal Conclusion Concerning the Evolutionary Nature of Man*. New York: Atheneum.

Ardrey, R. (1978) *Kari o suru saru—nikushoku kōdō kara hito o kangaeru*. Japanese translation of R. Ardrey (1976). K. Tokuda (trans.). Tokyo: Kawade Shobō Shinsha.

Arima, T. (2002) *Hobbusu "Rivaiasan" no ningenzō: riseiteki ningen no imēji* (The human image in Hobbes' *Leviathan*: The image of rational man). Tokyo: Kindaibungeisha.

Aristotle (1961) *Seijigaku* (Politics). M. Yamamoto (trans.). Tokyo: Iwanami Bunko.

Aristotle (2001) *Seijigaku* (Politics). N. Ushida (trans.). Kyoto: Kyoto University Press.

Asmarom, L. (1973) *Gada: Three Approaches to the Study of African Society.* New York: The Free Press.

Aureli, F., C. M. Schaffner, C. Boesch, S. K. Bearder, J. Call, C. A. Chapman, R. Connor, A. Di Fiore, R. I. M. Dunbar, S. P. Henzi, K. Holekamp, A. H. Korstjens, R. Layton, P. Lee, J. Lehmann, J. H. Manson, G. Ramos-Fernandez, K. B. Strier and C. P. van Schaik (2008) "Fission-fusion dynamics: New research framework". *Current Anthropology*, 49: 627–654.

Baba, H. (1999) "Hito no kōgekisei to shokujin" (Human aggression and cannibalism). In K. Fukui and H. Harunari (eds), *Tatakai no shinka to kokka no seisei* (The evolution of disputes and the creation of the state). Tokyo: Tōyō Shorin, 33–56.

Balikci, A. (1989) *The Netsilik Eskimos.* Garden City: Natural History Press.

Bamford, S. (2004) "Conceiving relatedness: Non-substantial relations among the Kamea of Papua New Guinea". *The Journal of the Royal Anthropological Institute*, (N.S.) 10: 287–306.

Barnes, J. A. (1954) "Class and committees in a Norwegian island Parish". *Human Relations*, 7: 39–58.

Barton, R. A. (2000) "Socioecology of baboons: The interaction of male and female strategies". In P. M. Kappeler (ed.), *Primate Males*. Cambridge: Cambridge University Press, 97–107.

Baudrillard, J. (1990) *Seduction.* B. Singer (trans.). Oxford: Macmillan Education.

Beckerman, S. and P. Valentine (eds) (2002) *Cultures of Multiple Fathers: The Theory and Practice of Partible Paternity in Lowland South America.* Miami: University Press of Florida.

Bodenhorn, B. (1989) *The Animals Come to Me, They Know I Share—Iñupiaq Kinship, Changing Economic Relations and Enduring World Views on Alaska's North Slope.* PhD thesis: Cambridge University.

Bodenhorn, B. (1990) "I am not the great hunter, my wife is". *Études/Inuit/Studies*, 14(2): 55–74.

Boissevain, J. (1974) *Friends of Friends: Networks, Manipulators and Coalitions.* London: Basil Blackwell.

Bott, E. (1955) "Urban families: Conjugal roles and social networks". *Human Relations*, 8: 345–384.

Bradley, B. J., D. M. Doran-Sheehy, D. Lukas and C. Boesch (2004) "Disposed male networks in western gorillas". *Current Biology*, 14: 510–513.

Bradley, B. J., M. M. Robbins, E. A. Williamson, H. D. Steklis, N. G. Steklis, N. Eckhardt, C. Boesch and L. Vigilant (2005) "Mountain gorilla tug-of-war:

Silverbacks have limited control over reproduction in multi-male groups". *Proceedings of the Natural Academy of Science*, 102: 9418–9423.

Bygott, J. D. (1979) "Agonistic behavior and dominance among wild chimpanzees". In D. Hamburg and E. McCown (eds), *The Great Apes*. Menlo Park: Benjamin/ Cummings, 405–427.

Caillois, R. (1939) *L'homme et le sacre*. Paris: Leroux.

Caillois, R. (1963) *Bellone ou la pente de la guerre*. Paris: Niset.

Caillois, R. (1974) *Sensōron*. Japanese translation of R. Cailliois (1963). S. Akieda (trans.). Tokyo: Hōsei University Press.

Caillois, R. (1994) *Ningen to sei naru mono*. Japanese translation of R. Caillois (1939). F. Tsukahara et al. (trans.). Tokyo: Serika Shobō.

Campbell, J. R. (2006) "Who are the Luo? Oral tradition and disciplinary practices in anthropology and history". *Journal of African Cultural Studies*, 18(1): 73–87.

Carsten, J. (1995) "The substance of kinship and the heat of the hearth: Feeding, personhood and relatedness among the Malays in Pulau Langkawi". *American Ethnologist*, 22: 223–241.

Carsten, J. (2000) "Introduction". In J. Carsten (ed.), *Cultures of Relatedness: New Approaches to the Study of Kinship*. Cambridge: Cambridge University Press, 1–36.

Cartmill, M. (1993) *A View to a Death in the Morning: Hunting and Nature through History*. Cambridge: Harvard University Press.

Clastres, P. (1974) *La Société contre l'"Etat: Recherches d'"anthropologie politique* (Society against the state: Studies in political anthropology). Paris: Editions de minuit.

Clastres, P. (1977) *Archeologie de la violence: la guerre dans les societes primitives*. Paris: Editions de l'Aube.

Clastres, P. (1989) *Kokka ni kōsuru shakai: Seiji jinruigaku kenkyū*. Japanese translation of P. Clastres (1974). K. Watanabe (trans.). Tokyo: Suiseisha.

Clastres, P. (1994) *Archaeology of Violence*. J. Herman (trans.). New York: SEMIOTEXT(E).

Clastres, P. (2003) *Bōryoku no kōkogaku—mikai shakai ni okeru sensō*. Japanese translation of P. Clastres (1977). M. Marimo (trans.). Tokyo: Gendai Kikakushitsu.

Clemens, Z., B. Merker and M. Ujhelyi (2008) "Observations on paternal care in a captive family of white-handed gibbons (*Hylobates lar*)". *Gibbon Journal*, 4: 46–50.

Copeland, S. R., M. Sponheimer, D. J. de Ruiter, J. A. Lee-Thorp, D. Cordon, P. J. le Roux, V. Grimes and M. P. Richards (2011) "Strontium isotope evidence for landscape use by early hominids". *Nature*, 474: 76–78.

Damas, D. (1972) "Central Eskimo systems of food sharing". *Ethnology*, 11(3): 220–240.

Darwin, C. (1872) *The Expression of Emotions in Man and Animal*. London: John Murray.

Darwin, C. (1921) *Ningen oyobi dōbutsu no hyōjō*. Japanese translation of C. Darwin (1872). G. Andō and A. Okamoto (trans.). Tokyo: Nihon Hyōronsha.

Dawkins, R. (1976) *The Selfish Gene*. Oxford: Oxford University Press.

Dawkins, R. (1982) *The Extended Phenotype—The Long Reach of the Gene*. Oxford: Oxford University Press.

Dawkins, R. (1987) *Enchō sareta hyōgen-kei*. Japanese translation of R. Dawkins (1982). T. Hidaka et al. (trans.). Tokyo: Kinokuniya Shoten.

Dawkins, R. (1991) *Rikoteki na idenshi*. Japanese translation of R. Dawkins (1976). T. Hidaka et al. (trans.). Tokyo: Kinokuniya Shoten.

De Certeau, M. (1987) *Nichijōteki jissen no poietīku* (*L'Invention du quotidien, 1, Art de Faire*. Paris: Union Generale d'Editions. 1980.). T. Yamada (trans.). Tokyo: Kōkubunsha.

De Waal, F. (1982) *Chimpanzee Politics: Power and Sex among Apes*. Maryland: The John Hopkins University Press.

De Waal, F. (1989) *Peacemaking among Primates*. Cambridge: Harvard University Press.

De Waal, F. (1993) *Nakanaori senjutsu—reichōrui wa heiwana kurashi o dono yōni jitsugen shite iru ka*. Japanese translation of F. De Waal (1989). T. Nishida and T. Enomoto (trans.). Tokyo: Dōbutsusha.

De Waal, F. (1998) *Seiji o suru saru: Chinpanjī no kenryoku to sei*. Japanese translation of F. De Waal (1982). T. Nishida (trans.). Tokyo: Dōbutsusha.

De Waal, F. (2005a) *Our Inner Ape: A Leading Primatologist Explains Why We Are Who We Are*. Riverhead Books: New York.

De Waal F. (2005b) *Anatano naka no saru*. Japanese translation of F. De Waal (2005a). R. Fujii (trans.). Tokyo: Hayakawa Shobō.

Devereux, G. (1967) *From Anxiety to Method in Behavioral Sciences*. The Hague: Mouton.

Di Fiore, A. (2003) "Molecular genetic approaches to the study of primate behavior, social organization, and reproduction". *Yearbook of Physical Anthropology*, 46: 62–99.

Di Fiore, A. (2009) "Genetic approaches to the study of dispersal and kinship in New World primates". In P. A. Garber, A. Estrada, J. C. Bicca-Marques, E. W. Heyman and K. B. Strier (eds), *South American Primates*. Berlin: Springer-Verlag, 211–250.

Di Fiore, A. and D. Rendall (1994) "Evolution of social organization: A reappraisal for primates by using phylogenetic methods". *Proceedings of the Natural Academy of Science*, 91: 9941–9945.

Di Fiore, A., A. Link, C. A. Schmitt and S. N. Spehar (2009) "Dispersal patterns in sympatric woolly and spider monkeys: Integrating molecular and observational data". *Behaviour*, 164: 437–470.

Diamond, J. (1992) *The Third Chimpanzee: The Evolution and Future of the Human Animal*. New York: Perrenial Library.

Diamond, J. (1993) *Ningen wa doko made chinpanjī ka?—jinrui shinka no eikō to kageri*. Japanese translation of J. Diamond (1992). M. Hasegawa and T. Hasegawa (trans.). Tokyo: Shinyōsha.

Diamond, J. (1997) *Guns, Germs and Steel: The Fates of Human Societies*. New York: W. W. Norton.

Diamond, J. (2000) *Jū, byōgenkin, tetsu—1 man 3000 nen ni wataru jinrui shi no nazo jōgekan* (Vols. 1 and 2). Japanese translation of J. Diamond (1997). A. Kurahone (trans.). Tokyo: Sōshisha.

Dobson, F. S. (1982) "Competition for mates and predominant juvenile male dispersal in mammals". *Animal Behaviour*, 30: 1183–1192.

Douadi, M. I., S. Gatti, F. Levero, G. Duhamel, M. Bermejo, D. Vallet, N. Menard and E. J. Petit (2007) "Sex-biased dispersal in western lowland gorillas (Gorilla gorilla gorilla)". *Molecular Ecology*, 16: 2247–2259.

Durham, W. (1992) *Coevolution: Genes, Culture, and Human Diversity*. Stanford: Stanford University Press.

Durkheim, E. (1915) *The Elementary Forms of Religious Life: A Study in Religious Sociology*. J. W. Swain (trans.). London: Allen and Unwin.

Ekejiuba, F. (1995) "Down to fundamentals: Women-centred hearth-holds in rural West Africa". In F. D. Bryceson (ed.) *Women Wielding the Hoe: Lessons from Rural Africa for Feminist Theory and Development Practice*. Oxford: Berg, 47–61.

Elias, N. and E. Danning (1966) *Sport et civilisation*. Paris: Fayard.

Elias, N. and E. Danning (1995) *Supōtsu to bunmeika*. Japanese translation of N. Elias and E. Danning (1966). A. Ōhira (trans.). Tokyo: Hōsei University Press.

Ellanna, L. (1991) "More than subsistence—The harvest of wild resources and Alaskan native cultural and social identity". *International Symposium on Dietary and Housing Aspect of Northern Peoples—Proceedings of the 6th International Abashiri Symposium*. Abashiri: Hokkaido Museum of Northern Peoples, 28–37.

Ellsworth, J. A. (2000) "Molecular evolution, social structure, and phylogeography of the mantled howler monkey (Alouatta palliate)". PhD thesis: University of Nevada.

Epstein, A. L. (1969) "Network and urban society organization". In J. C. Mitchell (ed.) *Social Networks in Urban Situations: Analyses of Personal Relationships in Central African Towns.* Manchester: Manchester University Press, 77–116.

Eriksson, J., H. Siedel, D. Lukas, M. Kayser, A. Erler, C. Hashimoto, G. Hohmann, C. Boesch and L. Vigilant (2006) "Y-chromosome analysis confirms highly sex-biased dispersal and suggests a low male effective population size in bonobos (*Pan paniscus*)". *Molecular Ecology*, 15: 939–949.

Evans-Pritchard, E. E. (1940) *The Nuer: A Description of the Modes of Livelihood and Political Institutions of a Nilotic People.* Oxford: Oxford University Press.

Evans-Pritchard, E. E. (1965 [1949]) "Luo tribes and clans". In E. E. Evans-Pritchard (ed.) *The Position of Women in Primitive Societies and Other Essays in Social Anthropology.* London: Faber and Faber, 205–227.

Fienup-Riordan, A. (1983) *The Nelson Island Eskimo.* Anchorage: Alaska Pacific University Press.

Fienup-Riordan, A. (1990) *Eskimo Essays.* New Brunswick: Rutgers University Press.

Fietz, J. and K. H. Dausmann (2003) "Costs and potential benefits of parental care in the nocturnal fat-tailed dwarf lemur (*Cheirogaleus medius*)". *Folia Primatologica*, 74: 246–258.

Fietz, J., H. Zischler, C. Schwiegk, J. Tomiuk and K. H. Dausmann (2000) "High rates of extra-pair young in the pair-living fat-tailed dwarf lemur, Cheirogaleus medius". *Behavioral Ecology and Sociobiology*, 49: 8–17.

Fox, E. A. (2002) "Female tactics to reduce sexual harassment in the Sumatran orangutan (*Pongo pygmaeus abelii*)". *Behavioral Ecology and Sociobiology*, 52: 93–101.

Frank, A. (1991) "For a sociology of the body: An analytical review". In M. Featherstone, M. Hepworth and B. S. Turner (ed.) *The Body: Social Process and Cultural Theory.* London: Sage Publications, 36–102.

Fredsted, T., C. Pertoldi, J. M. Olesen, M. Eberle and P. M. Kappeler (2004) "Microgeographic heterogeneity in spatial distribution and mtDNA variability of gray mouse lemurs (*Microcebus murinus*, Primates: Cheirogaleidae)". *Behavioral Ecology and Sociobiology*, 56: 393–403.

Fredsted, T., M. H. Schierup, L. F. Groeneveld and P. M. Kappeler (2007) "Genetic structure, lack of sex-biased dispersal and behavioral flexibility in the pair-living fat-tailed dwarf lemur, *Cheirogaleus medius*". *Behavioral Ecology and Sociobiology*, 61: 943–954.

Freud, S. (1918) *Totem and Taboo: Resemblances Between the Psychic Lives of Savages and Neurotics.* A. A. Brill (trans.). New York: MoffatYard and Company.

Freud, S. (1940) "Medusenhaupt". *Int. Z. Psychoanal. Imago*, 25.

Freud, S. (1997) "Medūsa no kubi: sōkō". Japanese translation of S. Freud (1940). In *Erosu ronshū* (Collection of Eros theory). G. Nakayama (trans.). Tokyo: Chikuma Gakugei Bunko, 275–279.

Fuentes, A. (2002) "Patterns and trends in primate pair bonds". *International Journal of Primatology*, 23: 953–978.

Fukui, K. (1979) "Cattle color symbolism and inter-tribal homicide among Bodi". In K. Fukui and D. Turton (eds) *Warfare among East African Herders: Senri Ethnological Studies 3*. Osaka: National Museum of Ethnology.

Fukui, K. (1984) "Tatakai kara mita buzoku kankei: Higashi Afurika ni okeru ushi bokuchikumin Bodi Meken o chūshin ni" (Tribal relationships from the perspective of battle: Focusing on the East African pastoralists of Bodi Meken). *Minzokugaku Kenkyū* (Ethnographic studies), 48(4): 471–480.

Fukui, K. (1996) "Tatakai no hajimari to shinka" (The beginning of battle and evolution). In National Museum of Japanese History (ed.) *Wakoku Midaru* (The country of Japan in disorder). Tokyo: Asahi Shinbun, 132–136.

Fukui, K. (1999) "Tatakai no shinka to minzoku no seizon senryaku" (The evolution of battle and the tribal survival strategy). In K. Fukui and H. Harunari (eds) *Jinrui ni totte no tatakai towa* (1) *Tatakai no shinka to kokka no seisei* (What battle means for humankind (1): The evolution of battle and the formation of the state). Tokyo: Tōyō Shorin, 161–184.

Fukui, K. et al. (2004) "Tokushū: Hito wa naze tatakau no ka" (Special feature: Why do humans fight). In K. Fukui (ed.) *Kikan minzokugaku* (Quarterly ethnology), 109: 4–62.

Funabiki, T. (1981) "On pigs of the Mbotgote in Malekula". In M. Allen (ed.), *Vanuatu: Politics, Economics and Ritual in Island Melanesia*. New York: Academic Press, 173–188.

Funabiki, T. (1982) *Mbotgote Ritual: A Study on the Ritual Life and Social Organization of the Mbotgote in the Interior Region of Malekula Island, Vanuatu*. PhD thesis: University of Cambridge.

Furuichi, T. (1991) "Fukei shakai o gyūjiru mesu tachi" (The patrilineal society controlled by females). In T. Nishida (ed.) *Saru no bunkashi* (Monographs of primate cultures. Tokyo: Heibonsha, 561–581.

Furuichi, T. (1999) *Sei no shinka, hito no shinka—Ruijinen Bonobo no kansatsu kara* (The evolution of sex, the evolution of humans: From the observation of bonobo apes). Tokyo: Asahi Shinbunsha.

Furuichi, T. (2002) "Hito jōka no shakai kōzō no shinka no saikentō—Shokumotsu no bunpu to hatsujōsēhi ni chakumoku shite" (Evolution of social structure of Hominoids: Reappraisal from food distribution and estrus sex ratio). *Reichōrui Kenkyū* (Primate Research), 18: 187–201.

Gerloff, U., B. Hartung, G. Fruth, G. Hohmann and D. Tautz (1999) "Intra-community relationships, dispersal pattern, and paternity success in a wild living community of bonobos (Pan paniscus) determined from DNA analysis of faecal samples". *Proceedings of the Royal Society of London, Series B, Biological Sciences*, 266: 1189–1195.

Girard, R. (1965) *Deceit, Desire, and the NOVEL: Self and Other in Literary Structure.* Y. Freccero (trans.). Baltimore: The Johns Hopkins University Press.

Girard, R. (1972) *La Violence et le Sacré.* Paris: Editions Bernard Grasset.

Girard, R. (1977) *Violence and the Sacred.* P. Gregory (trans.). Baltimore: The Johns Hopkins University Press.

Girard, R. (1982) *Bōryoku to seinaru mono.* Japanese translation of R. Girard (1972). Y. Furuta (trans.). Tokyo: Hōsei Daigaku Shuppan-kyoku.

Girard, R. (1987) *Things Hidden Since the Foundation of the World.* S. Bann and M. Metteer (trans.). London: The Athlone Press.

Goldberg, T. L. and R. W. Wrangham (1997) "Genetic correlates of social behavior in wild chimpanzees: Evidence from mitochondrial DNA". *Animal Behaviour*, 54: 559–570.

Goodall, J. (1971) *In the Shadow of Man.* Glasgow: William Collins Sons & Co. Ltd.

Goodall, J. (1973) *Mori no rinjin: Chinpanjī to watashi.* Japanese translation of J. Goodall (1971). H. Kawai (trans.). Tokyo: Heibonsha.

Goodall, J. (1986) *The Chimpanzees of Gombe: Patterns of Behavior.* Cambridge: Belknap Press.

Goodall, J. (1990a) *Through a Window: Thirty Years Observing the Gombe Chimpanzees.* London: Weidenfeld and Nicolson.

Goodall, J. (1990b) *Yasei chinpanjī no sekai.* Japanese translation of J. Goodall (1986). Yukimaru Sugiyama et al. (trans.). Kyoto: Minerva Shobō.

Goodall, J. (1994) *Kokoro no mado—Chinpanjī to no 30 nen.* Japanese translation of Goodall, J. (1990a). K. Takasaki, H. Takasaki and J. Itani (trans.). Tokyo: Dōbutsusha.

Goossens, B., J. M. Setchell, S. S. James, S. M. Funk, L. Chikhi, A. Abulani, M. Ancrenaz, I. Lackman-Ancrenaz and M. W. Bruford (2006) "Philopatry and reproductive success in Bornean orangutans (Pongo pygmaeus)". *Molecular Ecology*, 15: 2577–2588.

Graeber, D. (2001) *Toward an Anthropological Theory of Value: The False Coin of Our Own Dreams*. New York: Palgrave Macmillan.

Greenwood, P. J. (1980) "Mating systems, philopatry and dispersal in birds and mammals". *Animal Behaviour*, 28: 1140–1162.

Gursky, S. (2007) "Tarsiformes". In C. J. Campbell, A. Fuentes, K. C. Mackinnon, M. Panger and S. K. Bearder (eds) *Primates in Perspective*. New York: Oxford University Press, 73–85.

Hamada, Y. (1999) "Reichōrui no keitō bunrui" (Phylogenetic classification of primates). In T. Nishida and S. Uehara (eds) *Reichōruigaku o manabu hito no tameni* (For apprentices of primatology). Kyoto: Sekai Shisōsha, 25–49.

Hammond, R. L., L. J. L. Handley, B. J. Winney, M. W. Bruford and N. Perrin (2006) "Genetic evidence for female-biased dispersal and gene flow in a polygynous primate". *Proceedings of the Royal Society*, 273: 479–484.

Handley, L. J. and N. Perrin (2007) "Advances in our understanding of mammalian gender-biased dispersal". *Molecular Ecology*, 16: 1559–1578.

Hapke, A., D. Zinner and H. Zischler (2001) "Mitochondrial DNA variation in Eritrean hamadryas baboons (*Papio hamadryas hamadryas*): Life history influences population genetic structure". *Behavioral Ecology and Sociobiology*, 50: 483–492.

Harako, R. (1976) "The Mbuti as hunters: A study of ecological anthropology of the Mbuti pygmies (1)". *Kyoto University African Studies*, 10: 37–99.

Haraway, D. (1989) *Primate Visions: Gender, Race, and Nature in the World of Modern Science*. New York: Routledge.

Hart, D. and R. W. Sussman (2005) *Man, the Hunted: Primates, Predators, and Human Evolution*. New York: Westview Press.

Hart, D. and R. W. Sussman (2007) *Hito wa taberarete shinka shita*. Japanese translation of D. Hart and R. W. Sussman (2005). N. Itō (trans.). Kyoto: Kagaku Dōjin.

Hasegawa, M. and T. Hasegawa (2000) *Shinka to ningen kōdō* (Evolution and human behaviour). Tokyo: University of Tokyo Press.

Henry, M. (2000) *Incarnation: Une philosophie de la chair*. Paris: Seuil.

Henry, M. (2007) *Juniku: "Niku" no tetsugaku*. Japanese translation of M. Henry (2000). Y. Naka (trans.) Tokyo: Hōsei University Press.

Hill, R. A. and F. Lee (1998) "Predation risk as an influence on group size in cercopithecoid primates: Implications for social structure". *Journal of Zoology*, 245: 447–456.

Hobbes, T. (1651) *Leviathan*.

Hobbes, T. (1914) *Leviathan 1*. London: The Aldine Press.

Hobbes, T. (1992) *Ribaiasan*. Japanese translation of T. Hobbes (1914). H. Mizuta
(trans.) (1992). Tokyo: Iwanami Bunko.

Hosaka, K. (2002) "Shuryō, nikushoku kōdō" (Hunting and predation). In T. Nishida,
S. Uehara and K. Kawanaka (eds) *Mahale no chinpanjī—Pansuroporojī no 37
nen* (Chimpanzees of Mahale: Thirty-seven years in panthropology). Kyoto:
Kyoto University Press, 219–260.

Hrdy, S. B. (1977) "Infanticide as a primate reproductive strategy". *American Scientists*, 65: 40–49.

Huck, M., P. Lottker, U.-R. Bohle and E. W. Heyman (2005) "Paternity and kinship
patterns in polyandrous moustached tamarins (Saguinus mystax)". *American
Journal of Physical Anthropology*, 127: 449–464.

Hussein, T. (2006) *Historical and Current Perspectives on Inter-ethnic Conflicts in
Northern Kenya*. Masters thesis: Norwegian University of Science and Technology.

Hutchinson, S. (1996) *Nuer Dilemmas: Coping with Money, War and the State*. Berkeley: University of California Press.

Ichikawa, M. (1978) "The residential group of the Mbuti pygmies". *Senri Ethnological
Studies*, 1: 131–88.

Ichikawa, M. (1982) *Mori no shuryō min: Mubuti pigumī no seikatsu* (Hunters of the
forest: The life of the Mbuti Pygmy). Kyoto: Jinbun Shoin.

Ichikawa, M. (1986) "Afurika shuryō saishū shakai no kasosei" (The plasticity of
African hunter-gatherer societies). In J. Itani and J. Tanaka (eds) *Shizen Shakai
no Jinruigaku—Afurika ni ikiru* (Anthropology of natural societies: Living in
Africa). Kyoto: Akademia Shuppankai, 279–311.

Idani, G. (1990) "Relations between unit-groups of bonobos at Wamba, Zaire: Encounters and temporary fusions". *African Study Monographs*, 11: 153–186.

Ikegami, Y. (2004) "'Kizuna—kyōdōsei o toinaosu' joron" (An introduction to
"Bonds—questioning communality"). *Iwanami kōza shūkyō 6 kizuna* (Iwanami
religion series, vol. 6, bonds). Tokyo: Iwanami Shoten, 1–24.

Imamura, H. (1982) *Bōryoku no ontorogī* (The ontology of violence). Tokyo: Keisō
Shobō.

Imamura, H. (1989a) "Ideorogī to purakutisu" (Ideology and practice). In S. Tanabe
(ed.) *Jinruigaku teki ninshiki no bōken: Ideorogī to purakutisu* (The adventure of
the anthropological understanding: Ideology and practice). Tokyo: Dōbunkan,
123–147.

Imamura, H. (1989b) *Haijo no kōzō* (The structure of exclusion), Tokyo: Seidosha.

Imamura, H. (1993) *Tai de kangaeru* (Philosophizing in Thailand). Tokyo: Seidosha.

Imamura, H. (1994a) *Chūgoku de kangaeru* (Philosophizing in China). Tokyo: Sei-dosha.

Imamura, H. (1994b) *Kindaisei no kōzō: "Kuwadate" kara "kokoromi" e* (The structure of modernity: From "attempt" to "trial"). Tokyo: Kōdansha.

Imamura, H. (2000) *Kōeki suru ningen, homo komunikansu: Zōyo to kōkan no nin-gengaku* (Humans who trade: The anthropology of gifts and exchange). Tokyo: Kōdansha.

Imamura, H. (2005a) *Kōsō suru ningen* (Disputatious humans). Tokyo: Kōdansha.

Imamura, H. (2005b) *Marukusu nyūmon* (Introduction to Marxism). Tokyo: Chi-kuma Shinsho.

Imamura, H. (2007) *Shakaisei no tetsugaku* (The philosophy of sociality). Tokyo: Iwanami Shoten

Imamura, H. (ed.) (1988) *Gendai shisō o yomu jiten* (Encyclopedia of contemporary thought). Tokyo: Kōdansha Gendai Shinsho.

Imamura, H. and S. Imamura (2007) *Girei no ontorogī* (Ontology of rituals). Tokyo: Kōdansha.

Imamura, K. (1992) "Sentoraru Karahari San ni okeru saishū katsudō" (Gathering activities of Central Kalahari San). *Afurika Kenkyū* (African studies), 41: 47–73.

Imamura, K. (1993), "San no kyōdo to bunpai: Josei no seigyō katsudō no shiten kara" (Cooperation and distribution of the San: From the perspective of women's livelihood activities). *Afurika Kenkyū* (Africa studies), 42: 1–25.

Imamura, K. (1996) "Dōchō kōdō no shosō: Busshuman no nichijō seikatsu kara" (Various phases of acts of conformity: From the day-to-day lives of the Bush-men)". In K. Sugawara and M. Nomura (eds) *Komyunikēshon to shite no karada* (Body as communication). Sōsho: Karada to bunka (Series: Body and culture) (vol. 2), Taishūkan Shoten, 71–91.

Imanishi, K. (1949) "Seibutsu shakai no ronri" (Biotic communities). Tokyo: Mainichi Shinbunsha.

Imanishi, K. (1957) "Nihonzaru kenkyū no genjō to kadai" (The present condition of and challenges facing Japanese macaque research). *Primates* 1: 1–29.

Imanishi, K. (1960) "Tori, saru, ningen—aidentifikēshon o sasaeru ippan riron ga kanō darōka" (Birds, monkeys, humans: Is a general theory of identification possible?). *Jinbun gakuhō* (Journal of humanities), 12: 1–24.

Imanishi, K. (1971) *Seibutsu shakai no ronri* (The theory of societies of living things). Tokyo: Shisakusha.

Imanishi, K. (1976) "Reichōrui shakai no shomondai" (Issues in primate society). In *Bessatsu saiensu tokushū dōbutsu shakaigaku—saru kara hito e* (Supplemental issue of *Science* "Animal sociology: From monkeys to humans"). Tokyo: Nikkei Shinbunsha, 9–21.

Imanishi, K. (1987) "Mure seikatsusha tachi" (Group dwellers). *Kikan Jinruigaku* (Anthropology quarterly), 181: 82–96.

Imanishi, K. (2002) *Seibutsu no sekai* (The world of organisms and others). Chūkō Classics. Tokyo: Chūō Kōron Shinsha.

Ingold, T. (2000) *The Perception of the Environment*. London: Routledge.

Inoue, E., M. Inoue-Murayama, L. Vigilant, O. Takenaka and T. Nishida (2008) "Relatedness in wild chimpanzees: Influence of paternity, male philopatry, and demographic factors". *American Journal of Physical Anthropology*, 137: 256–262.

Isbell, L. A. and D. van Vuren (1996) "Differential costs of locational and social dispersal and philopatry and their consequences for female group-living primates". *Behaviour*, 133: 1–36.

Isbell, L. A. and T. P. Young (2002) "Ecological models of female social relationships in primates: Similarities, disparities, and some directions for future clarity". *Behaviour*, 139: 177–202.

Ishihara, Y. (2005) *Ishihara Yoshirō shibunshū* (Collected poems and essays of Yoshirō Ishihara). Tokyo: Kōdansha.

Itani, J. (1972) *Reichōrui no Shakai Kōzō* (Social structure of primates). Tokyo: Kyōritsu Shuppan.

Itani, J. (1977) *Chinpanjī ki* (On chimpanzees). Tokyo: Kōdansha.

Itani, J. (1985) "The evolution of primate social structure". *Man*, 20: 593–611.

Itani, J. (1986a) "Reichōrui shakai kōzō no shinka" (Evolution of primate social structures). *Sōzō no sekai* (The world of creativity), 58: 160–185.

Itani, J. (1986b) "Ningen byōdō kigen ron" (Thesis of the origin and bases of equality among men). *Shizen shakai no jinruigaku: Afurika ni ikiru* (Anthropology on natural society: Living in Africa). Kyoto: Academia Shuppankai, 349–389.

Itani, J. (1987a) "Reichōrui no shakai kōzō" (Primate social structure). In *Reichōrui shakai no shinka* (The evolution of primate society). Tokyo: Heibonsha, 33–208.

Itani, J. (1987b) *Reichōrui shakai no shinka* (The evolution of primate society). Tokyo: Heibonsha

Itani, J. (1987c) "Reichōrui shakai kōzō no shinka" (Evolution of primate social structure). In *Reichōrui shakai no shinka* (The evolution of primate society). Tokyo: Heibonsha, 297–325.

Itani, J. (1987d) "Shakai kōzō o tsukuru kōdō" (Behaviors underlie social structures). In *Reichōrui shakai no shinka* (The evolution of primate society). Tokyo: Heibonsha, 223–245.

Itani, J. (1988) "The origin of human equality". In M. R. A. Chance (ed.) *Social Fabrics of the Mind*. Hove: Lawrence Erlbaum Associations Ltd., 137–156.

Itani, J. (1991a) "Shakai no kōzō to hikōzō—Saru, tori, hito" (Structure and anti-structure in society: Monkeys, birds, humans). *Illume*, 32: 41–56.

Itani, J. (1991b) "Karuchā no gainen—aidentifikēshon ron sono go" (The concept of monkey culture: After identification theory). In T. Nishida, K. Izawa and T. Kanō (eds) *Saru no bunkashi* (Primatological study of culture in non-human primates). Tokyo: Heibonsha, 269–277.

Itani, J. (1993) *Yasei no ronri* (Logic of the wild). Tokyo: Heibonsha.

Itani, J. (1995) "Kongun—Nite hinaru mono eno sokohatonai kanshin" (Mixed species associations: Indeterminate interest in others that are similar, but not the same). *Iden*. 4912: 2–3.

Itani, J. (2008) *Itani Junichirō chosakushū* (Collection of Junichirō Itani's work). Vols. 1–4. Tokyo: Heibonsha.

Itani, J. (2009) *Itani Junichirō chosakushū* (Collection of Junichirō Itani's work). Vol. 5. Tokyo: Heibonsha.

Itani, J. and R. Harako (1977) *Jinrui no shizenshi* (A natural history of humans). Tokyo: Yūzankaku.

Itani, J. and J. Tanaka (ed.) (1986) *Shizen shukai no jinruigaku—Afurika ni ikiru* (Anthropology of natural societies: Living in Africa). Kyoto: Akademia Shuppankai.

Itani, J. and T. Yoneyama (ed.) (1984) *Afurika Bunka no Kenkyū* (Studies of African culture). Kyoto: Akademia Shuppankai.

Itoh, N. (2003) "Matomaru koto no mekanizumu" (Mechanism of assembling). In M. Nishida, K. Kitamura and J. Yamagiwa (eds) *Ningensei no kigen to shinka* (Origin and evolution of human nature). Kyoto: Shōwadō, 233–262.

Itoh, N. and T. Nishida (2007) "Chimpanzee grouping patterns and food availability in Mahale Mountains National Park, Tanzania". *Primates*, 48(2): 87–96.

Kaganoi, S. (2009) *Meruro-Ponti: Shokuhatsu suru shisō* (Merleau-Ponty: Inspiring ideology). *Tetsugaku no gendai o yomu* 8 (Reading the present of philosophy 8). Tokyo: Hakusuisha.

Kakeya, M. (1983) "Netami no seitaigaku" (Ecology of jealousy). In R. Otsuka (ed.) *Gendai no jinruigaku, seitai jinruigaku* (Anthropology of today, ecological anthropology). Tokyo: Shibundō, 229–241.

Kakeya, M. (1991) "Byōdōsei to fubyōdōsei no hazama: Tongwe shakai no Muwami seido" (Between equality and inequality: *Mwami* system in the Tongwe society). In J. Tanaka and M. Kakeya (eds) *Hito no shizenshi* (Natural history of humanity). Tokyo: Heibonsha, 59–88.

Kamilian, J. A. (2007) "Survey of feuding families and clans in selected provinces in Mindanao". In W. M. Torres III (ed.) *Rido, Clan Feuding and Conflict Management in Mindanao*. Manila: Asia Foundation.

Kano, T. (1991) "Ningen shakai no genkei ka" (Is this the prototype of human society?). In T. Nishida (ed.) *Saru no bunkashi* (Monographs of primate culture). Tokyo: Heibonsha, 241–266.

Kano, T. (1992) *The Last Ape: Pygmy Chimpanzee Behavior and Ecology*. E. O. Vineberg (trans.). Ann Arbor: University Microfilms International.

Kappeler, P. M., B. Wimmer, D. Zinner and D. Tautz (2002) "The hidden matrilineal structure of a solitary lemur: Implications for primate social evolution". *Proceedings of the Royal Society of London, Series B, Biological Sciences*, 269: 1755–1763.

Kawada, J. (ed.) (2006) *Hito no zentaizō o motomete—21 seiki hitogaku no kadai* (In search of a holistic view of humans: Issues in the study of 21st century man). Tokyo: Fujiwara Shoten

Kawai, K. (2002a) "'Teki' no jittaika katei: Dodosu ni okeru reidingu to tasha hyōshō" (The process of substantialization of the "enemy": Raiding and representation of others in Dodoth). *Afurika Repōto* (Africa report), 35: 3–8.

Kawai, K. (2002b) "'Chimei' to iu chishiki: Dodosu no kankyō ninshiki ron, josetsu" (The knowledge of "geographical names": The environmental epistemology of Dodoth, preface). In Shun Satō (ed.) *Kōza seitai jinruigaku 4: Yūbokumin no sekai* (Ecological anthropology series 4: The world of pastoralists). Kyoto: Kyoto University Press, 17–85.

Kawai, K. (2004) "Dodosu ni okeru kachiku no ryakudatsu to rinsetsu shūdan kan no kankei" (The looting of livestock and relationship with neighboring groups in Dodoth). In J. Tanaka, S. Sato, K. Sugawara and I. Ohta (eds) *Yūbokumin, nomado: Afurika no genya ni ikiru* (Pastoralists, nomads: Living in the wilderness of Africa). Kyoto: Shōwadō, 542–566.

Kawai, K. (2006) "Kyanpu idō to chō uranai" (Camp transfer and intestine reading). In R. Nishii and S. Tanabe (eds) *Shakai kukan no jinruigaku: Materiaritī, shutai, modanitī* (Anthropology of social space: Materiality, identity, modernity). Kyoti: Sekaishisōsha, 175–202.

Kawai, K. (2007) "Dodosu no chō uranai: Bokuchikumin no yūdō ni kakawaru jōhō to chishiki shigen no keisei o megutte" (Intestine reading of Dodoth: Centering on information relating to nomadism of pastoralists and formation of intellectual resources). In C. Daniels (ed.) *Chishiki shigen no in to yō* (The ying and yang of intellectual resources). Tokyo: Kōbundō, 29–71.

Kawai, M. (1964) *Nihonzaru no seitai* (An ecology of Japanese macaques). Tokyo: Kawade Shobō.

Kawai, M. (1977) *Gorira tankenki* (Exploring gorillas). Tokyo: Kōdansha.

Kawamoto, H. (2000) *Ōtopoiesisu 2001—hibi aratani mezameru tameni* (Autopoiesis 2001: So as to awake again each day). Tokyo: Shinyōsha.

Keen, I. (2006) "Constraints on the development of enduring inequalities in Late Holocene Australia". *Current Anthropology*, 47(1): 7–38.

Keesing, R. M. (1975) *Kin Groups and Social Structure*. New York: Holt, Rinehart, and Winston.

Keesing, R. M. (1982) *Shinzoku shūdan to shakai kōzō*. Japanese translation of R. M. Keesing (1975). M. Ogawa et al. (trans.). Tokyo: Miraisha.

Kelly, R. L. (1995) *The Foraging Spectrum*. Washington: Smithsonian Institution Press.

Kiefer, T. (1985) "Folk Islam and the supernatural". In A. Ibrahim, S. Siddique and Y. Hussan (eds) *Readings on Islam in Southeast Asia*. Singapore: Institute of Southeast Asian Studies, 323–325.

Kiefer, T. (1986) *The Tausug: Violence and Law in a Philippine Moslem Society*. Illinois: Waveland Press.

Kimura, D. (2003) *Kyōzai kankaku: Afurika no futatsu no shakai ni okeru gengo teki sōgo kōi kara* (Sense of coexistence: Verbal interaction in two African societies). Kyoto: Kyoto University Press.

Kishigami, N. (1995) "Extended family and food sharing practices among the contemporary Netsilik Inuit: A case study of Pelly Bay". *Bulletin of Hokkaido University of Education Part 1B*, 45(2): 1–9.

Kishigami, N. (1996) "Kanada kyokuhoku chiiki ni okeru shakai henka no tokushitsu ni tsuite" (The characteristics of social change in the Canadian Arctic). In H. Stewart (ed.) *Saishū shuryōmin no genzai* (The gather-hunters today: Changes in and reproduction of subsistence culture). Tokyo: Gensōsha, 13–52.

Kishigami, N. (1998) *Kyokuhoku no tami: Kanada Inuitto* (People of the Arctic: The Canadian Inuit). Tokyo: Kōbundō.

Kishigami, N. (2003) "Shuryōsaishū shakai ni okeru shokumotsu bunpai—shokenkyū no shōkai to hihanteki kentō" (The distribution of food in hunter-gatherer societies: An introduction to and critical examination of a variety of research). In *Kokuritsu Minzokugaku Hakubutsukan Kenkyū Hōkoku* (National Museum of Ethnology Research Reports), 27(4): 725–752.

Kishigami, N. (2007) *Kanada Inuitto no shoku-bunka to shakai henka* (Food culture of the Canadian Inuit and social change). Kyoto: Sekaishisōsha.

Kishigami, N. and H. Stewart (1994) "Gendai netsuriku inuitto shakai ni okeru shakai kankei ni tsuite" (Indigenous social relations in a contemporary Canadian Inuit society: A case study from Pelly Bay, Northwest Territories, Canada). *Bulletin of the National Museum of Ethnology*, 19(3): 405–448.

Kitamura, K. (1983) "Pygmy chimpanzee association patterns in ranging". *Primates*, 24(1): 1–12.

Kitamura, K. (2004) "'Hikaku' ni yoru bunka no tayōsei to dokujisei no rikai: Bokuchikumin Turukana no ninshikiron, episutemorojī" (The understanding of the diversity and originality of culture by "comparison": The epistemology of the pastoralist people of Turkana). In J. Tanaka, S. Satō, K. Sugawara and I. Ohta (eds) *Yūbokumin, nomado: Afurika no genya ni ikiru* (Pastoralists, nomads: Living in the wilderness of Africa). Kyoto: Shōwadō, 466–491.

Kitamura, K. (2007) "'Sekai to chokusetsu deau'—to iu ikikata: Higashi Afurika bokuchikumin-teki dokujisei ni tsuite no kōsatsu" (Meeting the world head on—a way of life: Consideration of the uniqueness of East African pastoralists). In K. Kawai (ed.) *Ikiru ba no jinruigaku: Tochi to shizen no ninshiki, jissen, hyōshō katei* (The anthropology of living space: The process from recognition of land and nature to practice and representation). Kyoto: Kyoto University Press, 25–57.

Kitamura, K. (2008) "What is 'Social'?: Undecidability and a creative coping mechanism in the process of making relations with others". *Primate Research*, 24(2): 109–120.

Koenig, A. (2002) "Competition for resources and its behavioral consequences among female primates". *International Journal of Primatology*, 23: 759–783.

Kummer, H. (1971) *Primate Societies: Group Techniques of Ecological Adaptation*. Chicago: Aldine-Atherton.

Kummer, H. (1978) *Reichōrui no shakai*. Japanese translation of H. Kummer (1971). Tokyo: Shakai Shisōsha.

Kurimoto, E. (1996) *Minzoku funsō o ikiru hitobito: Gendai Afurika no kokka to mainoritī* (People living the ethnic conflict: The nation and the minorities in modern Africa). Kyoto: Sekaishisōsha.

Kurimoto, E. (1999) *Mikai no sensō, Gendai no sensō* (Primitive war, modern war). Tokyo: Iwanami Shoten.

Kuroda, S. (1979) "Grouping of the pygmy chimpanzees". *Primates*, 20(2): 161–183.

Kuroda, S. (1980) "Social behavior of the pygmy chimpanzees". *Primates*, 21: 181–197.

Kuroda, S. (1982) *Pigumī chinpanjī: Michi no ruijinen* (The pygmy chimpanzee: The unknown ape). Tokyo: Chikuma Shobō.

Kuroda, S. (1984) "Interaction over food among pygmy chimpanzees in Wamba,

Zaire". In R. L. Susman (ed.) *The Pygmy Chimpanzees: Evolutionary Biology and Behavior*. New York: Plenum Press, 301–324.

Kuroda, S. (1999) *Jinrui shinka saikō: Shakai seisei no kōkogaku* (Reconsideration of human evolution: Archeology of the emergence of the hominid society). Tokyo: Ibunsha.

Langergraber, K. E., H. Siedel, J. C. Mitani, R. W. Wrangham, V. Reynolds, K. Hunt and L. Vigilant (2007) "The genetic signature of gender-biased migration in patrilocal chimpanzees and humans". *Plos One*, 10: 1–7.

Lappan, S. (2007) "Patterns of dispersal in Sumatran siamangs (Symphalangus syndactylus): Preliminary mtDNA evidence suggests more frequent male than female dispersal to adjacent groups". *American Journal of Primatology*, 69: 692–698.

Lawler, R. R., A. F. Richard and M. A. Riley (2003) "Genetic population structure of the white sifaka (Propithecus vereauxi verreauxi). At Beza Mahafaly Special Reserve, southwest Madagascar (1992–2001)". *Molecular Ecology*, 12: 2307–2317.

Le Galliard, J-F., G. Gundersen, H. P. Andreassen and N. C. Stenseth (2006) "Natal dispersal, interactions among siblings and intra-sexual competition". *Behavioral Ecology*, 17: 733–740.

Lee, R. B. (1979) *The !Kung San: Men, Women and Work in a Foraging Society*. Cambridge: Cambridge University Press.

Lee, R. B. and I. De Vore (1968) *Man the Hunter*. Chicago: Aldine Publishing Company.

Lee, R. B. and R. Daly (eds) (1999) *The Cambridge Encyclopedia of Hunters and Gatherers*. Cambridge: Cambridge University Press.

Lehmann, J., A. H. Korstjens and R. I. M. Dunbar (2007) "Fission-fusion social systems as a strategy for coping with ecological constraints: A primate case". *Evolutionary Ecology*, 21: 613–634.

Leus, T. (2006) *Aadaa Boraanaa: A Dictionary of Borana Culture*. Addis Ababa: Shama Books.

Lévi-Strauss, C. (1949) *Les structures élémentaires de la parenté*. Paris: Presses Universitaires de France.

Lévi-Strauss, C. (1971) *The Elementary Structures of Kinship*. Boston: Beacon Press.

Lévi-Strauss, C. (2001) *Shinzoku no kihon kōzō*. Japanese translation of C. Lévi-Strauss (1949). K. Fukui (trans.). Tokyo: Seikyūsha.

Luhmann, N. (1984) *Soziale systeme: Grundriß einer allgemeinen Theorie*. Frankfurt: Suhrkamp.

Luhmann, N. (1993) *Shakai shisutemu riron*. Japanese translation of N. Luhmann (1984). T. Satō (trans.). Tokyo: Kōseisha-Kōseikaku.

Majima, I. et al. (2006) "Tokushū: chūkan shūdan no mondaikei" (Special edition: Issues of middle groups). *Bunka Jinruigaku* (Cultural anthropology), (71)1: 22–118.

Marcus, G. and M. Fischer (1986) *Anthropology as Cultural Critique: An Experimental Moment in the Human Sciences*. Illinois: University of Chicago Press.

Matsuba, S. (2008) "'Niku no kyōdōtai' no kanōsei" (Possibility of the "community of flesh"). *Shisō* (Thought), 1015: 85–101.

Matsuzawa, T., M. Takai and H. Hirai (2007) "Reichōruigaku e no shōtai" (An introduction to primatology). In Primate Research Institute, Kyoto University (ed.), *The Science of Primate Evolution*. Kyoto: Kyoto University Press, 1–10.

Merleau-Ponty, M. (1959) *The Philosopher and His Shadow, Signes*. R. C. McCleary (trans.). Evanston: Northwestern University Press.

Merleau-Ponty, M. (1968) *The Visible and the Invisible*. C. Lefort (ed.), A. Lingis (trans.). Evanston: Northwestern University Press.

Mitani, J. C., A. Merriwether and C. Zhang (2000) "Male affiliation, cooperation and kinship in wild chimpanzees". *American Journal of Physical Anthropology*, 109: 439–454.

Mitchell, J. C. (ed.) (1969) *Social Networks in Urban Situations: Analyses of Personal Relationships in Central African Towns*. Manchester: Manchester University Press.

Mithen, S. (1996) *The Prehistory of the Mind: A Search for the Origins of Art, Religion and Science*. London: Thames and Hudson.

Mithen, S. (1998) *Kokoro no senshi jidai*. Japanese translation of S. Mithen (1996). S. Matsuura and M. Makino (trans.). Tokyo: Seidosha.

Mito, S. (1971) *Kōshima no saru* (Monkeys of the Koshima Island), Tokyo: Populasha.

Mori, A. (1990) "Ruijinen shakai no sosenkei no fukugen kara jinrui shakai e" (Emergence of human society: Examined through reconstruction of the ancestral society of Pongids). *Afurika Kenkyū* (Journal of African Studies), 69–85.

Morin, P. A., J. J. Moore, R. Chakraborty, R. Jin, J. Goodall and D. S. Woodruff (1994) "Kin selection, social structure, gene flow, and the evolution of chimpanzees". *Science*, 265: 1193–1201.

Morrogh-Bernard, H. C., N. V. Morf, D. J. Chivers and M. Krutzen (2011) "Dispersal patterns of Orangutans (Pongo spp.) in a Bornean peat-swamp forest". *International Journal of Primatology*, 32: 362–376.

Morse, M. (1923–4) "Essai sur le don—Forme et raison de l'échange dans les sociétés archaïques". *L'Année Sociologique*.

Morse, M. (1973) *Shakaigaku to jinruigaku (Sociologie et anthropologie.* 1re éd. Paris: Presses universitaires de France 1950). T. Yuji, S. Ito and T. Yamaguchi (trans.). Tokyo: Kōbundō.

Morton, J. (1999) "The Arrernte of Central Australia". In R. B. Lee and R. Daly (eds) *The Cambridge Encyclopedia of Hunters and Gatherers.* Cambridge: Cambridge University Press, 329–334.

Munshi-South, J. (2008) "Female-biased dispersal and gene flow in a behaviorally monogamous mammal, the large tree shrew *(Tupaia tana)".* *PLos ONE,* 3: 1–6.

Myers, F. R. (1986) *Pintupi Country, Pintupi Self: Sentiment, Place, and Politics among Western Dessert Aborigines.* Berkeley: The University of California Press.

Naitō, N. (2004) "Bokuchikumin ariāru no fukugōteki na aidentiti keisei—dōitsu keiken no kyōyū ni motozuku kizoku ishiki keisei no jirei kara" (Shared experiences and the reconstruction of social categories: A case study of complex ethnic identity among the Ariaal pastoralists.) In J. Tanaka, S. Satō, K. Sugawara and I. Ohta (eds) *Yūdōmin Nomaddo—Afurika no Genya ni Ikiru.* (Nomad: Life in the wild in Africa). Kyoto: Shōwadō, 567–592.

Naka, Y. (2008) "Shintai no jiko shokuhatsu: Meruro-Ponti, Anri, Biran" (Self-incitement of the body: Merleau-Ponty, Henry, Biran). *Shisō* (Thought), 1015: 102–120.

Nakagawa, N. (1999) "Shoku wa shakai o tsukuru—Shakai seitaigakuteki apurōchi" (Food makes the society: A socioecological approach). In T. Nishida and S. Uehara (eds) *Reichoruigaku o manabu hito no tameni* (For apprentices of primatology). Kyoto: Sekai Shisōsha, 50–92.

Nakagawa, N. (2007) *Sabanna o kakeru saru—Patasu monkī no seitai* (The monkey that runs the savannah: The ecology and society of the patas monkey). Kyoto: Kyoto University Press.

Nakagawa, N. and K. Okamoto (2003) "Van Schaik no shakaiseitaigaku moderu: Tsumikasanete kitamono to tsuminokosarete kitamono" (Van Schaik's socioecological model: Developments and problems). *Reichōrui Kenkyū* (Primate Research), 19: 243–264.

Nakajima, T. (2007) *Zankyō no Chūgoku Tetsugaku* (The reverberation of Chinese philosophy). Tokyo: University of Tokyo Press.

Ndaywel, e N. (1999) "The political system of the Luba and Lunda: Its emergence and expansion". In B. A. Ogot (ed.), *General History of Africa V: Africa from the Sixteenth to the Eighteenth Century (Unesco General History of Africa).* California: James Curry, 290–299.

Nietlisbach, P. and M. Krutzen (2010) "Male-specific markers reveal sex-biased dispersal in orangutans (Pongo spp.)". *Primate Research* (supplement), 26: 154.

Nievergelt, C. M., L. J. Digby, U. Ramakrishnan and D. S. Woodruff (2000) "Genetic analysis of group composition and breeding system in a wild common marmoset (*Callithrix jacchus*) population". *International Journal of Primatology*, 21: 1–20.

Nievergelt, C. M., T. Mutschler, A. T. C. Feistner and D. S. Woodruff (2002) "Social system of the Alatran gentle lemur (*Happalemur griseus alatrensis*): Genetic characterization of group composition and mating system". *American Journal of Primatology*, 57: 157–176.

Nimmo, H. A. (1972) *The Sea People of Sulu*. San Francisco: Chandler Press.

Nishida, M. (2003) "Shakai shinka to heiwaryoku" (Social evolution and the power of peace). In M. Nishida, K. Kitamura and J. Yamagiwa (eds) *Ningensei no kigen to shinka* (Origin and evolution of human nature). Kyoto: Shōwadō, 63–96.

Nishida, M., K. Kitamura and J. Yamagiwa (eds) (2003) *Ningensei no kigen to shinka*. Kyoto: Shōwadō.

Nishida, T. (1968) "The social group of wild chimpanzees in Mahali Mountains". *Primates*, 9: 167–224.

Nishida, T. (1977) "Mahare sankai no chinpanjī (I): Seitai to tan'i shūdan kōdō" (Chimpanzees in the Mahale mountains I: The ecology and unit-group structure). In J. Itani (ed.) *Chinpanjī ki* (Monographs of chimpanzees). Tokyo: Kōdansha, 543–638.

Nishida, T. (1981) *Yasei chinpanjī kansatsu ki* (Observation record of wild chimpanzees). Tokyo: Chūkō Shinsho.

Nishida, T. (1994) *Chinpanjī omoshiro kansatsu ki* (Fun observation record of chimpanzees). Tokyo: Kinokuniya Shoten.

Nishida, T. (2007 [1999]) *Ningensei wa doko kara kita ka—Sarugaku kara no apurōchi* (The origins of humanity: Approach from the studies of monkeys). Kyoto: Kyoto University Press.

Nishida, T. (2001) "Kyōtsū sosen no shakai" (The society of a common ancestor). In T. Nishida (ed.) *Hominizēshon* (Hominization). Kyoto: Kyoto University Press, 9–32.

Nishida, T. (2012) *Chimpanzees of the Lakeshore*. Tokyo: Tokyo University Press.

Nishida, T., M. Hiraiwa-Hasegawa, T. Hasegawa and Y. Takahata (1985) "Group extinction and female transfer in wild chimpanzees in the Mahale National Park, Tanzania". *Z. Tierpsychol*, 67: 284–301.

Nishii, M. (1998) "Kokusaihō ni okeru kaizoku" (Pirates in international law). In *Heibonsha sekai daihyakka jiten* (Heibonsha world encyclopedia), CD-ROM edition. Tokyo: Heibonsha.

Nishii, R. and S. Tanabe (eds) (2006) *Shakai kūkan no jinruigaku—Materiariti, shutai, modaniti* (Anthropology of social space: Materiality, identity and modernity). Kyoto: Sekai Shisōsha.

Niwa, F. (1993) *Nihonteki shizenkan no hōhō—Imanishi seitaigaku no imi suru mono* (Methodology of the Japanese view of nature: The meaning of Imanishi's ecology). Tokyo: Nōson Gyoson Bunka Kyōkai (Rural Culture Association Japan).

Nuttall, M. (1992) *Arctic Homeland*. Toronto: University of Toronto Press.

Nyambedha, E. O. and J. Aagaard-Hansen (2007) "Practices of relatedness and the re-invention of Duol as a network of care for orphans and widows in Western Kenya". *Africa*, 77(4): 517–534.

Oda, M. (1989) *Kōzōshugi no paradokkusu: Yasei no keijijōgaku no tameni* (Paradox of structuralism: To metaphysics in the wild). Tokyo: Keisō Shobō.

Oda, M. (1994) *Kōzōshugi no fīrudo* (The field of structural anthropology). Kyoto: Sekaishisōsha.

Oda, M. et al. (2004) "Tokushū kyōdōtai to iu gainen no datsu/saikōchiku". *Bunka Jinruigaku*, (69)2: 236–312.

Omura, K. (2002a) "Construction of *Inuinnaqtun* (Real Inuit-way): Self-image and Everyday Practices in Inuit Society." In H. Stewart, A. Barnard and K. Omura (eds), *Self and Other Images of Hunter-Gatherers* (Senri Ethnological Studies No. 60). Osaka: National Museum of Ethnology, 101–111.

Omura, K. (2002b) "Kanada kyokuhokuken ni okeru chishiki wo meguru funsō" (Conflict between Inuit traditional knowledge and scientific knowledge: A case of the wildlife co-management regime in the Canadian Arctic). In N. Kishigami and T. Akimichi (eds) *Funsō no umi* (Conflict over marine resources). Kyoto: Jimbun Shoin, 149–167.

Omura, K. (2005a) "Science against modern science: The socio-political construction of otherness in Inuit TEK (traditional ecological knowledge)". In N. Kishigami and J. Savelle (eds), *Indigenous Use and Management of Marine Resources* (Senri Ethnological Studies 67). Osaka: National Museum of Ethnology, 323–344.

Omura, K. (2005b) "Repetition of different things: The mechanism of memory in traditional ecological knowledge of the Canadian Inuit". In K. Sugawara (ed.), *Construction and Distribution of Body Resources: Correlations between Ecological, Symbolic and Medical Systems*. Tokyo: Research Institute for Language and Culture of Asia and Africa. Tokyo University of Foreign Studies, 79–107.

Omura, K. (2007a) "Seikatsusekai no shigen toshiteno shintai—Kanada inuitto no seigyo ni miru shintaishigen no kōchiku to kyōyū" (The body as the resource for the construction of the life-world: The construction and sharing of bodily experience in the subsistence activities of the Canadian Inuit). In K. Sugawara (ed.) *Shintaishigen no Kōchiku to Kyōyū* (The construction and sharing of physical resources, resource anthropology 9). Tokyo: Kōbundō, 59–88.

Omura, K. (2007b) "From knowledge to poetics". *Japanese Review of Cultural Anthropology*, 7: 27–50.

Omura, K. (2008) "Kakawariaukoto no yorokobi—Kanada inuitto no kankyō no shirikata to tsukiaikata" (The pleasures of getting involved: The Canadian Inuit's ways of knowing and ways of associating with the environment). In Y. Yama, A. Furukawa and M. Kawada (eds) *Kankyō minzokugaku—Atarashii fīrudogaku e* (Environmental folklore: New fields of study). Kyoto: Shōwadō, 34–57.

Omura, K. (2009) "Ikirukoto no uta: Kanada inuitto no hanga no miryoku" (The songs of lives: The art prints of the Canadian Inuit). In R. Saito, N. Kishigami and K. Omura (eds) *Kyokuhoku to shinrin no kioku: Inuitto to hokuseikaigan indian no āto* (Memories of life in the Arctic and forest: The art of the Inuit and Northwest Coast People). Kyoto: Shōwadō, 18–23.

Ōoka, S. (1994 [1951]) "Nobi" (Fires on the plain). Ōoka Shōhei zenshū 3 (Collected works of Shōhei Ōoka 3). Tokyo: Chikuma Shobō.

Parkin, D. (1978) *The Cultural Definition of Political Response: Lineal Destiny among the Luo*. London: Academic Press.

Perrin, N. and L. Lehmann (2001) "Is sociality driven by the costs of dispersal or the benefits of philopatry? A role for kin-discrimination mechanisms". *American Naturalist*, 154: 282–292.

Phiri, K. M. (1999) "The Northern Zambezia: Lake Malawi region". In B. A. Ogot (ed.) *General History of Africa V: Africa from the Sixteenth to the Eighteenth Century (Unesco General History of Africa)*. California: James Curry, 300–314.

Pope, T. R. (2000) "The evolution of male philopatry in neotropical monkeys". In P. M. Kappeler (ed.) *Primate Males*. Cambridge: Cambridge University Press, 219–235.

Prentice, I. C. and I. Kuijt (eds) (2004) *Complex Hunter-Gatherers: Evolution and Organization of Prehistoric Communities on the Plateau of Northwestern North America*. Salt Lake City: The University of Utah Press.

Purvis, A. (1995) "A composite estimate of primate phylogeny". *Philosophical Transactions of the Royal Society, Services B*, 348: 405–421.

Radcliffe-Brown, A. R. (1930/31) "The social organization of Australian tribes". *Oceania*, 1: 34–63.

Radespiel, U., S. M. Funk, E. Zimmermann and M. W. Bruford (2001) "Isolation and characterization of microsatellite loci in the grey mouse lemur (Microcebus murinus) and their amplification in the family Cheirogaleidae". *Molecular Ecology*, 1: 16–18.

Ray, B. (2006) "Elementary schools help to nurture 830 Melanesian languages". *Papua New Guinea Yearbook 2006*, 144–148.

Reichard, U. H. and C. Barelli (2008) "Life history and reproductive strategies of Khao Yai Hylobates lar: Implications for social evolution in Apes". *International Journal of Primatology*, 29: 823–844.

Reno, P. L., R. S. Meindl, M. A. McCollum and C. O. Lovejoy (2003) "Sexual dimorphism in Australopithecus afarensis was similar to that of modern humans". *Proceedings of the Natural Academy of Science*, 100: 9404–9409.

Reynolds, V. and F. Reynolds (1965) "Chimpanzees of the Budongo Forest". In I. DeVore (ed.) *Primate Behaviour: Field Studies of Monkeys and Apes*. New York: Holt, Rinehart and Winston, 368–424.

Richards, A. I. (1939) *Land, Labour and Diet in Northeastern Rhodesia*. London: Oxford University Press.

Roberts, A. (1974) *History of the Bemba*. New York: Longman.

Roberts, A. (1976) *A History of Zambia*. New York: Africana Publishing Company.

Robinson, P. (1985) *Gabbra Nomadic Pastoralism in Nineteenth and Twentieth Century Northern Kenya: Strategies for Survival in a Marginal Environment*. PhD thesis: Northwestern University.

Roos, C. and T. Geissmann (2001) "Molecular phylogeny of the major hylobatid divisions". *Molecular Phylogeny and Evolution*, 19: 486–494.

Rousseau, J. J. (1755), *Discours sur l'origine et les fondements de l'inégalité parmi les homes* (A discourse on the origin and foundation of inequality among mankind).

Rousseau, J. J. (1972) *Ningen fubyōdō kigen ron*. Japanese translation of J. J. Rousseau (1755). N. Hiraoka and K. Honda (trans.). Tokyo: Iwanami Bunko.

Rousseau, J. J. (1978) *Ningen fubyōdō kigen ron*. Japanese translation of J. J. Rousseau (1755). In Y. Hara (trans.), *Rusō zenshū dai 4-kan* (Complete works of Rousseau, vol. 4). Tokyo: Hakusuisha.

Saitō, N. (2007) "Ruijin'en to ningen wa dorehodo chikaika" (How close are apes and humans?). In J. Yamagiwa (ed.) *Hito wa dono yōni shite tsukuraretaka* (How were humans made?), *Science of Man* series. Tokyo: Iwanami Shoten, 81–93.

Sakurai, T. (2011) *Imamura Hitoshi no shakai tetsugaku, nyūmon: mezameru tame ni* (Introduction to the social philosophy of Hitoshi Imamura for awakening). Tokyo: Kōdansha.

Santos-Granero, F. (1991) *The Power of Love: The Moral Use of Knowledge amongst the Amuesha of Central Peru*. London: Athlone Press.

Sasaki, M. (1991) *Afōdansu—Atarashii Ninchi no Riron* (Affordance: New theory of cognition). Tokyo: Iwanami Shoten

Sather, C. (1997) *The Bajau Laut*. New York: Oxford University Press.

Schutz, A. (1970) *On Phenomenology and Social Relations*. H. R. Wagner (ed.). Chicago: University of Chicago Press.

Sedgwick, E. K. (1985) *Between Men: English Literature and Homosocial Desire*. New York: Columbia University Press.

Service, E. and M. Sahlins (1960) *Evolution and Culture*. Ann Arbor: University of Michigan Press.

Service, E. and M. Sahlins (1976) *Shinka to bunka*. Japanese translation of E. Service and M. Sahlins (1960). R. Yamada (trans.). Tokyo: Shinsensha.

Service, E. E. R. (1971) *Primitive Social Organisiation: An Evolutionary Perspective*. New York: Random House.

Service, E. E. R. (1979) *Mikai no shakai soshiki*. Japanese translation of E. E. R. Service (1971). M. Matsuzono (trans.). Tokyo: Kōbundō.

Shiino, W. (2007a) "*Dala*, the life and residence space of the Kenya Luo: Its formation and change in symbolic meaning". In K. Kawai (ed.) *Anthropological Study on the Arena of Human Life: Perception, Practice and Representation Process on the Land and Nature*. Kyoto: Kyoto University Press, 331–362.

Shiino, W. (2007b) "Malo Malo!—The Luo are rising higher and higher". In Tama Africa Center (ed.) *The Lure of African Pop Music*. Yokohama: Shumpusha Publishing.

Shima, T. (2004) *Saru no shakai to hito no shakai—kogoroshi o fusegu shakai kōzō* (Monkey society and human society: Social structures that prevent infanticide). Tokyo: Taishūkan.

Shimizu, A. (1992) "Rekishi, minzoku, shinzoku, soshite jūjutsu" (History, ethny, kinship and magic). Osaka, National Museum of Ethnology. *Minpaku Tsushin*, 58: 84–92.

Soga, T. (2006) "Changes in knowledge of time among Gabra Miigo Pastoralists of Southern Ethiopia". *Nilo-Ethiopian Studies*, 10: 23–44.

Soga, T. (2007) "'*Kishō shigen' o meguru kyōgō to iu shinwa: Shigen o meguru minzoku kankei no fukuzatsusei o megutte*" (The myths of competition over "scarce resources": On the complexity of relations among ethnic groups surrounding resources). In K. Matsui (ed.) *Shizen no shigenka* (Turning nature into resource), *Shigen jinruigaku 6* (Resource anthropology, vol. 6). Tokyo: Kōbundō, 205–249.

Southall, A. (1952) *Lineage Formation among the Luo*. London: Oxford University Press.

Sprague, D. (2004) *Saru no shōgai, hito no shōgai: Jinsei keikaku no seibutsugaku* (Monkey lives, human lives: The biology of life planning). Kyoto: Kyoto University Press.

Stairs, A. and G. Wenzel (1992) "I am I and the environment". *Journal of Indigenous Studies*, 3(1): 1–12.

Sterck, E. H. M., D. P. Watts and C. P. van Schaik (1997) "The evolution of female social relationships in nonhuman primates". *Behavioral Ecology and Sociobiology*, 41: 291–309.

Steward, J. (1955) *Theory of Culture Change: The Methodology of Multi-linear Evolution*. Chamaign: University of Illinois Press.

Steward, J. (1979) *Bunka henka no riron*. Japanese translation of J. Steward (1955). T. Yoneyama (trans.). Tokyo: Kōbundō.

Stewart, H. (1990) "Shokuryō, josei, sekaikan—Chūbu kyokuhoku kanada no dentō inuitto shakai ni okeru shokuryō no kakutoku to bunpai" (Food, women, cosmology: Acquisition and sharing of food in traditional Inuit society). *International Symposium on Dietary and Housing Aspect of Northern Peoples—Proceedings of the 4th International Abashiri Symposium*. Abashiri: Hokkaido Museum of Northern Peoples, 46–50.

Stewart, H. (1991) "Shokuryō bunpai ni okeru danjyo no yakuwari buntan ni tsuite—Netsiriku inuitto shakai ni okeru emono, bunpai, sekaikan" (The allotment of male and female roles in food distribution: Game, distribution and world view in Netsilik Inuit society). *Shakai Jinruigaku Nenpō* (Annual report of social anthropology), 17: 45–56.

Stewart, H. (1992) "Teijū to seigyō—netsiriku inuitto no dentōteki seigyōkatsudō to shokuseikatsu ni miru keishō to henka" (Sedentarism and subsistence activities: Continuity and change of traditional Netsilik Inuit subsistence and food ways). *Sedentary and / or Migratory Life in the North—Proceedings of the 6th International Abashiri Symposium*. Abashiri: Hokkaido Museum of Northern Peoples, 75–87.

Stewart, H. (1995) "Gendai no netsiriku inuitto shakai ni okeru seigyōkatsudō" (Subsistence activities in modern-day Netsilik Inuit society). *People and Cultures of the Tundra—Proceedings of the 9th International Abashiri Symposium*. Abashiri: Hokkaido Museum of Northern Peoples, 36–67.

Stewart, H. (1996) "Genzai no shuryōsaishyūmin ni totteno seigyōkatsudō no igi" (The meaning of subsistence activities for modern-day hunters and gatherers).

In H. Stewart (ed.) *Saishū-shuryōmin no genzai* (The gather-hunters today: Changes in and reproduction of subsistence culture). Tokyo: Gensōsha, 125–154.

Strier, K. B. (1990) "New world primates, new frontiers: Insights from the wooly spider monkey, or Muriqui (*Brachyteles arachnoids*)". *International Journal of Primatology*, 11: 7–19.

Strier, K. B. (1999) "Why is female kin bonding so rare? Comparative sociality of neotropical primates". In P. C. Lee (ed.) *Comparative Primate Socioecology*. Cambridge: Cambridge University Press, 300–319.

Strier, K. B. (2006) *Primate Behavioral Ecology* (3rd edition). Boston: Allyn, P. & B. Bacon.

Sugawara, K. (2002) *Kanjō no enjin = hito* (Emotional hominids = humans). Tokyo: Kōbundō.

Sugiyama, Yukimaru. (1965) "On the social change of Hanuman langurs (Presbytis entellus) in their natural conditions". *Primates*, 6: 381–418.

Sugiyama, Yukimaru. (1981) *Yasei chinpanjī no shakai* (The society of wild chimpanzees). Tokyo: Kōdansha.

Sugiyama, Yukimaru. (1987) *"Reichōruishakai no tayōsei o tekiō no kanten kara kangaeru"* (Adaptation and variability of primate social structure). *Kikan Jinruigaku* (Quarterly anthropology), 18: 3–49.

Sugiyama, Yukimaru. (1990) *Saru wa naze mureru ka* (Why monkeys herd together?). Tokyo: Chuō Kōronsha.

Sugiyama, Yukimaru. (1993 [1980]) *Kogoroshi no kōdōgaku* (The behavioral science of infanticide). Tokyo: Hokuto Shuppan (Kōdansha).

Sugiyama, Yukimaru. (ed.) (1996) *Saru no hyakka* (Encyclopedia of monkeys). Tokyo: Data House.

Sugiyama, Yukimaru. (ed.) (2000) *Reichōrui seitaigaku—kankyō to kōdō no dainamizumu*. Kyoto: Kyoto University Press.

Sugiyama, Yukimaru. (2002) "Chinpanjī to ningen ni okeru bunsansuru sei ni tsuite" (On the dispersing sex of chimpanzees and humans). *Reichōrui Kenkyū* (Primate research), 18: 19–33.

Sugiyama, Yukimaru. (2007) *Bunka no tanjō—hito ga hito ni naru mae* (The birth of culture: Before hominids became humans). Kyoto: Kyoto University Press.

Sugiyama, Yuko. (1987) "Usu o kashite kudasai: seikatsu yōgu no shoyū to shiyō o meguru Bemba josei no maikuro poritikusu" (May I borrow your mortar?: Micro politics of the Bemba women on the ownership and usage of daily utensils). *Afurika Kenkyū* (African studies), 30: 9–69.

Sugiyama, Yuko. (2004) "Kieta mura, saisei suru mura: Bemba no nōson ni okeru noroi jiken no kaishaku to ken'i no seitōsei" (Village that disappears, village that revives: The interpretation of curse incidents in the rural villages of Bemba and the legitimacy of authority). In H. Terashima (ed.) *Byōdō to fubyōdō o meguru jinruigakuteki kenkyū* (Anthropological study on equality and inequality). Kyoto: Nakanishiya Shuppan, 134–171.

Sugiyama, Yuko. (2006) "On *Sansamukeni* (Be happy/Feel comfortable), or the power that controls space and the feeling of co-presence". In T. Imada, K. Hiramatsu and K. Torigoe (eds), *The West Meets the East in Acoustic Ecology*. Hirosaki: Japanese Association for Sound Ecology and Hirosaki University International Music Centre, 16–21.

Sugiyama, Yuko. (2007a) "'Miombo-rin nara doko e demo' to iu shinnen ni tsuite: Yakihata nōkōmin Bemba no idōsei ni kansuru kōsatsu" (On the faith of "going anywhere as long as there are Miombo woodland": A study on the mobility of the Bemba shifting cultivators). In K. Kawai (ed.) *Ikiru ba no jinruigaku: Tochi to shizen no ninshiki, jissen, hyōshō katei* (Anthropology of the land we live on: Recognition, practice and symbolization process of the land and nature). Kyoto: Kyoto University Press, 239–270.

Sugiyama, Yuko. (2007b) "Shokumotsu no michi, okane no michi, keii no michi—Genkin, shokumotsu, Miombo-rin: Afurika yakihata nōkomin ni okeru shigen to datsu-shigenka no purosesu" (The path of food, the path of money, the path of respect: Cash, food, Miombo woodland: Resource and de-recycling process of the shifting cultivators of Africa). In N. Kasuga (ed.) *Kahei to shigen* (Currency and resource). Tokyo: Kōbundō, 147–188.

Suwa, G. (2006) "Kaseki kara mita hito no shinka" (The evolution of homo sapiens as seen in fossils). In H. Ishikawa et al. (eds) *Hito no shinka* (The evolution of homo sapiens). Tokyo: Iwanami Shoten, 13–64.

Suwa, G., R. T. Kono, S. W. Simpson, B. Asfaw, C. O. Lovejoy and T. D. White (2009) "Paleobiological implications of the Ardipithecus ramidus dentition". *Science*, 69: 94–99.

Tachikawa, K. (1991) *Yūwakuron* (Enticement theory). Tokyo: Shinyōsha.

Tachimoto, N. (1996) *Chiiki kenkyū no mondai to hōhō: Shakai bunka seitai rikigaku no kokoromi* (Issues and methodology of area studies: An attempt for a socio-cultural ecodynamics). *Chiiki kenkyū sōsho 3*, (Area studies series 3). Kyoto: Kyoto University Press.

Tachimoto, N. (2001) *Kyōsei no shisutemu o motomete—Nusantara sekai kara no teigen* (In search of a system of coexistence: Proposal from the Nusantara world). *Gendai shakai no chikaku hendō o yomu 3* (Reading the cataclysm of the contemporary society 3). Tokyo: Kōbundō.

Takacs, Z., J. C. Morales, T. Geissmann and D. J. Melnick (2005) "A complete species-level phylogeny of the Hylobatidae based on mitochondrial ND 3-ND 4 gene sequences". *Molecular Phylogenetics and Evolution*, 36: 456–467.

Takiura, S. (1974) "Niku no sonzai ron: Meruro-Ponti ni okeru genshōgaku no kisū" (Ontology of the flesh: The outcome of the phenomenology of Merleau-Ponty). *Gendai Shisō* (Contemporary thought), 28: 112–126.

Tanabe, S. (2002) "Joshō nichijōteki jissen no *esunogurafī*—katari, comyuniti, aiditi (An introduction to the ethnography of everyday practice". In S. Tanabe and M. Matsuda (eds) *Nichijōteki Jissen no Esunogurafī* (The ethnography of everyday practice). Kyoto: Sekai Shisōsha, 1–38.

Tanabe, S. and M. Matsuda (eds) (2002) *Nichijōteki Jissen no Esunogurafī—katari, comyuniti, aidentiti* (The ethnography of everyday practice: Narratives, communities and identities). Kyoto: Sekai Shisōsha.

Tanaka, J. (1971) *Busshuman—Seitai jinruigakuteki kenkyū* (Bushmen: Research in ecological anthropology). Tokyo: Shisakusha.

Tanaka, J., M. Kakeya, M. Ichikawa and I. Ohta (eds) (1996) *Zoku: Shizen shakai no jinruigaku—henbō suru Afurika* (Anthropology of natural societies—Changing Africa: Sequel). Kyoto: Akademia Shuppankai.

Tanaka, M. (2006) "Mikuro jinruigaku no kadai" (The challenge for micro-anthropology). In M. Tanaka and M. Matsuda (eds) *Mikuro jinruigaku no jissen: Eijenshī, shintai, nettowāku* (The practice of micro-anthropology: Agencies, bodies and networks). Kyoto: Sekai Shisōsha, 1–37.

Tanaka, M. (2009a) "Shūkyōgaku wa yūwakusuru" (Religious studies and seduction). *Shūkyō Kenkyū* (Religious studies), 359: 35–57.

Tanaka, M. (2009b) "Joshō: Fetishizumu kenkyū no kadai to tenbō" (Introductory chapter: The challenge and perspectives of fetishism studies). In M. Tanaka (ed.), *Fetishizumu kenkyū 1: Fetishizumuron no keifu to tenbō* (Fetishism studies 1: The genealogy and perspectives of fetishism theory). Kyoto: Kyoto University Press, 3–38.

Tanaka, M. and M. Matsuda (eds) (2006) *Mikuro jinruigaku no jissen—Eijenshī, nettowāku, shintai* (The practice of micro-anthropology: Agencies, bodies and networks). Kyoto: Sekai Shisōsha.

Tanno, T. (1976) "The Mbuti net-hunters in the Ituri Forest, Eastern Zaire: Their hunting activities and band composition". *Kyoto University African Studies*, 10: 101–35.

Terashima, H. (1984) "Mbuti āchā no bando kōzō" (Band composition of the Mbuti archers). In J. Itani and T. Yoneyama (eds) *Afurika bunka no kenkyū* (Research on African cultures). Kyoto: Akademia Shuppankai, 3–41.

Terashima, H. (1985) "Variation and composition principles of the residential group (band) of the Mbuti Pygmies: Beyond a typical/atypical dichotomy". *African Study Monographs*, Supplementary Issue 4: 103–120.

Terashima, H. (2004) "Ningen wa naze byōdō ni kodawaru no ka" (Why are humans obsessive about equality?). In H. Terashima (ed.), *Byōdō to fubyōdō o meguru jinruigaku-teki kenkyū* (Anthropological research on equality and inequality). Kyoto: Nakanishiya Shuppan, 3–52.

Tokoro, I. (1996) "Ekkyō no minzokushi" (Border-crossing ethnography). In S. Yamashita (ed.) *Iwanami kōza bunka jinruigaku dai7kan: Idō no minzokushi* (Iwanamai Koza cultural anthropology, vol. 7: The ethnography of migration). Tokyo: Iwanami Shoten, 159–186.

Tokoro, I. (1999) *Ekkyō—Sūrū kaiiki sekai kara* (Border transgressions: From the world of the Sulu maritime world). Tokyo: Iwanami Shoten.

Tokoro, I. (2003) "Transformation of shamanic rituals among the Sama of Tabawan Island, Sulu Archipelago". In S. Yamashita and J. S. Eades (eds) *Globalization in Southeast Asia: Local, National, and Transnational Perspectives*. New York: Berghahn Books, 165–178.

Tokoro, I. (2006) "Border crossing and the politics of religion in Sulu". *Militant Islam in Southeast Asia (Asian Cultural Studies* 15: Special Issue), 121–136.

Tournier, M. (1967) *Vendredi ou les Limbes du Pacifique*. Paris: Gallimard.

Tournier, M. (1996) *Furaidē aruiwa Taiheiyō no meikai* (Friday or the limbo of the Pacific). Japanese translation of M. Tournier (1967). K. Sakakibara (trans.). Tokyo: Iwanami Shoten.

Treves, A. and C. Chapmann (1996) "Conspecific threat, predation avoidance, and resource defense: Implications for grouping in langurs". *Behavioral Ecology and Sociobiology*, 39: 43–53.

Tsukahara, T. (1993) "Lions eat chimpanzees: The first evidence of predation by lions on wild chimpanzees". *American Journal of Primatology*, 29: 1–11.

Turner, V. (1972[1957]) *Schism and Continuity in an African Society: A Study of Ndembu Village Life*. Manchester: Manchester University Press.

Turner, V. (1996) *Girei no katei*. Tokyo: Shin Shisakusha. M. Tomikura (trans.) from *The Ritual Process: Structure and Anti-Structure* (1969). Aldine Transaction.

Turton, D. (1997) "Introduction: War and ethnicity". In D. Turton (ed.) *War and Ethnicity: Global Connections and Local Violence*. Rochester: University of Rochester Press, 1–45.

Uchibori, M. (1989) "Minzokuron memorandamu" (Memorandum on the notion of ethnos). In S. Tanabe (ed.) *Jinruigakuteki ninshiki no bōken—ideorogī to purakutisu* (The adventures of anthropological episteme: Ideology and practice). Tokyo: Dōbunkan, 27–48.

Uchibori, M. (2006) "Henbō ka kaibō ka: Afurika no seitai jinruigaku ni yosete (Transformation or dissection?: To the ecological anthropology of Africa)" *Afurika Kenkyū* (Journal of African Studies), 69: 143–145.

Uchibori, M. (2007) "Rinia na kūkan: Iban no kōdō kankyō ni okeru senkei hyōshō ni mukete no josetsu" (Linear space: An introduction to linear representations in the behavioral environment of the Iban). In K. Kawai (ed.) *Ikiru ba no jinruigaku: tochi to shizen no ninshiki, jissen, hyōshō katei* (The anthropology of living space: The process from awareness of land and nature to practice and representation). Kyoto: Kyoto University Press, 119–140.

Uexküll, J. von, and G. Kriszat (1934) *Streifzüge durch die Umwelten von Tieren und Menschen: Ein Bilderbuch unsichtbarer Welten* (Stroll through the Umwelten of animals and humans). Berlin: J. Springer. T. Hidaka and S. Haneda (trans.) *Seibutsu Kara Mita Sekai* (1973). Tokyo: Sinsisakusya.

Umezaki, M. and R. Ohtsuka (2002) "Changing migration patterns of the Huli in the Papua New Guinea Highlands: A genealogical-demographic analysis". *Mountain Research and Development*, 22: 256–262.

Utami, S. S., B. Goossens, M. W. Bruford, J. de Ruiter and J. A. R. A. M. van Hooff (2002) "Male bimaturism and reproductive success in Sumatran orangutans". *Behavioral Ecology*, 13: 643–652.

Valderrama Aramayo, M. X. C. (2002) "Reproductive success and genetic population structure in Wedge-Capuchin monkeys". PhD thesis: Columbia University.

van Schaik, C. P. (1989) "The ecology of social relationships amongst female primates". In V. Standen and R. A. Foley (eds) *Comparative Socioecology*. Oxford: Blackwell, 195–218.

van Schaik, C. P. (1996) "Social evolution in primates: The role of ecological factors and male behaviour". In W. G. Runciman, J. Maynard-Smith and R. I. M. Dunbar (eds), *Evolution of Social Behaviour Patterns in Primates and Man*. Oxford: Oxford University Press, 9–31.

van Schaik, C. P. and P. M. Kappeler (1997) "Infanticide risk and the evolution of male-female association in primates". *Proceedings of the Royal Society of London, Series B*, 264: 1687–1694.

van Schaik, C. P. and P. M. Kappeler (2003) "The evolution of social monogamy in primates". In U. H. Reichard and C. Boesch (eds) *Monogamy: Mating Strategies and Partnerships in Birds, Humans and Other Mammals.* Cambridge: Cambridge University Press, 59–80.

van Schaik, C. P. and R. I. M. Dunbar (1990) "The evolution of monogamy in large primates: A new hypothesis and some crucial tests". *Behaviour*, 115: 30–61.

Viegas, S. de Matos (2003) "Eating with your favourite mother: Time and sociality in a Brazilian Amerindian community". *The Journal of the Royal Anthropological Institute* (N. S.), 9: 21–37.

Warren, J. F. (1981) *The Sulu Zone 1768–1898: The Dynamics of External Trade, Slavery and Ethnicity in the Transformation of a Southeast Asian Maritime State,* Singapore: Singapore University Press.

Warren, J. F. (1986) "Who were the Balangingi Samal? Slave raiding and ethnogenesis in nineteenth century Sulu". *Journal of Asian Studies*, 37(3): 477–490.

Watts, D. and J. C. Mitani (2002) "Hunting and meat sharing by chimpanzees at Ngogo, Kibale National Park, Uganda". In C. Boesch, G. Hohmann and L. Marchant (eds) *Behavioral Diversity in Chimpanzees and Bonobos.* Edinburgh: Cambridge University Press, 244–255.

Watts, D., J. C. Mitani and H. Sherrow (2002) "New cases of inter-community infanticide by male chimpanzees at Ngogo, Kibale National Park, Uganda". *Primates*, 43: 263–270.

Wenzel, G. (1991) *Animal Rights, Human Rights.* Toronto: University of Toronto Press.

White, L. A. (1959) *The Evolution of Culture.* New York: McGraw-Hill.

Wilkins, J. F. and F. W. Marlowe (2006) "Sex-biased migration in humans: What should we expect from genetic data?". *BioEssays*, 28: 290–300.

Williams, J. M., W. Oehlert, J. V. Carlis and A. E. Pusey (2004) "Why do male chimpanzees defend a group range?". *Animal Behaviour*, 68: 523–532.

Wilson, M. L., and R. W. Wrangham (2003) "Intergroup relations in chimpanzees". In W. H. Durham (ed.) *Annual Review of Anthropology*, vol. 32. Palo Alto: Annual Reviews, 363–392.

Wimmer, B. D. and P. M. Kappeler (2002) "The effects of sexual selection and life history on the genetic structure of red-fronted lemur, Eulemur fulvus rufus, groups". *Animal Behaviour*, 64: 557–568.

Wimmer, B., D. Tautz and P. M. Kappeler (2002) "The genetic population structure of the gray mouse lemur (Microcebus murinus), a basal primate from Madagascar". *Behavioral Ecology and Sociobiology*, 52: 166–175.

Wolff, J. O. (1992) "Parents suppress reproduction and stimulate dispersal in opposite-sex juvenile white-footed mice". *Nature*, 359: 409–410.

Wolff, J. O. (1994) "More on juvenile dispersal in mammals". *Oikos*, 71: 349–353.

Woodburn, J. (1982) "Egalitarian societies". *Man* (N. S.), 17(3): 431–51.

Wrangham, R. W. (1980) "An ecological model of female-bonded primate groups". *Behaviour*, 75: 262–300.

Yamagiwa, J. (1984) *Gorira—mori ni kagayaku hakugin no se* (Gorillas: Silver backs shining in forests). Tokyo: Heibonsha.

Yamagiwa, J. (1994) *Kazoku no kigen* (The origin of families). Tokyo: University of Tokyo Press.

Yamagiwa, J. (2005) *Gorira* (Gorilla). Tokyo: University of Tokyo Press.

Yamagiwa, J. (2006) "Gorira no fīrudo idengaku" (Field genetics in gorillas). In M. Murayama, K. Watanabe and A. Takenaka (eds) *Idenshi no mado kara mita dōbutsutachi—fīrudo to jikkenshitsu* (Animals seen from the window of genetics: Field and laboratory). Kyoto: Kyoto University Press, 267–280.

Yamagiwa, J. (2008) *Jinrui shinkaron: Reichōruigaku karano tenkai* (Human evolution: Perspective from primatology). Tokyo: Shōkabō.

Yamazumi, K. and Y. Engeström (ed.) (2008) *Nottowākingu: Musubiau ningen katsudō no sōzō e* (Knot working: For the creation of human activity that connects one another). Tokyo: Shinyōsha.

Name Index

Subject Index